Old Electrical Wiring
Maintenance and Retrofit

Other McGraw-Hill Construction Titles

American Electrician's Handbook, 13/e, by Croft & Summers
Electrical Contracting Forms and Procedures Manual
by Johnson & Whitson
Rapid Electrical Estimating and Pricing, 5/e, by Kolstad & Kohnert
The Lineman's and Cableman's Handbook by Kurtz
Low Voltage Wiring Handbook by Maybin
McGraw-Hill's National Electrical Code Handbook, 22/e, by McPartland
Handbook of Practical Electrical Design by McPartland
McGraw-Hill's Handbook of Electrical Construction Calculations,
revised edition, by McPartland
Yearbook Supplement to the McGraw-Hill National Electrical Code
Handbook by McPartland
How to Start and Operate an Electrical Contracting Business, 2/e, by Ray
Practical Electrical Wiring, 17/e, by Richter & Schwan
Handbook of Electrical Design Details by Traister
Security/Fire Alarm Systems: Design, Installation, and Maintenance 2/e,
by Traister
McGraw-Hill's Illustrated Pocket Guide to the 1996 NEC Tables
by Traister
McGraw-Hill's Illustrated Index to the 1996 NEC Code by Traister
Industrial Electrical Wiring by Traister
The Electrician's Troubleshooting Pocket Guide by Traister
The Electrician's Pocket Reference by Traister
Handbook of Electrical Construction Tools and Materials by Whitson

Old Electrical Wiring Maintenance and Retrofit

David E. Shapiro

McGraw-Hill

New York   San Francisco   Washington, D.C.   Auckland   Bogotá
Caracas   Lisbon   London   Madrid   Mexico City   Milan
Montreal   New Delhi   San Juan   Singapore
Sydney   Tokyo   Toronto

Library of Congress Cataloging-in-Publication Data

Shapiro, D. E. (David E.)
  Old electrical wiring maintenance and retrofit / David E. Shapiro.
    p. cm.
  ISBN 0-07-057879-6
  1. Electrical wiring, Interior—Maintenance and repair.   2. Historic buildings—
Maintenance and repair.   I. Title.
  TK3271.S46   1998
  621.319'24'0288—dc21   97-48869
                                                                               CIP

McGraw-Hill

A Division of The **McGraw·Hill** Companies

1 2 3 4 5 6 7 8 9 0   DOC/DOC   9 0 3 2 1 0 9 8

ISBN 0-07-057879-6

The sponsoring editor for this book was Zoe G. Foundotos, the editing supervisor was Sally Glover, and the production supervisor was Claire Stanley. It was set in ITC Century Light per the EL3 Specs by Michele M. Bettermann and Paul Scozzari of McGraw-Hill's Professional Book Group composition unit, Hightstown, N.J.

Printed and bound by R. R. Donnelley & Sons Company.

McGraw-Hill books are available at special quantity discounts to use as premiums and sales promotions, or for use in corporate training programs. For more information, please write to the Director of Special Sales, McGraw-Hill, 11 West 19th Street, New York, NY 10011. Or contact your local bookstore.

 This book is printed on recycled, acid-free paper containing a minimum of 50 percent recycled, de-inked fiber.

Many people helped this book come into being by bringing me to where I could write it. Some helped me develop as a writer, some as an electrician, and most as a person. They include the following.

My parents:

> Lillian Shapiro-Michael, an example of integrity and of working to make a positive difference; and
>
> Louis Leonard Shapiro, who showed me integrity, and the pleasure and value to be derived from physical work. He also eased my way to join the IBEW.

Mr. Brown, a grade-school English teacher.

Dr. Leslie Horst. Her coaching made possible my first success as a writer of substantive material—my master's thesis.

Arthur Hesse and Ronald Kotula, chief inspectors whose professionalism, courtesy, knowledge, and good will helped make me a competent master electrician.

Creighton Schwan, my mentor, not the least as a writer for the electrical trade; an urbane, erudite, ever-civil, ever-helpful columnist and code authority.

Various trade magazines which provided learning through their articles and responsa and enabled me to get my feet wet writing about the industry.

The editors for whom I wrote columns, including Walt Albro, Nancy Bailey, and Jeanne LaBella; they helped me gain experience and credibility writing about safety.

IBEW Local 3, which gave me a start in the electrical trade, and taught a wonderful attitude of workmanship and pride.

Bill Zinsser, author of *On Writing Well*, who helped me value clear language.

Wilbert J. McKeachie, whose writing gave me some understanding of effective teaching.

My customers, who have trusted me and supported me for so many years and whose electrical systems provided many of the lessons conveyed in this book.

The International Association of Electrical Inspectors, IAEI— a place where persons of good will in this industry come together to further electrical safety.

And finally, my friends and lovers over the years, including most especially Mary Jo Hlavaty, Judith Oarr, and Sheila Wolinsky, who supported me through the twists and turns of my career path with trust, advice, and more.

I offer deep thanks to the following people who contributed more directly and immediately to this book.

M J Hlavaty provided major help with copyediting.

Charley Forsberg provided information on PVC.

Dr. Jesse Aronstein reviewed the material on aluminum, early on.

Joseph Tedesco, a man with strong knowledge and strong opinions, reviewed an early version of the entire book and offered feedback.

Sam Levinrad provided exhaustively detailed feedback, giving the author, and readers, the benefit of his very wide knowledge and experience in the electrical industry.

Creighton Schwan provided equally helpful and careful comments. He also provided the entre to Hal Crawford, which smoothed the way for the making of this book.

Dave Noonan at Bryant offered useful comment on the glossary and other bits.

Arthur Hesse looked through an early version of the text and confirmed the reasonableness of the suggestions in Chapter 12, "Inspection issues," from the perspective of a universally respected chief inspector with decades of experience and the highest reputation for integrity.

Contents

10 Historic Buildings   207

11 Commercial Settings   217

12 Inspection Issues   247

Part IV: Supplementary Material   259

13 Dating and History   261

Foreword

THE ELECTRICAL CONSTRUCTION INDUSTRY IS SO FAST moving that it takes a lot of effort on the part of building owners, electricians, electrical contractors, inspectors, apprentices, and others to keep up with new methods, equipment, wiring materials, and tools being introduced to the marketplace in a steady stream, not to mention absorbing the revisions to the National Electrical Code, of which there is a new, large batch every three years.

Well, you can relax now, for this book looks back to the way it used to be done, and there is no urgency for you to absorb it all at once. As with most technical books, *Old Electrical Wiring Maintenance and Retrofit* does not lend itself to being read through in one session. Rather, it is a book you will find to be valuable reference work. A lot of the information is applicable to your current work: safety practices, customer relations, inspector relations, etc., but keep the book handy for when you run into something you don't recognize. An installation completed last week is "old work," and sometimes such a job can be as poorly done as a really old job, but in general what is covered in *Old Electrical Wiring Maintenance and Retrofit* is wiring done years ago. The trick is to try to figure out what was in the mind of the electrician at the time the job was done originally. Just because the work is old does not mean it was not well done, nor is it assured that a job done yesterday is necessarily first-rate. *Old Electrical Wiring Maintenance and Retrofit* will help you make the judgments necessary in sorting out the good from the bad.

David Shapiro has done a fine job of researching the material for this book. The use of electricity in buildings is over 100 years old, so there are no persons alive who have seen it all. You will find that reading *Old Electrical Wiring Maintenance and Retrofit* is just like talking to an "old timer," a friendly, talkative one at that, and one who speaks your language.

I notice the mention of illegal three-way hookups referred to as the Carter System. Over 50 years ago, when I was an apprentice in San Francisco, we referred to these illegal and unsafe three-way circuits as "Los Angeles three-ways." I later learned that electricians in Los Angeles referred to them as "San Francisco three-ways." In any case, pointing the finger does not accomplish anything. It does not matter who messed it up—it only matters that you know how to fix it, and *Old Electrical Wiring Maintenance and Retrofit* will help you do that.

Insulation deteriorates with age, so one of your major tasks is knowing when you can salvage something, or whether it must be replaced. This book will help you make those decisions.

Experience is of great value when troubleshooting an old job, but if you don't have the experience yourself, *Old Electrical Wiring Maintenance and Retrofit* gives you an insight into the experiences of uncounted others to help you. By the way, when troubleshooting, be sure to carefully listen to the owner or occupant who is trying to let you know what happened. Even though you may not speak the same language, technically, if you listen very carefully you will probably get a helpful clue.

I wish this book had been around years ago—it would have been most useful to me, and I know it will be so for you.

W. Creighton Schwan
Hayward, California
September 17, 1997

Old Electrical Wiring
Maintenance and Retrofit

Overview and Introduction

ALL THINGS GET OLD AND CROTCHETY AND EVENTUALLY die: electrical systems are no exception. Power wiring has not been around forever; William A. Wittbecker's 1902 handbook, *Domestic Electrical Work*, took it for granted that readers were wiring from batteries. Things changed rapidly over the next quarter-century. By the time of the Great Depression, two-thirds of the homes and businesses in this country had been electrified. Quite a few, perhaps even a majority, of those buildings are still around, and most still rely on their original wiring. Close to 40% of single-family homes in use in 1991 were built before 1940, according to NAHB statistics. Millions upon millions of dollars will be spent to repair and upgrade their systems. Even much of the wiring installed during the building boom following World War II has decaying insulation. If you are a contractor, you could work on these old systems nonstop till you retire. If you are an inspector, as long as you work you will be walking onto jobs where you have to make those hard calls about what can be grandfathered and what has got to go.

There are many, many problems in working with old wiring. Expertise in dealing with those problems will make your work more valuable and set you off from your competitors. To make this your specialty, you need plenty of patience with idiosyncrasy and frailty. You also need some background. So-called old work demands different knowledge and skills than does new work; and that is even more true of work on genuinely old wiring. No book, and no author, has the knowledge, and the space, to walk you through everything you could encounter in old work, in detail. This text will get you started, give you a leg up. But, for instance, when you encounter equipment whose labels, not to mention installation instructions, are long gone, nothing will substitute for experience to teach you that a particular fixture is probably suited for no more than 25-watt or 60-watt or 100-watt lamps, or that a terminal probably is intended for the attachment of more than a single conductor.

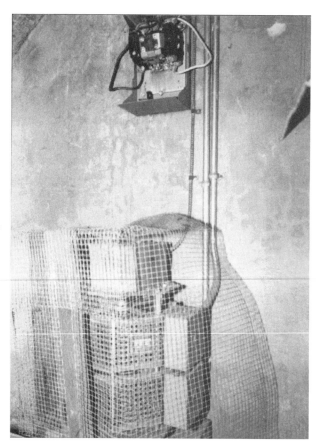

■ **1-1** *This is an old dimmer! Looking inside the protective cage, the two stacked units on the bottom are variable autotransformers. The smaller unit on top is a motor (operated by a switch in the same room as the two lighting circuits being dimmed) that is used to change the transformers' setting. There's a coverless contractor high above. You will find no further description. It would be quite impossible to discuss every queer duck that you might encounter in old buildings.*

If you are a rank novice, you would do well to start with a book such as *Practical Electrical Wiring*, which is referenced in Chapter 14, "Resources." Nonelectricians certainly can learn from this volume. It does, however, make the necessary assumption that readers who choose to apply the information hands-on have the needed training and experience, or sufficient professional supervision.

Whatever your background, please recognize that when the text mentions a tool or test that is unfamiliar to you, you need to seek instruction, rather than blundering through.

If you are a practical person who just wants to know how to handle the job in front of you, you may choose to skip around. There are bits of philosophy and history scattered through the book, context to help you understand how the types of installations you encounter came to be, and why particular fixes are suggested. There are also at least mentions of very unusual types of installation, which you may never ever encounter. There is no requirement that you read through every word. Still, you may be glad that information is available. On the other hand, if you are an old-timer, much will be old hat to you unless you have spent your entire career on new construction. You still will come across new bits, interesting information that will make this book worth reading.

Layout

It would be irresponsible to start off with Chapter 2, "Troubleshooting," without first alluding to the dangers. You would do well to flip right now to Chapter 17, "Dangers and Benefits." That is way back in Part 3 of the book, along with other material that on the one hand all readers would do well to read, but on the other hand not all readers will choose to go through. Part 1, in contrast, teaches basic information about dealing with old wiring. Part 2 covers topics that you definitely will need if, but only if, you encounter certain types of old wiring or deal with certain aspects of old wiring work.

Part 1: Getting into old wiring

Chapter 2

"Troubleshooting" starts getting into the nitty-gritty, but steps back frequently to discuss safety, history, and design issues. There is a "presenting problem"—a light is dead or flickers, a receptacle does not always work. The customer says, "I think I may have a short in my switch." The chapter looks at how to uncover and solve the actual, technical problem that led to a complaint. It addresses in detail the challenge of repairing damaged insulation.

Chapter 3

"Design Changes" continues the mix of hands-on, what's-in-this-ceiling troubleshooting with theory and history begun in Chapter 2.

Chapter 3 addresses two problems. The first is untangling the cumulative changes in a house's electrical layout. This requires you to guess at the logic that guided decades and generations of electricians and home repairers or maintenance workers. The second problem is the painful necessity of sometimes saying, "This is no longer legal; not only is the design unsafe, but I could jeopardize my license if I restored it." Specific issues related to these problems include closet lights and overfusing.

Chapter 4

"Switching Layouts" is short and straightforward. It focuses on a specific example of no-longer-acceptable wiring, based on changes in Chapter 3 of the NEC. This chapter's example underscores the difference between troubleshooting in a case where you know how a system was installed and in a case where "anything goes" may have been the rule. Chapter 4 introduces and examines Carter System (lazy neutral) switching, and describes creative solutions to that dangerous situation.

Chapter 5

"Accept, Adapt, or Uproot" continues on a similar theme. This chapter addresses a number of problems, such as being asked to make changes that would bring a system into violation, and dealing with crowded enclosures. It also explores one problem specific (though not exclusive) to installation of fans and heavy chandeliers: providing mechanical support after-the-fact.

Chapter 6

"Safety Surveys: The Big Picture" pulls away from troubleshooting. This chapter introduces a comprehensive safety survey and deals with educating customers concerning the difference between "shiny" and "safe." Ignorant or unscrupulous contractors who are assuredly not old wiring experts will sometimes do just enough to make a system look superficially good. In this chapter, you will learn to sell customers an inspection that goes way beyond what they can get elsewhere. A certified home inspector normally does not have the knowledge; a municipal or county inspector does not have the time, even if he or she has the knowledge and authority.

Chapter 7

"Setting Limits and Avoiding Snares" is about setting boundaries that will protect your license, your receivables, and your customers. It is easy to feel sorry for a customer who has rotten wiring, and who perhaps has been taken by previous electricians or home inspectors who did not know how to cope with such a system. Carelessly accommodating all sorts of requests can cost you dearly by saddling you with unanticipated liability. On the other hand, flexibility on your part can increase your business and offer a genuine service to customers who are themselves flexible.

Chapter 8

"Tearing It Out and Doing It Right" is about running largely or completely new wiring in old buildings. Even when you rewire, you benefit from understanding old wiring systems. For instance, when the safety survey you learned about in Chapter 6 leads you to conclude, "This is deadly," and the customers (and the customers' bankers) accept that conclusion, you will be in a much better position to rewire than some strange contractor. Not only will you have just gone through the structure, but you will be more knowledgeable about old structures and about what will need to be removed.

This chapter goes into options for major rewiring, including fishing, surface raceway, and "in-and-out." It also addresses coordination with other repair and renovation efforts.

Part 2, Advanced topics

5

After that, the text gets to the advanced and even oddball topics of Part 2. At this point, if you are not trying to get a broad education in old wiring, it is more reasonable to pick and choose, looking for information to help solve the problems you face.

Chapter 9

"Relatively Rare Situations" covers low-voltage switching, fuse-boxes in ceilings, aluminum wiring, bypassed meters, and other challenges. This volume cannot possibly address all the systems, designs, and equipment you will encounter. Still, this chapter does try to at least touch on some of the relatively unusual systems that are no longer installed, or not commonly installed. You may run across them.

Chapter 10

"Historic Buildings" is much less focused on wiring than are most of the other chapters. Like Chapters 6 and 7, Chapter 10 is about people and laws at least as much as it is about circuits. Questions addressed include the following: What kinds of building features do people cherish? What kind of changes can be restricted by covenant or law because of a building's special circumstances? How can you deal with such limitations when the wiring is shot?

Chapter 11

"Commercial Settings" brings you back to some special electrical situations. While most of what has been covered in earlier chapters applies to commercial installations as well as residential, the information in this chapter is more specific to commercial, institutional, and industrial work, and less relevant to homes. It focuses on single-phase/three-phase issues such as getting rid of phase converters and motor generators, on updating to more efficient lighting, and on complying with modern safety requirements in nonmodern workplaces.

Part 3, Supplementary material

Finally come the extras. These chapters are more than just added thoughts. Depending on your background, you may be referring to them from the very beginning.

Chapter 12

Inspectors, and contractors concerned about what the inspector will say, can benefit from most of the material in this book. This chapter picks out some special concerns that old wiring raises for jurisdictional authorities and for electricians dealing with those authorities.

Chapter 13

"Dating and history" will help you estimate the ages of systems and devices. This is essential to determining what may be grandfathered and what cannot possibly have been legal even when first installed. It includes the sort of information you need to become an expert, a scholar even, in your eyes and those of your customers.

Chapter 14

"Resources" offers suggestions of where to find obsolete breakers, where to find information on historic buildings, and where to find special equipment—both obsolete parts and devices that can solve problems posed by old systems. It includes sources for tools that can be particularly useful in troubleshooting or repairing old systems.

Chapter 15

"Communication" is an essential chapter for any reader who is not experienced in working closely with lay customers. Communicating with the untrained is a far more important part of rewiring than of new work. Not only are you more likely to deal with laypersons in selling the job, but you often will have to confer with them, again

6

and again, as you uncover kinks in the old wiring that require decisions. Furthermore, you may be working in occupied buildings—chopping and drilling around customers' businesses, or stepping carefully around their home lives.

Chapter 16

Certain hard-wired products may be dangerous because of manufacturing defects. This chapter, "Recalls," contains a list of products to watch for.

Chapter 17

The last chapter, "Hazards and Benefits," describes some of the dangers associated with electrical work, emphasizing those particularly threatening practitioners of this old-wiring specialty. The discussion that follows details the benefits that expertise in electrical renovation and the repair of old wiring can provide to all electricians, whether or not they specialize in such work.

Glossary

As you read through this book, you will find unfamiliar terms. Most of them will be in the glossary. You would do well to read its brief introduction and use guide the first time you look up a word, rather than treating it like a familiar dictionary.

The glossary is not quite a must-read, but it is more than just something to serve you when you are puzzled. If you choose to leaf through the glossary, you are bound to learn. The word list is more extensive and goes into far greater detail than the lists you will find in other wiring books. It emphasizes outdated language and slang, including many terms that are not even used in the body of the text. Besides translating terms, the glossary explains concepts and rules. Like the other chapters, it discusses the significance of various changes from old procedures and equipment to new—and legitimate and illegitimate ways to deal with those changes.

Appendices

There are three appendices. The first two contain material that belongs early in the book, perhaps in Chapter 2, "Troubleshooting," or Chapter 3, "Design Changes," and the third belongs with Chapter 15, "Communication." They have been relegated to the rear simply because of their length; placing them earlier would have broken the flow.

Familiarizing Yourself with Old Wiring

Troubleshooting

THIS CHAPTER EMPLOYS TROUBLESHOOTING AS A FRAMEwork for illustrating various issues that you will face in working on older wiring.

Safety has to come first. That material is planted right in front of the how-to, rather than tucked away in a preface, because it is dead critical and absolutely warrants "overlearning"—reading even though you know all about it. Then comes a discussion of the best ways to approach a troubleshooting puzzle, and next procedures to use in taking outlets apart for examination and testing. After you track down problems, the chapter talks about how to repair them and closes with advice on bonding and grounding.

To avoid breaking the flow of the chapter proper, two subjects that you should look at before you proceed to the next chapter are covered in appendices. Appendix 1 discusses special issues associated with older fluorescent fixtures; Appendix 2 talks about old splices.

A "heads-up" on safety

Before you dive into troubleshooting, here are two important points about safety. Safety begins with language, because safe thinking leads to safe practice. To drive home the point, consider what this book does to avoid sloppy language with regards to what folks tend to call "the neutral," even in two-wire circuits.

You will find a fair amount of slang in this book, but it rarely uses the term, "the neutral," because, especially in very old systems, what is called "the neutral" in 120-volt wiring is not always, technically, a neutral. And sloppy language can mislead you into making dangerous assumptions. In old wiring, even when the workmanship shows that the original installer was a professional, you dare not assume that he or she followed modern rules. "The return conductor" is more precise. The more formal "identified conductor" is another way of talking about the same thing, but using it can mislead you into making another dangerous assumption. In old wiring, even more than in other wiring, you cannot rely on the return conductor to be an identified, or white (or, theoretically, natural gray), conductor. You will be reminded of that repeatedly by use of the even more cumbersome phrase, "grounded conductor."

Another thing to remember is that you need to rely on both your tool training and your senses. Sniff for burned bits. Listen for crackles. Be ready for sparks or arcs waiting to zap your eyes or body. "Be ready" means that if you are working around wires that could arc with significant energy, you need both suitable goggles and sufficient clearance. Sensory alertness is most important for your self-protection. Aside from safety considerations, looking and listening and sniffing are also important tools for tracking down evasive problems.

This is illustrated by a customer, Nan, who took a tip from her electrician in using the nose as a test instrument. The electrician was hired to track down the reason a circuit was tripping, and one of the first things he did was to sniff at the various outlets and switch boxes on the circuit. A burnt smell led him to the ground-faulted switch, and the evidence of charring was confirmed when he removed box and cable from the wall. Two months later, Nan called to say that she needed him again. She had smelled the same "electrical burnt smell" at another switch box, one that in fact had checked out okay on the earlier visit. "Could it be hazardous? Should we have you over again?" Yes, and yes again. Smart woman. It seemed that the leak had reappeared, after a heavy rain, and traveled to a new part of the basement. (Of course, the problem disappeared by the time the electrician arrived, leaving no clues. That happens.)

Start troubleshooting by sizing up the problem

It is time to give you a sense of what a savvy electrician will do, and more importantly of how he or she will think, in working on old wiring. The following example will introduce the common ways that cables and fixtures deteriorate. Before actually delving into the wiring to solve the problem, there will be a break to discuss why you should tend to suspect certain trouble spots.

A customer in an old house complains about problems with a light. At first, this might look quite similar to troubleshooting at a pretty new house.

The first thing to do is to ask the customer to describe his or her observations; complaints are often heavy on amateur diagnosis but light on detail. You may need to ask follow-up questions in order to learn what the customer actually sees, hears, or smells. An appropriate answer to the request for symptoms might be, "The light dims sometimes" or "All the lights dim sometimes." Those two answers point in different directions.

Or the answer might be, "The lights sometimes take a long time to come on and don't always come on to full brightness." It might be, "I have to jiggle the switch to make the light come on, or stay on." It could be, "The light went out recently, and it couldn't have been the bulb, because it wasn't an old bulb." The latter throws in a little amateur diagnosis along with the behavioral description. At least this amateur diagnosis adds more information, unlike "The light's shorted," a "description" devoid of observational data, meaning that it tells you nothing except that the customer is not happy with the light.

Next, gather data firsthand. To start, attempt to reproduce the symptoms the customers reported. So if they say a light has stopped working, try the switch. If nothing happens when you flick the switch, check the lamp. This may seem ungracious, but it is far less foolish than spending expensive time hunting for something more exotic when in fact the customer has burned out a light bulb, or, more charitably, bought a case of bad light bulbs.

Incidentally, this scenario can be easily extrapolated to dealing with the complaint of a bad receptacle. In that case, an equivalent test would be to make sure that the (portable) appliance that failed to operate works, by plugging it in somewhere else, or to try plugging something else in the suspect outlet. Another useful step is making sure the receptacle is not switch-controlled. When the presenting problem is an errant light, the equivalent step is checking for three-way switching, or for the presence of both a wall switch and a pull chain.

You may wonder at the reason for talk about trying appliances rather than inserting the probes of a voltmeter. The reason is simple. You can suggest each of the steps described above to customers over the phone, if they sound as though they will be comfortable. Replacing fuses, even though they look fine, is far cheaper than having you over to diagnose a blown fuse. Flicking every circuit breaker off and then on again costs nothing. The tests may seem silly, but even if a possibility is obvious to a person working in the electrical industry, the customer may not have thought of it.

Even though you may lack a great deal of old-wiring expertise, your common sense can take you a little further. Besides the silly oversights listed above for elimination as the first step in checking for simple problems with lights and receptacles, there are other "usual suspects," as likely to be found in older wiring as they are in newer.

☐ Has someone tried to work on the system recently? (If someone has recently worked on the malfunctioning system, first of all look for careless workmanship, including work that might endanger you as you investigate. Also look for misapplication, such as buying the wrong ballast for an HID fixture, or putting the wrong leads together.)

☐ Has there been any reason to suspect water damage, such as a plumbing leak or storm infiltration? Has there been, or is there, smoke or a burnt smell? You might think that customers would mention such clues without prompting, but do not rely on them to do so.

Even when some of the "usual suspects" are present, avoid rushing to a premature conclusion. Many problems in old wiring have had nothing to do with recent amateur work. For instance, an electrician had a customer ask, "Was this weak ballast caused by the bathtub overflow from upstairs?" (His appropriately respectful answer was, "No, ma'am, it doesn't work that way. Didn't hurt to ask me, though.") The flickering fluorescent fixture had been re-lamped by the customer's adult son about a month before. His amateur efforts were futile but had not caused the problem: he had relamped with the correct lamps, and moreover had inserted them correctly. However, a month of operation on a dying ballast had nearly destroyed them—so after replacing the ballast the electrician had to relamp again.

Once you have eliminated the obvious, you need to put on your thinking cap: what clues do you have? You will proceed in different directions depending on whether the light takes a while to come on (if it is a fluorescent, perhaps it has a dying starter) or it flickers when you touch the switch (that is an easy one—the problem is almost certainly at the switch) or it sometimes will not come on for days. (The latter sometimes means that the light is controlled by multipoint switching, and that one of the seldom-used switches is bad. The light will not come on until that bad switch is flipped again into its conducting position.) The worst case is where there is a defective—in the very worst case, an intermittent—splice or termination, perhaps with a broken wire. This is the worst case for two reasons. First, often no damage is visible. Second, when you think you have isolated the problem, it may disappear.

Fatigue is often the villain

Besides any specific clues you come across, there are general rules that can guide your search. What is most likely to have deteriorated? Deterioration, whether of insulation, conductors, or mechanical components, is generally the result of fatigue, which results in mechanical fracturing. Fatigue is the result of repeated movement. Heat is one common cause. Thermal expansion and contraction are the most important reasons that heat causes deterioration, although heat also accelerates chemical change, cooking substances such as insulation and thus making them brittle. Alternately or simultaneously, heat can soften materials such as some insulations to the point where they flow away from any pressure against them, or simply flow down with the pull of gravity. Changes in materials themselves due to heat, however, are not the most likely source of an old-wiring problem—unless there has been severe scorching. Fatigue caused by thermal or other movement is the most common culprit—at least the most common after bad workmanship.

Switches and receptacles are likely suspects, due to movement

If fatigue is the result of movement, the first place to look is anywhere something moves. Switches are an obvious choice. Any switch is designed for so many, and only so many, operations. At some point, its springs break; its contacts pit. Old mercury switches are a long-lived exception. Because the contacts do not draw arcs the same way that those in snap switches do, they do not need springs and they do not burn contacts. That does not mean they never fail—but they do sink toward the bottom of the list of suspects.

The concern with switches is not just that the devices themselves may die. Each time the handle is flicked, there is a bit of jiggling. Therefore, attached wires suffer more mechanical disturbance at a switch than at a more passive device such as a smoke detector or an LED indicator, devices that suffer very modest thermal stress as current flow waxes and wanes.

In many cases, that slight movement caused by repeated flickings of the switch handle has nudged device terminals close enough to grounded enclosures that they arced. In many others, twist-on connectors apparently were loosened by repeated small movements to the point where they fell off.

Similar if perhaps lesser disturbance takes place at receptacles, as cord connectors are repeatedly plugged in and pulled out.

Troubleshooters need to keep in mind an additional problem associated with receptacles. Because receptacles generally have two terminal screws for hot conductors and two for grounded conductors, installers often save themselves some splicing by daisy-chaining, carrying downstream power through upstream receptacles. Therefore, if a connection loosens at one receptacle, that defect can affect switches, receptacles, and lighting outlets elsewhere on the circuit. "The case of the phantom switch" was a friend's name for one such intermittent, "Now you see it, now you don't" situation.

The same problem can be found, albeit less commonly, with switches that have been used as part of the current path for the hot conductor. With relatively modern switches, daisy-chaining is usually accomplished by using a screw terminal to connect one hot wire and inserting another in a "backwiring" hole—a legal, but less-than-optimal, way of doing things. An older, and currently uncommon, alternative is to loop a wire around one device's terminal screw (after skinning off less than an inch of insulation) and continue with the same wire on to the next device.

Four problems are commonly found with that system, and a fifth one more rarely.

☐ If too much insulation was stripped, there is that much extra, potentially hazardous, bare conductor in the switch box.

☐ If too little insulation was stripped, the terminal screw may pinch insulation, which serves as a spacer, keeping the screw away from making contact with bare copper.

☐ This may have been sometimes used as a solution when too little insulation was stripped: the conductor may not be wound a full $\frac{2}{3}$ to $\frac{3}{4}$ of the way around the terminal screw.

☐ Unless the installer was unusually generous in pulling wire into the box, bridging from switch to switch may have caused him to eat into the required six inches of free conductor. Furthermore, most commonly, very little wire is allotted to the span between one switch and its neighbor. The result of these constraints is that it can be difficult to pull the switches out to test or to work on them.

☐ In some of these installations, the electrician saved time by stripping enough insulation so that he could wrap the wire completely around a switch terminal, rather than carefully forming it into a partial loop. The result is that the wire overlaps itself, at one point. Therefore, the terminal screw makes solid contact with the wire only at that one overlapped point where its thickness is doubled.

Lighting outlets are also trouble spots

Lampholders for incandescent lights are Suspect #2. There are easily five reasons to suspect them.

☐ There is gross physical movement every time a lamp is replaced.

☐ Contact is commonly made through rivets, which can loosen or corrode.

☐ The switched lead usually connects to a vulnerable center contact. That contact is either spring metal, which flattens over time as it loses its resilience, or a solid lump (button, if you prefer), which can pit to such an extent that its corrosion offers very high resistance.

☐ Lampholders are subjected to long hours of cooking over 60, 100, and more watts of electricity. More precisely, they cook over whatever percentage of those watts the lamp—a rather inefficient device—fails to radiate away.

☐ In old houses, the boxes above ceiling lights often serve as splice boxes and are quite crowded. For that especial reason, lighting outlets—both those giving trouble and others upstream— often warrant investigation.

This brings up a bit of history. When houses were first wired, lighting was the only use foreseen for electricity. Therefore, unless the preexisting wall sconces of a previously gaslit house were being retrofitted, power was brought to one point in a room—a pull-chain lampholder at the center of the ceiling. In some cases, wiring was pulled into the gas pipe itself. In others, a gas fixture was taken down or modified, and an electric fixture was installed in its place, with the outlet box supported from the gas pipe.

Later, when receptacles were added, they were added by branching out from that one poor, often already-crowded box in the center of the ceiling. The ceiling box was originally designed either as an end-of-run, albeit rarely, or, worse, used to feed another ceiling outlet or outlets, without real consideration of the need for additional volume. So it has been crowded for a long, long time.

Now look again at fatigue associated with heat at lighting outlets. Anything close to a source of heat will expand and contract more than its fellow that has not been subjected to such cooking.

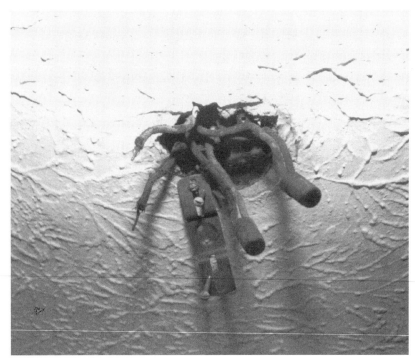

■ **2-1** *Crowded ceiling box.*

Anything crowded and unventilated also has extra reason to suffer from the effects of heat. Therefore, although the first suspect is always the silly one—a burned-out lamp—and the next is the switch, after that you should look at the lampholder and even the wiring inside the outlet serving that lampholder.

Because heat rises, wiring located above a lampholder gets especially toasty. For that additional reason, other factors being equal, you should be more likely to suspect a ceiling outlet than a wall sconce.

The effects of deterioration

Simple deterioration due to heating and cooling is less likely to cause a short circuit than to cause an open contact or connection. That includes the possibility of an intermittently open or even an arcing connection. Air is such a fine insulator! When insulation dries, it does not crumble unless it is moved. Even cracked, it still keeps conductors apart. If it does crack and crumble, the conductors that were protected by the insulation have to be moved against a grounded surface or each other—or damned close—before they will short. That is a lot of movement to result from heat expansion, or even from screwing and unscrewing lamps from a lampholder, so long as the lampholder itself does not twist. Yes, natural building movement, as well as vibration from activity on the floor above, could nudge wiring. However, neither seems nearly as likely to cause brittle insulation to fall off and wires to then touch as the possibility that someone has opened the enclosure and handled them, directly or indirectly.

When thermal expansion causes enough movement for arcing to result, GFCIs might seem to be useful in detecting resultant ground fault currents. Alas, according to Earl Roberts, an authority on such matters, tests show that GFCIs do not prevent overheated fixtures recessed in insulated ceilings from causing fires. Damaged wiring can cause a fire well before there is sufficient arcing to ground to operate a GFCI.

There is a very straightforward exception to the idea that deteriorated wiring at lampholders is more likely to result in open circuits than in shorts. The exception requires two events to occur. First, something has to push the wires against each other. Second, insulation damage due to aging, or to further twisting or pushing, causes them to short. If the lampholder is not secured solidly to the fixture, the lampholder may rotate every time a lamp is replaced, twisting the wires behind it. That greatly increases the probability of eventual shorting, making a short circuit about as likely as a broken conductor. Without something like this scenario, an open is the far more likely result of deterioration than is a short.

The type of lampholder most commonly presenting this problem is the canopy-mounted type. Two porcelain components screw together and are held from rotating by engaging with a bump in the thin metal canopy. Sometimes this is not much of a bump, and thus

■ **2-2** *A box (sold in 1918) that was "born crowded." If the box in Fig. 2-1 goes back that far, it may very well be grandfathered.*

not much of a restriction. Similar lampholders are also commonly used in outline lighting, as in theater marquees, and dressing mirror or vanity lights.

Live testing procedures

That is a fair amount of general information. In a moment, this discussion will come back to the specific example of troubleshooting a nonfunctioning light fixture. But first, it will do you no harm to scan a refresher on a few basics, before you start pulling wires apart.

To test live circuits, you need to be familiar with protective gear and proper tool-handling procedures; and in many cases, you should have a safety person available to rescue you in case of accident. If you are experienced enough to feel safe testing live electrical systems, you probably know that the logical, orderly way to troubleshoot is to start at the source of power and work your way downstream. It is wise to have an extension cord with you, offering a proven hot and grounded conductor. That way, you will not be misled by broken grounded conductors and unreliable grounds into thinking that an outlet is dead when it is live—perhaps live on both hot and return sides.

Back to the nonfunctioning light fixture that constituted the "presenting problem," quite a ways back. Suppose you test a lampholder for absence of the hot (presumably switched) or the grounded conductor and find both present. Why does the light remain dark? There could be a contact problem, caused by corrosion or loosening, that allows your voltmeter to show a complete circuit but does not pass power through to lamps.

Now suppose that you eliminate those possibilities and recheck; and the light works fine. Why was it not working? You do not get paid for a diagnosis of "You've got gremlins." There could be intermittency in the switch, even though it tested out okay (assuming you tested it). There could be intermittent contact further upstream, even in the fuseholder. There could be a problem that was temporarily reversed by the twisting associated with unscrewing the lamp. You are unlikely to find the answer from outside the outlet box.

Play it safe as you pull things apart

The next step in troubleshooting is to remove the fixture or device, or at least to separate it from the outlet box. It is strongly recom-

mended that you kill power before doing so, indeed before you undertake any step that involves pulling apart wiring. Turn it back on
again when your hands can be free.

Sure, everyone and their old pappies have dispensed with that step
at one time or another. Just recognize that when you do so, you are
taking an unnecessary risk. Consider the case of an electrician who
was blown away by operating a circuit breaker. The probable cause
is not important for your purposes. From the fact that this colleague sustained a bad injury from such a safe-appearing activity,
just extrapolate to the risk you take pulling apart rotten old wiring
without killing power. Incidentally, when in doubt, a cautious electrician will operate an overcurrent device from the side rather than
in front of it, and with the panelboard cover in place.

Ensure that power is off

Old wiring is notorious for inadequate labeling. Suppose you do
not know what circuit is involved. If there is power at the switch,
you are ahead of the game. It is a reasonably safe bet that going to
the loadcenter and turning off the power feeding that switch will
interrupt the power trying to get through to the light. On many a
job, however, the switch the electrician thought controlled the
light, indeed the switch he or she was told controlled the light, was
not the right switch. You can sometimes start troubleshooting at
the switch, before taking things apart at the outlet.

Warning: there still could be other circuits feeding the outlet box
to which the light is mounted, using the outlet as a junction box. Old
self-contained fixtures, such as continuous rows of fluorescents
that serve as their own wiring channels, also could have been
used as raceways for additional circuits. Receptacle outlets too
can host extraneous circuits; nothing in the NEC forbids sharing
enclosures with circuits of the same voltage and system, so long
as box-fill restrictions are observed. In addition, be wary of receptacles that have been split-wired—whether with two independent circuits, multiwire circuits, or simply one outlet switched
and the other not.

Even worse than the presence of additional circuits is this hazardous scenario. You check for voltage between the center contact
and the shell and find nothing. You confirm that the outlet is dead
by checking for voltage both between the center contact and the
outlet box and between the shell and the outlet box. Nothing. Just
to be safe, you triple-check by testing for voltage between the extension cord's grounded conductor and the lampholder's center
contact. Again, nothing.

How could this be hazardous? Hazard lurks when the return path is interrupted, and the shell is the hot or switched hot contact—especially when the outlet box is not well grounded. There is a similar problem if the center contact is the hot contact, the outlet box is a bad ground, and the shell is backfed from another circuit. Here, the extension cord test will save you. In either case, making the assumption that the outlet is dead can mean getting zapped. How? A bad ground is not the same as no ground. A little pressure on the box, just the wrong amount, can cause it to make better contact with an armored cable connector. Or starting to loosen an 8-32 canopy or cover plate screw can cut it free from the paint that has insulated it. Then, suddenly, it is well-enough grounded to shock you when you touch it and something live. This is why an extension cord is so useful for testing—but it is not sufficient to use your voltmeter, with the cord, to check only the center contact for voltage.

Backfeeds that make the shell live even when the fixture is correctly wired can have various causes. One involves a grounded conductor that is broken upstream. The power is fed into it from a load on a different outlet on this circuit—or a circuit properly or improperly sharing its grounded conductor. Another, which you will see in Chapter 4, comes from a switching system that totally ignores the "center contact = hot contact" convention.

The obvious, if seemingly radical, counter to this danger is to kill all power to the building. If you choose to do so, warn the customer of the need to reset clocks and timers (including those associated with stoves, setback thermostats, and computers) and answering machines that lack battery backup. Make sure that the people paying your bill are not unduly inconvenienced. See to the following:

☐ Computers are turned off, unless they have uninterruptible power supplies with ample batteries;

☐ Important processes—from fine machining to laundry drudgery—are not stopped where they should not be;

☐ Refrigerated perishables are not put at risk;

☐ No one is endangered by being caught in the dark.

Unfortunately, even operating the main disconnect is no guarantee of safety. In some instances, even pulling the meter is not enough. Later chapters will talk more about some of the ways in which old buildings can trick you. Suffice it to say that with old wiring, you can be even less sure than usual that when you think you have killed a circuit, it is dead, or that when you think you have shut

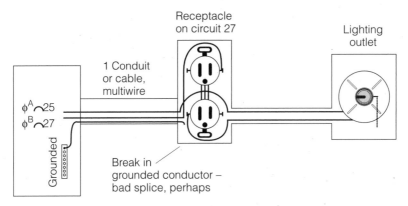

The lighting outlet on circuit 25 actually could operate on up to twice the normal voltage;because of the break in the grounded conductor, it can only complete its circuit through the receptacle and circuit 27, dividing phase to phase (or hot leg to hot leg) voltage with whatever load is plugged into the receptacle.If this were not a legitimate multiwire circuit – if #25 and #27 came off the same leg or phase, it could not operate without a good grounded conductor.

■ **2-3** *Here is how backfeeding can result in double or zero voltage between your two conductors—because they are both hot.*

down everything in a building, all power is off. Whether or not you are knowingly taking chances, it is always wise to assume you are dealing with live wires.

Despite the fact that there is no absolute guarantee that you can kill all power, you need to locate the circuit on which you are working, both for safety and to ensure that the problem is not one of a tripped overcurrent device. There are clues that can help you shortcut the search for the right circuit. Most old buildings, alas, have miserable circuit directories—when circuit directories are present at all. "Plugs and lights" may be the legend on five out of six circuits, and even when that information is correct, it provides little assistance. You can gain some tentative clues by considering the equipment that continues to work. If you are troubleshooting a light, as in the present example, throw breakers or pull fuses to find out whether adjoining rooms' lights are all on the same circuit, or whether the lights and receptacles in a room tend to be on the same circuit. If so, the function or lack of function of adjoining lights, or nearby receptacles, lets you make a good guess about where to start.

Now use that extension cord. First, plug it into a known live outlet, preferably an outlet that is on a different circuit altogether (but ideally one coming from the same busbar), and test the other end of the cord with your voltmeter to confirm that you find both hot and grounded conductors. It is relatively rare for both halves of the circuit to go simultaneously, even at a much-abused ceiling outlet. Therefore, your extension cord will let you trace the circuit using either the grounded conductor or the hot one from the outlet itself, and the other conductor from the cord. Then you can look for the originating loadcenter and circuit.

If power comes up to the light through the switch, many experienced folks are inclined to test there first, for two reasons. One is the general principle that you start troubleshooting at, or closer to, the source of power. The other is that it makes more sense to work at arm height rather than from a ladder, other factors being equal.

If all the outlet offers that you can trace back is a grounded conductor, disconnect it at the loadcenter. With a 50' extension cord and a 25' drop light, you will often be able to power a light or a radio from a combination of the suspect outlet plus a good source, and monitor the light or radio from the loadcenter. You know when you have disconnected the identified conductor associated with the right circuit, because the light or radio goes off. If you are thus able to confirm the identity of the cable or raceway, you can disconnect or pull the fuse or flip the circuit breaker feeding the hot wire or wires.

In really large buildings, teams have used telephones for this monitoring. For instance, you might plug a radio into the outlet you are checking, or in a pigtailed receptacle combining the one good lead from that outlet with a lead going into your extension cord (which is plugged into a working outlet), and use a telephone near that location to dial to a portable telephone (connected to a different number), or to a telephone number represented by a jack located near the panelboard. Then, as you pull things apart at the panelboard, the cue that lets you know you have been successful in locating the offending circuit is that your radio goes silent.

Patent circuit tracers come in handy, too. They are discussed in Chapter 14, "Resources." Ultimately, however, the best tool is the intelligent, informed mind—which is not to deny that accessories are useful.

In some cases, you may be unable to find a single grounded conductor that seems to correspond to the return wire you found at the outlet. This could mean someone indiscriminately spliced together the grounded conductors of various circuits at some junction box. If that is the case, it is probably wise for you to trace all of those grounded conductors involved, after disconnecting them. Make a note, mental or otherwise, that they need to be untangled as soon as you have a chance. Once you have disconnected them, you then have the arguable joy of checking out all the hot conductors associated with the wires you have disconnected, to figure out which might serve the nonfunctioning outlet.

A possibility more dangerous than that of commingled grounded conductors is that the installer may have used a ground return. In that case, you may be stuck with working live, or with killing power to the entire building—if kill it you can.

Be alert to nonelectrical hazards

Now that you are really ready to remove or separate the fixture from the outlet, protect yourself! OSHA rules aside, there may rarely be a need for a hard hat on a construction job where no one is working above you, but deciding that does not mean you should scorn protection—goggles, respirators, and, yes, sometimes protective clothing (including headgear). In old houses, anything can rain down. Okay, "anything" is an overstatement. Try this list:

dust;

plaster in various forms including powder, chunks, and slabs;

sawdust;

termite castings;

cockroach castings and eggs;

mouse feces;

rockwool;

asbestos;

old newspaper;

crumbled wire insulation;

nails;

tools;

former food;

sand;

stripped cable armor;

dead light bulbs;

broken glass;

pens and pencils;

scraps of sheet metal;

used tissues;

and pieces of brick.

What is that trash doing between an electrical box and a fixture? Most of it has no business there, but sometimes it is there nonetheless. Complain to the mayor. If it is not between the box and the fixture, it may be alongside the box, and drop down because a bit of ceiling comes loose when you disturb it. Protect yourself. Protect your eyes, protect your respiratory system, protect your head and body and hands. (Chapter 17, "Hazards and Benefits," has plenty more on the risks associated with this work.)

Recognize the fragility of age—the risk of damaging building structure

Speaking of that bit of ceiling that comes down when you disturb it, who is responsible? One answer is Father Time, but that answer may not satisfy a less-reasonable customer. Here again you might want to refer to Chapter 15, "Communication." It helps to point out the risk of damage in advance. Old plaster is more brittle than new, and certainly falls apart more readily than plasterboard. Plaster on wood lath is more apt to come loose in chunks than plaster on wire lath, because the keys are fewer. And old wood is more brittle, more apt to crack, than is fresher wood.

You can minimize the risk of unnecessary damage by using the same tricks you apply when working on newer buildings. Use a sharp knife to score or cut paint; that is one example. Some swear by a putty knife, instead. A putty knife will pry, without the risk of gashing. On the other hand, a utility knife is less likely to gouge or rip. Banging equipment to free it can work well with, say, a cover plate glued to drywall with excess joint compound; however, doing so carries greater risk to older, more fragile wall and ceiling surfaces. Older fixtures, similarly, are often more fragile than new ones.

Here are just a few of the fragile bits you could discover, to your chagrin.

☐ The separator between the actual socket and a metal lampholder, which is basically paper or light cardboard, has dried out and crumbles.

☐ A cracked, or at least crazed, glass fixture part such as a shade is ready to break apart when handled.

☐ Similarly, a wooden fixture part has dried out to the point where it is ready to disintegrate.

☐ And, of course, rubber and cloth parts turn to dust.

One more caution before you take that fixture apart. Be very careful in this part of the process, in those cases where you are not absolutely certain power is off. This is when you can so very easily turn a bad ground into one with sufficiently low resistance to draw an arc, right in your face. This is when you can so very easily twist or pull a conductor so the wire breaks, or the insulation breaks away and you can get shocked, can even be grabbed by current exceeding your "let-go" threshold. Okay, it is not a 480 volt industrial-quality shock, but most shocks that kill are 120 volts. And just a strong tingle or a minor zap can knock a person off a ladder or jerk handheld pliers right into his or her smile. Be especially wary in old buildings with metal ceilings.

There is one more category of risk. Besides damaging the building as you take things apart, and besides damaging yourself, there is the risk that you may leave things in more dangerous shape than they were when you started. This is not a discussion of what to do if wire insulation falls apart in your hands; the present concern is more subtle. The January-February 1997 issue of *NFPA Journal* described a Texas church fire, causing well over a million dollars' worth of damage, that resulted from minor carelessness on the part of electricians performing repairs. They worked on a spotlight, mounted on a swivel base, located near a wood-shingle roof. After they left, the light rotated down, so the lamp aimed at the roof from a few inches away. After two nights, the shingles ignited.

Without investigating the situation, it would be plausible to speculate that a locknut had been insufficiently tightened. That is all it would take. A safety-minded contractor is careful to do the job right. Going further, he or she asks, "What could go wrong?" The vulnerable parts are then double-checked.

Proceeding inside an old outlet box

Consider these issues as you unscrew the screws holding the fixture to the ceiling or to the outlet box, or unscrew the bezel ring to get the canopy down.

Okay, the fixture is down. Now you go a-hunting. The easy case is where the fixture comes away, leaving the wire ends in the box. Yes, they do get that rotten, and many is the time an electrician has wondered how a fixture managed to conduct power to and from lamps as long as it did. The answer is usually that the break did not matter so much, because the broken ends were still in contact. Mind you, it was a high-resistance contact, and could have caused a fire, but by golly the light lit!

In this easiest case, you found the wires broken off right at the lampholder. Sometimes the problem is not so evident that it will yield to a quick visual inspection.

Here is the good news: it is unheard of for a wire to simply break inside a cable, in the middle of a run, buried somewhere within the wall. Oh, there are cases where a cable gets pierced by a nail or screw, which cuts and sometimes opens or shorts a conductor—but those are rare exceptions. Wires almost always fail at or near connections. That means, in most legal wiring, that they die inside enclosures—or at least at accessible locations.

Why do wires die, so conveniently, inside enclosures? Sometimes they were nicked in the original process of stripping off the insulation. Sometimes they were shoved back and forth, back and forth, inside the enclosure. Sometimes they were twisted and untwisted as splices were made and remade. At outlets, as opposed to junction boxes, they almost always are subjected to the heat of being close to where electricity flows across a resistance—the resistance of adjacent splices; the resistance of receptacle terminals, prongs, and cord connectors; the resistance of lampholders plus lamp filaments. Switch boxes are intermediate in the heat stress they put on connections. All this explains why, even if the problem is not immediately evident, you are probably looking in the right sort of place.

Before you assume that there is a hidden problem, it is worth your while to test again now that you have pulled things apart. Practically speaking, that means putting power back on if you have it off. If possible, you will turn it off again before digging your fingers into the wires as you proceed further. The reason to test again for power is that screws do get loose, and if the connection to the lampholder involves terminal screws, it would be awfully silly to let a little corrosion or loosening mislead you into getting hurt.

Troubleshooting old splices

To test for voltage (or for continuity), you may need to get into the splices. At this point you may discover a significant difference between old and modern materials. What is likely to go bad with old, soldered-and-taped splices? Very, very little. The conductor insulation may well be brittle, but that bad insulation, not the connection, will be the reason to consider rewiring. Splices have gone bad inside the tape—wires broken or making bad contact inside the splice. However, that is quite rare. Just about the only way for a twisted-and-soldered splice to go bad is for the installer to have done a bad soldering job. Of course, if the splice has been pushed back and forth, back and forth, fatigue might easily have broken off one of the conductors outside the splice—but just inside the tape! That is something you may come upon more than once. Similarly, wires have broken off just at the screw terminal of a lampholder, switch, receptacle or fuseholder. Old splices are discussed further in Appendix 2.

Broken wires

Before you turn power back on, do check for broken conductors elsewhere in the enclosure. On rare occasions, wires have broken off inside their original insulation. Just like thoroughly fractured or dislocated bones in a person's arm, these may manifest their presence by making sharp angles. Wires bend in curves unless you apply lots of precise force, say with your needlenose pliers. So if a conductor makes an angle, it is suspect. If nothing is quickly evident, and you are sure power is off, you can check the conductors for undue flexibility. With a broken wire inside, rubber-and-cloth (more precisely, rubber-and-cotton braid) insulation or, in high-heat locations, asbestos-and-cloth insulation, bends quite freely compared to rubber-and-cloth insulation with continuous #14 copper inside. It is easy to confirm the break—old insulation is not strong enough to keep you from pulling it apart unless there is a wire inside to give it some backbone.

One more way to check for breaks as well as for faulty splices, if you are confident that it is safe for you to do live testing, is to poke through the insulation with the sharp tip of your voltmeter probe. An alternative, if you would rather not work live, is to use your ohmmeter or continuity tester. That means probing through the insulation in two places, or, for conductors terminated at, say, the lampholder, probing through the insulation at the point where a conductor enters the enclosure, and again where it is terminated. Why unwrap the entire splice, and then have to redo it? The insulation has always appeared intact after the tester's pinpoint was withdrawn. At most, a wrap or two of tape has made up for any injury.

Example: no power at the outlet

It is time to return again from the general case to a specific example. Suppose the source of power does not come through the switch, which merely contains the feed and the switch leg. (You may know them by other names; what is meant is that the grounded conductor was not brought through the switch box.) You find no power at the switch. You find no power in the lamp socket, nor do you find power at the terminals at the back of the lampholder. (You did need to check that, because it is always possible that the installer, or somebody messing with the system subsequent to its original installation, arranged that the switch broke the return rather than the hot conductor. Chapter 4, "Some Esoteric Switching Layouts," will look at even stranger possibilities.)

This example will be relatively simple. Assume that the light you are examining is the only one controlled by its single-pole, single-throw switch, and that there are only three splices in the box at the lighting outlet, to which three cables are connected. (This assumes that, as is usually the case with old wiring, there are no insulated grounding conductors, or that these three splices are in addition to the splice combining the grounds.)

Here is what you find as you track the wires. One splice contains two conductors: one of those heads out of the box in what will be called Cable A, and the other goes to the lampholder—ideally to the center contact. Assume—but you always need to check such assumptions—that the first splice constitutes power returning from the switch, the "switched hot lead." The second splice contains three or more conductors; one goes to the lampholder, and the others head out of the box in Cables B and C. That second splice should contain the grounded conductors (another assumption). The third splice contains a number of conductors, none of which goes to the lampholder, but one of which seems to leave the box in Cable A, the one suspected of going to the switch. That certainly should be the splice containing the hot conductors coming from the source of power and continuing downstream. Even if it turns out not to be the right splice, it certainly is worth testing for voltage if you do any live testing. Given that you found no power at the lampholder, and given that the other two splices connect to the lampholder, where else could you look within this outlet? Since, unless you have encountered a Chapter 4 scenario, one splice seems to contain the switched conductor, and the other should be the return, the third ought to be unswitched power. Hence it is the furthest upstream point at the outlet that you can check for power.

Suppose you do not get a voltage reading at that third splice. This means either that there is no power at the box or that it was not wired the way one would expect. You need to check somewhere else. With solder-and-tape installations, probing each conductor going into the other splices can be at least as useful as probing the splices themselves. Unless you are pretty clear that the circuit is dead, or are pretty clear as to which outlet power is coming from (which outlet, in other words, is the next one upstream) and can check to see if that one is dead, your best bet is to assume that power is getting to the box, but not making it through some single conductor or connection. That working hypothesis means that the first thing to check for is a conductor that does have power. Then you look for the break.

Evaluating and dealing with bad wires

What should you do when you do find broken conductors or bad insulation? The problem is not all that different from the one you face when you are simply replacing an old fixture, and have to undo and redo an old splice. There, too, insulation cracks and crumbles. Wires are too short. Or wires are barely long enough, but need to be shortened because the conductors inside are damaged or the surrounding insulation is gone.

Those terms require definition. A damaged conductor is one that is nicked or in some other way reduced in diameter. When is insulation "gone"? You can pretty well rely upon this rigorous test of insulation condition. It is simple: fold a conductor double and straighten it. If this does not crack the insulation (or fatigue the conductor to the point that it breaks), you might figure it has a few years left. At least this section of it. Do recognize that any testing is a stressor, and will bring marginal equipment that much closer to failure. See Chapter 6, "Safety Surveys," for a more extensive discussion of evaluating wiring systems.

Healing old conductors Suppose the wires fail your test. What should you do? You could throw up your hands and say the whole place needs rewiring. See Chapter 7, "Setting Limits and Avoiding Snares," for discussion of the pitfalls inherent in taking halfway measures at correction, and Chapter 8, "Tearing It Out and Rewiring," for how to carry out a more radical choice. Here are a few other options. They may be less ideal than rewiring, but in some cases they are more realistic.

Some electricians talk of stripping the insulation off new wire out of the truck, and slipping it onto old wires. You may not find that approach very satisfactory. For one thing, you really still need to tape the transition between the new slipped-on plastic insulation and the old rubber or cloth-and-rubber.

Around the middle of the twentieth century, there was another clever system for repair of deteriorated wiring. What was commonly known as rag wiring has rubber insulation with a cloth, or braided cotton, layer over it. (Other fibers may have been used as well, and the discussion will henceforth use the term, "cloth," for convenience.) The latter was primarily for mechanical protection of the rubber, because although dry fabric has a high dielectric rating, the cotton or hemp cloth can absorb moisture, greatly reducing its insulating value. There used to be a product which looked a bit like that cloth covering, but was used as insulation repair. Woven of glass fiber, it expanded slightly as you pushed it over conductors, but tightened over them when you gave it a tug. It worked like the old children's plaything, the "Chinese Finger Trap." If you do encounter any installation using it, meaning wire that seems to be covered with just a cloth-like material, it is probably ancient enough that the insulation was crumbling by the middle of the twentieth century. Do not expect it to be repairable now, shrink tubing or no. On the other hand, a similar product was still being manufactured in the 1990s, for OEM use. It was supplied to and by Ruud Lighting for installers to apply to raise the temperature rating of conductors feeding some of Ruud's lighting fixtures.

Another old option, but one that is still available, is to heal the old insulation. Of course that is a misnomer. Insulation is not living; it cannot be healed. Fussing aside, insulating varnish can be applied over tired rubber, and that certainly will perform an adequate job of insulating. Take warning: the varnish is brittle when it dries, and sticky and messy to apply. It is also possible that the solvent is toxic, and perhaps intoxicating, to breathe in any quantity—not to mention flammable. Still, varnish is one of the best choices when you can not rewire. It is important to point out that there are a number of liability issues. First, do you trust that this will leave a safe job? Second, has the product you are using been investigated for this specific purpose? If you are not sure, look on the container, or contact the manufacturer. Supply house salespersons rarely can give you a definitive answer, but they can usually find you a phone number for the manufacturer's representative.

Heat-shrink tubing is another fine choice, but you do need to keep in mind its shortcomings. When you heat it, it shortens longitudinally as well as circumferentially. This means you have to make sure you cut a length that extends far enough towards the end of the bare wire and, at the other end, far enough over reasonably good old insulation, that the tubing does not creep away too far as it tightens. Also, you are applying considerable heat, which raises a concern about flammability. You are not very likely to have problems using heat guns and hair dryers; however, many have also used matches—and that is taking a risk.

Tape is a distant third choice. It is fine when you can fully reach in and apply it correctly, but it is not the equal of shrink tubing or even varnish, particularly where access is limited. If you do tape, please color code the conductors. It does not cost you any significant money to keep a roll of white and a roll of green tape in your toolbox as well as the black. Red and blue will do you no harm, either.

Taping old wires can be tricky. Make sure the tape does not pull off old insulation. Both ends of your tape must adhere to something solid. Usually this means sticking your tape to the conductor itself at one end, and to relatively resilient insulation at the other. If the insulation is dried out "all the way till morning," the situation calls for rewiring.

Just in case you did not have the benefit of the same fine apprentice training some enjoyed in this regard, here is a bit more detail about taping. Double-wrap taping is an extra-trustworthy system. It is not complicated, just careful. The tape is applied under enough tension to make it closely follow the contours of what you are taping. Each wrap of the tape overlaps the previous wrap by 50%. In consequence, you get a double thickness— which ensures that there are no gaps or voids even if the tape has minor defects.

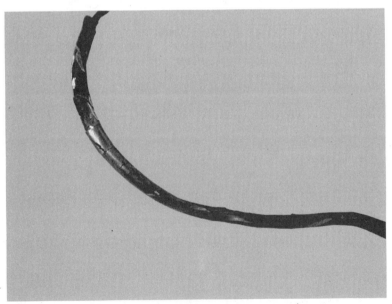

■ **2-4** *When the insulation is this bad, you are far, far better off not attempting to repair it.*

Those are the basic fixes. A little anticipatory healing can be a good idea, even if the insulation is not yet actually cracked. Not infrequently, you will find a rubber-and-cloth insulated conductor whose cloth protection is fairly shot, all loose, and in the way, but whose rubber insulation itself still has some life. That conductor too deserves some additional protection such as taping, because the rubber was not intended to do its job without the mechanical protection provided by the cloth covering. In that circumstance, cut any loose- and-useless cloth out of the way and apply tape or heat-shrink tubing over those conductors.

The very best fix is pulling fresh conductors into the box, despite the fact that you cannot rewire the whole occupancy or building. If you are dealing with pipe or tubing as opposed to cable, it is absurd to waste the time "putting band-aids on" dying or overly short conductors. Pull in new ones unless there is a strong reason not to do so.

What could be a strong reason not to pull new wires? Here is an example. A customer with just that kind of rotten-all-the-way wiring preferred taping rather than replacing conductors because he did not have major cash to spend right then. His crowded ceiling boxes had wires heading out here, there, and everywhere. If the electrician had started replacing conductors, he would have been going for days—just to clean up one outlet. Taping the cracked insulation at the one box where he was having trouble put at least a temporary lid on his problem, without disturbing the wires at four or five other boxes. Here are some of the other factors that went into the electrician's choice:

☐ The communication was absolutely clear; the customer knew that this was not a permanent, long-term fix, and was willing to take the risk. The electrician documented his concerns and advice.

☐ The occupancy was of fireproof construction.

☐ If the electrician had out-and-out refused to take any halfway measure, the customer would have employed a handyperson to try to get things working. While the threat of unethical or incompetent competition must never determine how you work, it is unquestionably true that you do not work in a vacuum.

☐ And the most important factor of all is that the electrician was able to tape the wires in such a way that he felt confident that they were unlikely to short out.

Another reasonable, competent electrician might well have insisted on all or nothing.

Unfortunately, when you are dealing with cable rather than conduit, rewiring is not as easy a matter as using the old wires as drag lines to pull in new from the reel. As will be discussed later, you usually can cut back on the cable sheath or armor if you tussle with it and haul more into the box—or if you replace the box with a bigger one, so the cables need not be quite so long. If you have not attempted this before, you will be surprised at how much better the conductor insulation usually appears just a couple of inches outside the enclosure, where it has not been subjected to the same heat.

Receptacle troubleshooting and replacement

So far, this has focused on troubleshooting old lights. As mentioned at the beginning of this chapter in the discussion of silly little preliminary tests, it is easy to extrapolate from old lights to old receptacles. Receptacles are like lampholders rather than switches in that they have no internal moving parts. (At this point, the discussion is ignoring lampholders and receptacles that incorporate switches.) Also like lampholders, receptacles' failure often results from incremental contact deterioration or loosening.

When you replace a receptacle in an old wiring system, there are a couple of things to watch for that are relatively rare with newer systems. One is a lack of color coding or polarization. Use your voltmeter, and if you feel particularly conscientious, add a little white tape to color-code the grounded conductors, turning them into properly identified conductors. Another is lack of grounding. Bonding (discussed below) is easy to achieve, but how much good does it do if there is no ground present?

Grounding is a major contributor to electrical safety. Given how common three-prong cords are, the need for grounding receptacles is quite evident. In the general consideration of grounding, NEC Section 250-42 (a) exempts equipment that is not within eight feet of grounded surfaces, and does not fall under one of the other subsections of 250-42. Nevertheless, NEC Section 410-18, in the area that focuses on lighting fixtures, requires the grounding of lights, and does not include a similar exemption. Ground them. For one thing, when a light has a brass pull chain, if that chain reaches to within eight feet of a grounded surface, judicious electricians consider that to fall within the NEC's intent for grounding under Article 250. For another, someone relamping an ungrounded fixture who comes in contact with some grounded surface is at risk of shock if the fixture has a short but no ground.

Consider a person who is standing on a washing machine to change a lamp—or someone standing on a chair to relamp, but brushing against a steam pipe.

There are a number of solutions to the problem of ungrounded receptacles. Before getting to them, here are two changes that you are required to make when replacing a receptacle. If the receptacle is on a multiwire circuit, NEC Section 300-13 (b) requires that you pigtail the grounded conductors rather than letting their continuity rely on the receptacle terminal screws. This requirement may be extended to all receptacles in the 1999 NEC. This is a case where you can holler all day long about how the design is preexisting and thus should be grandfathered, but you will get little sympathy. The reason is that the fix takes one solderless connector, a few inches of white wire, and a few minutes. This change is easily as important as the other change that supersedes any claim of grandfathering: the Section 210-7 (d)(2) requirement that any replacement receptacles in locations where GFCIs are now required be GFCI protected.

Bonding equipment on which you have worked

Before getting back to grounding, it is important to talk about bonding. Once you have dealt with a wiring system, you need to take what measures you can to make it safe. Bonding is not very difficult. The NEC used to consider a device adequately grounded and bonded when it was held by a 6-32 metal screw to a grounded metal enclosure. That changed. Now even two 8-32 metal screws are considered adequate to do the job only if so Listed or if they press the device's yoke against a metal enclosure that is surface-mounted. Most authorities having jurisdiction will consider NEC Section 250-74 Exception 1, permitting metal-to-metal contact to bond receptacles, and by extension, other devices, to have been met if the enclosure is merely protruding, rather than actually surface-mounted.

The best solution is to bond the receptacle to ground wires that are spliced together and secured to the enclosure if it is metal. If there are no ground wires, but the enclosure is grounded, the next best solution is bonding to the enclosure using a 10-32 or larger screw in a tapped hole. If there is no prepared grounding screw hole, you can drill and tap one. Or if there is an extra, unused, cable clamp, you are highly unlikely to get a complaint from an inspector if you remove that and use its screw (with a washer, if it seems necessary because of the shape of the screw's head) to bond the box.

One other solution is to use a self-grounding receptacle, which has a spring clip (Listed, although in some people's opinion flimsy) that ensures positive contact between the mounting screw and the yoke. Yet another solution is to use a bonding wire secured to a grounding clip, which you work onto the side of the box. Grounding clips can work okay: they are Listed devices. However, it is also easy to install them in such a way that they do not make solid contact, or in such a way that they are forced off the box as you button things up. This makes them pretty much a last choice.

Grounding

Grounding is a lot more complicated than bonding. Bonding simply ensures that a device yoke or body remains at the same potential as the enclosure. Grounding brings them to an approximation of ground potential so touching them while grounded would pose no hazard. If an old enclosure does not seem to be grounded, the solution will depend on the reason you find no ground.

There are three common reasons for an enclosure to test as ungrounded. One is poor contact because of interference. Perhaps your tester could not find ground, even though it is present, because of dirt, paint, or corrosion. The solution is to scrape away the interference—then bond to clean, shiny metal. If necessary, restore corrosion protection such as enamel (spray paint normally will suffice) after making your connection.

A second cause is loose contact between the enclosure and the metal system feeding it. For this reason, even if an enclosure appears on testing not to be grounded, go ahead and test the conduit or armor for ground, if one of those is the wiring method. Armor held by a loose clamp or connector can itself be grounded without grounding the enclosure. Similarly, conduit held on the inside of the enclosure by only a bushing rather than being secured by a bushing and paired locknuts, out and in, can make lousy contact. This can make you think you have a bigger problem than you do.

Then there are unprofessional installations pure and simple. The contact is poor because the installer was unconcerned or inept. In some of these illegal messes, the cable or conduit is merely stuck in through a knockout. Or armored cable is sort of secured with an NM/loom clamp. In other old jobs, someone who was ignorant even for his or her day tried to secure a locknut against the side of a round box. (This is not likely to be a problem with receptacles as much as with lights; but you may run into it, especially with surface-mounted equipment.)

There is no mystery to handling these problems. Tighten that locknut or setscrew. Add a second locknut. Install the correct connector, in an appropriate opening. Make sure the armor is squarely under that armored cable clamp, and carefully cinch it down. Keep a few two-hole, wraparound armored cable clamps in your truck, for replacing those that do not quite do the job. Keep a few nonsetscrew armored cable connectors around, too. Setscrews can push armor right through frail insulation.

Suppose you still find no ground. If everything seems secure at this end, and the wiring method, on the face of it, should provide a ground, find the other end of the cable or raceway. Somewhere along the circuit, grounding has been interrupted. If necessary, start at the panelboard and work forward. You usually will come upon a case of poor contact. If doing so is at all possible, correct that poor connection. This will be a far better—safer and more professional—job than one of the kludges (slapdash solutions) suggested below.

What about those cases where you just cannot find an interrupted ground, or where the wiring method itself does not allow for grounding? The latter is the third common cause for a missing ground. In the 1996 NEC, Article 336 still defines nonmetallic-sheathed cable as an assembly of two or more insulated conductors; the ground is optional even now. At one point it was uncommon. Until very recently, knob and tube wiring, traditionally two-wire, was still a code-recognized method.

In such cases, you face an array of not fully satisfactory choices. The one that is absolutely certain, as always, is to yank out what is in place and rewire. If that is not an option, the next best choice is to run a grounding conductor, adequately protected, to a reliable ground.

That was a loaded sentence. What is adequate protection for a grounding conductor? Fishing it in the walls is fine—unless some carpenter comes along and cuts into the wall, or a plumber fishes (meaning, in this case, rams) a pipe through the same space. The latter is a low-probability occurrence, so, all in all, fishing is probably the most reasonable choice.

The next question is what constitutes a reliable ground. Plumbing pipes used to be strongly recommended. They have fallen into disfavor because of the increasing popularity of nonmetallic plumbing. When a bad section of copper is cut away and replaced with

CPVC, the plumber has no way to know that he or she has disabled your life-safety grounding.

The most reliable ground is the grounding terminal in a properly installed loadcenter. Its ground is either the grounded service conductor, a feeder ground, or in the worst case, at least a grounding electrode conductor. These grounding paths are far less likely to be disrupted than those of the plumbing system, or even of other parts of the electrical system.

One practice which you should never engage in, and which you need to be on the lookout for, is the use of ground wires and bonding where there is no ground. Suppose that you extend a non-grounding circuit using a wiring method that incorporates a ground. Now suppose that the outlet from which you extended the circuit suffers a ground fault. It will not trip the overcurrent device, because there is no ground return path. Instead, it will cause that enclosure and cover plate to become live if metallic, and the same with the enclosure and cover plate you wired in downstream.

The danger may be even worse if the downstream outlet is three-prong, which can be legal provided that it is GFCI-protected—but it may be three-prong regardless of legality. If the upstream outlet is the one with the GFCI, and the ground fault is from its line side hot wire to its enclosure, the GFCI will have no reason to trip, and anything connected to those third prongs will become live! It will stay live until something draws enough current to operate the overcurrent device: for instance, 15 amperes for several hours, or 30 amperes for a couple of minutes. For that reason, when you install a GFCI device (as opposed to a GFCI circuit breaker) to permit you to use three-prong receptacles on an ungrounded circuit, take an extra step: tape over the line-side hot terminal after attaching the wire, even if doing so is not commonly your practice.

For further discussion of grounding and bonding, see the end of Chapter 5, "Accept, Adapt, or Uproot."

"Troubleshooting" is the longest chapter in the technical section of this book. Before moving on to the next few chapters, which highlight how differently electricians used to design installations, consider thumbing ahead to the first two appendices. They were pulled out from the middle of this chapter so that the text would flow smoothly and give you the sense of looking over an old-timer's shoulder as he troubleshoots an actual wiring installation. Appendices 1 and 2 contain basic information that may help you troubleshoot old wiring.

Design Changes

THIS CHAPTER ADDRESSES TWO PROBLEMS COMMONLY encountered in working on older structures. One is technical: the accumulated changes that accrue in a building's electrical layout. The other is social: the painful necessity of sometimes saying, "This is no longer legal; not only is the design unsafe, but I could jeopardize my contracting license if I restore it." (If you are an inspector, the equivalent statement is, "Sorry, that is not acceptable in this jurisdiction.")

Handling these problems carries forward the work of Chapter 2, where you started to tussle with an old installation you were troubleshooting. Handling the first problem, accumulated changes, means guessing the logic that guided decades and generations of electricians and home repairers or maintenance workers. The second takes you ahead to the territory of Chapters 6 and 7, "Safety Surveys" and "Setting Limits," as well as Chapter 15, "Communication." Of course, you have much less need to concern yourself with these matters if local rules require you to perform a complete upgrade.

Mysteries

This chapter starts off with a troubleshooting example. You will find, though, that much of the same exploration, intuition, and calculation are required when you are called in to make additions to old wiring, as opposed to repairs.

Puzzling switches

Here is what you are confronted with, in the customer's words. "There's a funny switch in this room that doesn't control anything" or "The light in my closet no longer works." "I want to know what this wire is" poses a similar challenge. Happily, in many of the latter cases one glance will tell you that the wire in question must serve a doorbell or a telephone jack. That lets you reassure the customer right quickly; unless someone was terribly creative in material selection, the wire or cable poses no hazard even if you are unable to trace where it leads. (Yes, this might not be true if it terminates in a hazardous location. If you are dealing with a gas station, or some other type of customer with classified zones, please keep that caveat in mind.)

Some simple possibilities

What could that funny switch be? Any number of things. It could be a switch that was bypassed by a previous electrician, to change a receptacle from switched to unswitched. It could be a three-way switch bypassed so that its mate acts like a plain snap switch.

Blanking it off would have been a better job. Now that you are the person on the spot, it is up to you to make that change. If it is a multigang switch, eliminating it is not quite the simple matter of replacing a single gang switch cover with a single gang blank cover. You have three or four choices. First, odd combination covers are manufactured. For instance, you can purchase a two-gang cover with one side blanked and the other having an opening for a single switch. Second, switch blanks or dummies are manufactured. They consist of a switch yoke with a plastic rectangle located where the switch handle would be. They are expensive for do-nothings, presumably because of the low volume sold; but they do let you use a standard multigang switch cover plate. Third, at least one company makes snap-together multigang cover plates, whose sections can be mixed and matched as needed. Fourth, where legal, you may be able to replace the switch with a receptacle.

If the "funny switch" is not one that was intentionally disconnected, what else could be the solution to your mystery? It could be a switch that has failed either "on" or "off." You know what to do then: replace it, test the replacement, and that is that! You will find worse cases, though, in older buildings. They represent examples of what happens when an electrical system is monkeyed with, and a building's layout is modified indiscriminately, over the years.

Sometimes you get to play the masterful magician and solve the customer's puzzle with a wave of your wit. A simple solution presents itself, and makes you look awfully smart, when you ask, "Is there any light or receptacle outlet—or even just half of a receptacle—that seems to work off and on, without rhyme or reason?" and the customer says "Yes." Even the customer can check, under your direction, to find out if the mystery switch controls that receptacle or lighting load. In some instances the light is outside, or on the other side of a stairwell door, so no one put together the idea of the switch and the light. Far goofier, and more connected with the checkered history of older homes, is the possibility that somebody installed a wall between the switch and the outlet it controls. How was anybody to even think that the dead receptacle in Joe's room was connected to the dummy switch in Sally's? Greater expertise will be needed in those cases where the switch operates unreliably, intermittently. In those cases, bypassing the switch, or even replacing it, may be the best way to confirm the diagnosis.

Similarly, if an outlet has been mysteriously dead, it could well be that the mystery switch controls it but failed in nonconducting mode so power could not make its way through to the outlet. That, too, is easy to test. Remember, though, that the test may require installing a fresh lamp, if what has been long dead is a lighting outlet.

Messier possibilities

Those are relatively easy situations to identify. The next couple of possibilities are real messes. The lighting outlet could well have been buried—whether the original fixture was ceiling or sconce. Yes, NEC Section 370-29 says that it is always illegal to bury an electrical box so that it is not accessible—not to mention burying it where its location is unknown. No one has gone to jail for doing so. And when the person doing so is a carpenter or home repairer or maintenance worker, he or she is not even putting an electrical license in disrepute.

How do you find a buried box? It can be tough. Where you notice a suspended ceiling, do be alert to the possibility that a light was "buried" there. If you see a suspicious dimple in wall or ceiling, that is worth investigating. If there is no outlet at the center of the ceiling, give that location a very close look, and perhaps tap against the wall or ceiling to listen for suspicious echoes. A patent circuit tracer can sometimes help you track cables and thus locate terminations.

If you do find a buried outlet, the proper response is to expose it. A buried cable end with no box is even more of a problem. Terminate the cable in a box, and blank it off, except in two circumstances. One is where you identify the circuit and can make use of the cable to serve an outlet. The other exception is where you can remove the cable entirely. One must, naturally, apply this rule with intelligence. If the cable is embedded in masonry, it is reasonable to insist that it is dead before you leave; it is unreasonable to demand that it be chopped free in order to terminate it in an accessible box.

You probably do not risk your license if you do not open up the sort of mess represented by a buried cable, but correcting it is certainly the more judicious course. And if your customer refuses to authorize that work, document your recommendation, and leave off that piece of troubleshooting. In Chapters 5, 6, and 7, you will find further discussion of "Last man in" doctrine—the idea that if you do not speak up, the customer may assume you are somehow conveying approval of what already exists.

Although this next situation is more rare, a dead light or receptacle outlet can mean a buried switch. The way this can come to be is that a "funny" switch, such as the one described above, is dismissed as not doing anything. So someone removes the switch and plasters over the box.

Somewhat more commonly, a multigang switch box has a blanked section, or a disconnected switch, that used to control the outlet that has you puzzled. Or maybe the switch you are looking for is not disconnected, but redirected. If you have reason to believe that another, switched, outlet is a relatively new innovation, it is quite possible that the switch that controls the new outlet is the self-same one that used to control the formerly switched outlet you are troubleshooting.

Problems created by ignorance about boxes and wires

The solution to the case of a removed switch can be as simple as replacing the missing switch. Rarely will you be that lucky. More often, a switch was removed because its space was needed for some other purpose. In that case you have several options. These options also apply on jobs where you are not troubleshooting but need room in the a box because you have been called in to add an outlet or switch at that location.

The simplest solutions entail some compromise of function or rely on combining devices. If it is acceptable to your customer that the wall sconce and the ceiling light come on together, you have eliminated the need for one switch. If it is acceptable to your customer that the switches operate horizontally rather than vertically, devices combining two switches (or a switch and a receptacle) on one yoke are readily available. Interchangeable, Three-in-one, or "Despard" devices are archaic. They are addressed later. It would be unwise to add them nowadays, because they will constitute a nuisance for the next electrician.

Finally, if unit switching is acceptable, that too can free up a switch for another purpose. Acceptability must take the NEC into account: beware of the customer who encourages you to put both paddle fan and light on one switch. It is too likely that the customer will on occasion want the fan on and the light off, and thus will use the light switch on the fixture itself. If the fan-light combination is the sole switch-controlled light in the room, the room will no longer comply with the requirement in NEC Section 210-70(a) that habitable rooms have lights controlled by wall switches.

Adding room to enclosures

Rather than compromise or consolidate, you can expand. If you are dealing with a sectional switch box, you can cut open the wall and by hook and by crook—that is, by wiggling and contorting your fingers and probably scraping a knuckle—add another

■ **3-1** *Adding a Surface Metal Raceway (e.g. Wiremold™) multi-gang box to extend a single-gang box*

section to the box, unless it extends from stud to stud. Of course, if the box in the wall is nonmetallic or simply nonsectional, this option is unavailable. A less attractive approach is to add a surface extension, using a surface metal raceway box, perhaps multigang.

If the box in the wall does not permit you to add sections, for instance if it is a square box, you are not stuck. You may be able to add a second box adjacent to it, using a chase nipple, or a short nipple with locknuts for spacing. When you use this option, you have to make sure you respect the rule in NEC 300-20(a) about not inducing currents. Also, beware of spacing the new and old enclosures in such a way that the cover plates overlap or line up unattractively. (Using an offset nipple can result in the latter problem.) See also the section called "Crowded boxes" in Chapter 5.

Suppose you find that an outlet is dead, but its enclosure contains both hot and grounded conductors. A switch leg—and only a switch leg, no third wire to carry a grounded conductor along— leaves for points unknown and does not carry return current. If you truly cannot find where the switch leg goes, it may be time to consider disconnecting it and capping the wires. (They still count in your box fill calculations.) If the load requires switching, one option is running a new line to a new switch—if and only if box fill calculations permit adding the additional two conductors to the enclosure. Box fill and box replacement are covered in detail in Chapter 5, "Accept, Adapt, or Uproot," and in NEC Article 370.

Grandfathering issues

Now it is time to shift topics back to a different kind of decision. You have seen some discussion of what to do about deterioration and about careless installations. In old buildings, changes in doctrine, in how things are done, frequently present at least as much difficulty as do bad or damaged work, whether or not those doctrinal changes are reflected in the NEC. This is addressed further in Chapter 12, "Inspection Issues."

Backwired devices

Here is a 1993 change that could result in some inconvenience. Backwiring of receptacles and switches is now legal only with #14 solid copper conductors. You may still have in stock devices marked for backwiring with #12 conductors, or still be able to purchase them, but that use is no longer permitted—even

46

if the switch or receptacle you are replacing was backwired with #12.

The possible inconvenience results from not being able to utilize one advantage backwiring can provide. When conductors are on the short side, backwiring sometimes allows you to make a good connection anyway. This rule change means you must pigtail those short conductors when they are #12 AWG. On the plus side, the result is a better job.

Closet lights

How about that closet light? You know that according to the NEC, most lights in closets would no longer be legal to install. Or do you really know that? The statement that they would no longer be legal is true only if most closets are relatively shallow clothes closets. That is only clothes closets. There is nothing but common sense to keep you from installing or repairing or replacing a bare porcelain lampholder right over the shelves in a pantry-type closet, a cleaning closet, even a linen closet! The NEC permits them, although there have been repeated proposals to close that loophole.

Not only are clothes closet lights restricted, they are not readily grandfathered. Most inspectors will not consider it okay to repair or replace a fixture located where its heat could conceivably set piled clothing on fire.

Solutions to the problem of closet lighting

Having said that, there are cases where a generous inspector will okay replacing a keyless or pull-chain porcelain in a clothes closet. What may make the choice acceptable is that you insert a PL adapter and tell the customer that only compact fluorescents are to be installed henceforth. Not all inspectors would go for that solution, and if you personally are uncomfortable with a fix that is permitted but less than optimal, by all means stick to your guns.

A bare lamp fixture in a clothes closet is an example of something quite easy to repair that a cautious electrician nevertheless usually will refuse to repair. It constitutes too great a fire hazard. Does that mean you should leave the customer stranded? That would not be very good business.

There are any number of safe, reliable solutions to the problem. A simple, if stern, approach is to say, "Such lights are no longer legal, and here is a reason you can understand. If you need a light

47

there, I can suggest possible solutions. If not, I should blank off the closet ceiling outlet."

The reason for blanking off the outlet is to reduce the incentive for some amateur to get it working, possibly with sloppy, high-resistance splices and terminations. A tidy job of blanking off an outlet, with a smooth, 5" cover, also reduces the likelihood that someone will remove the fixture and plaster over the outlet. Yes, you are drumming up a few more minutes' work. But in doing so, you are providing a genuine service to the customer.

Incidentally, a porcelain or plastic lampholder of the type that installs over and covers an octagon box or mud ring is not formally considered a fixture, and thus, according to one source, was never ever acceptable in a clothes closet. Other sources disagree and believe that the "lampholders" that cannot be used as free-standing fixtures are the brass-shell type that are installed as pendants, or as part of table lamps. Pendants, you may be surprised to learn, were legal in closets into the 1930s, and installation of other open incandescents was officially kosher until adoption of the 1990 NEC.

Here are some of the solutions you might suggest to customers who need light in a closet. A cheap, low-technology one for inelegant, occasional-use closets is a battery-operated light stuck to the wall. Sure, batteries die. Sure, this answer is not pretty. But if they just do not have the money, or do not want to spend the money, to hire you to do something more elaborate—something probably involving a permit—it is a dirt-simple fix.

The next solution, for those few closets that are shaped so as to accommodate it legally, is to replace the bare-lamp fixture with an enclosed incandescent, or perhaps a fluorescent. In many jurisdictions, such a straightforward swap also will not require a permit. If the wiring is in decent condition, the swap takes very little time.

The next solutions involve moving the fixture. These normally do call for permits and inspection; hence they are intrinsically more expensive. Many times (but not always) customers are far less concerned about unsightly equipment on a closet ceiling than out in the open. This means that the solution can sometimes be as simple as putting a surface extension over the outlet box, with a blank cover, and running cable from the extension box to a fluorescent strip over the closet door.

If the closet dimensions are such that even this is not legal, one solution remains. Put a light where it is unquestionably legal—

outside, shining in. Because you usually can work from inside the closet even though the outlet will be in the ceiling outside, you can still avoid an offensive mess that would require professional wall or ceiling repair.

Grandfathering: the bigger picture

There are a number of Code issues involved in grandfathering. The NEC is not automatically law. Adopting it or some alternative code, amending it, updating it, and above all interpreting it, are actions that rely on local authorities. Yet the NEC is all the law that can be addressed here. To find out how it applies in your jurisdiction, you have to check your local situation. In some jurisdictions, there are no authorities enforcing the NEC. Conscientious contractors practicing in those areas face unscrupulous competition with less backup than they deserve.

There are a number of criteria to consider when it comes to grandfathering. Was this installation ever legal? Is it dangerously deteriorated? Is it palpably unsafe for some other reason? How much do you need to change it to make it legal? How much work are you going to perform on it, and to how great an extent will the installation be modified? Does the NEC explicitly require updating the installation, as is the case when you replace outlets that nowadays require GFCI protection? Chapter 12, "Inspections," goes into greater detail on many of these issues. They certainly need to be resolved before you offer a customer a quote.

Defining your responsibilities with regards to what exists

Breaking the questions down, if an installation was never and will not ever be legal, your getting it to work can land you with responsibility for the whole mess in the event that anything ever goes wrong (depending on the lawyers). If an installation was otherwise legal but installed without a permit, you can similarly be landed with responsibility for it—unless you squawk. Some jurisdictions will allow you to apply for a permit, noting that someone else began the installation. They may send an inspector out before you begin, to tell you what changes need to be made and what you can let slide.

It is not your responsibility to engage in historical research to determine whether something preexisting originally passed inspection. Here are two considerations to guide you, in the form of questions.

☐ Can a reasonable person take the installation as having been professionally installed—perhaps, admittedly, by a competent but sloppy professional?

☐ Is it reasonable to assume that the original installer pulled a permit and passed inspection—albeit perhaps by an inspector whizzing through "on roller skates"?

If the answers are yes, you will not be faulted, by most reasonable people who apply the Code, for proceeding on that basis. If the answers are no, you would be wise to check with the authorities, unless you are planning to tear it out anyway. A very conservative lawyer might offer more stringent advice.

Clearly wrong installations do not require any detective work to evaluate. Suppose your jurisdiction has forbidden new knob-and-tube wiring since 1965. If you find an old light controlled by a relatively new wall switch that is fed by knob-and-tube thermoplastic-insulated wiring, you have found a violation. It is probably the work of an old-timer who never looked at a Code book after getting a license. You would be wise to at least have a word with the inspection department before you begin work on that mess.

An example: overfusing

Some installations you will encounter are technically illegal, but really not so bad. Is a 30-amp plug fuse okay on #14 branch circuit conductors (putting aside motor loads, covered under Article 430)? Maybe yes! How so? An oversized fuse is not a bad idea when used on the grounded conductor in an old, grandfathered fusebox that fuses both hot and grounded conductors. However, you should strongly discourage customers from leaving extra 30-ampere fuses around, so they do not succumb to the temptation to use those to "protect" the hot conductors when they blow fuses. Of course, grandfathered or no, if the old fuse box is service equipment, with a main bonding jumper, and there is sufficient room on the ground or neutral bar, you would be far better off moving all those grounded conductors into spare terminals on that bar. There is a sound reason for the rule in NEC Section 230-90(b) against fusing grounded conductors .

While on the subject of fuses, if there is any evidence that the customers have overfused, take it as your responsibility to install Type S or Fustat adapters, except in two circumstances. One is where the fuseholders are shot. If any internal connections in the loadcenter appear to be damaged or otherwise high-resistance, you have the responsibility to tell the customer they cannot be used. Usually, but not always, this means the loadcenter needs to be replaced.

The other situation where you would not install a no-tamper adapter to prevent overfusing is where the customer has installed a minibreaker. Minibreakers require Edison-base sockets, but a 15-amp minibreaker is also relatively unlikely to be unwittingly replaced with an oversized fuse.

More Specialized Issues and Situations

An Esoteric Switching Layout: the Carter System

THIS CHAPTER IS DEVOTED TO VARIATIONS ON ONE OLD, long-illegal but long-employed, wiring layout—multipoint control that economized by alternately switching the hot and the return conductor. Carter System or "Lazy Neutral" switching (it goes by other names as well) is an example of something that went from being legal and commonplace, to many decades of being illegal (violating, for example, NEC Section 380-2) and commonplace, to the present, when it is quite unfamiliar and outrageous-looking to newly trained electricians. Yet at one time it was very attractive because it offered an awfully clever solution to the challenge of doing the most with the least. It certainly was used for decades past the time when it became illegal. By 1920, the NEC required that all wires of a circuit be in the same cable, at least when run in armored cable.

A modern, legal parallel to Carter System wiring is the practice of saving time and materials by running multiwire circuits. Multiwire circuits can be fully code-compliant but do not constitute quite the tip-top job that separate two-wire branch circuits produce. Of course, even the latter are not quite as flexible as a Rolls Royce of an installation, consisting of individual branch circuits, or at least radial wiring, to each outlet. Every installer must draw the line somewhere. An illegal, and even more dangerous, parallel to Carter System wiring is the use of a ground return.

What Carter System switching accomplished

Here is one problem that Carter System switching solved, and solved very nicely. An old-time electrician found a pull-chain light fixture in the ceiling, or no fixture at all, and noticed receptacles scattered around the walls or baseboards. The customer wanted to control that light (or, if there was no light, wanted to install a new light and control it), from each of two room entrances. Of course the electrician wanted to economize by using the least cable possible. Carter System switching allowed the electrician to do the job using one conductor less in each run than otherwise would have been required. How that worked will be described a little later.

Here is another problem the system solved. That old-timer found a switch-controlled light, with power—both hot and grounded conductors—coming through the switch box rather than merely running in a switch leg down from the light. Wanting to add control from a second switch, the electrician figured that it would be awfully nice not to have to rerun a three-conductor cable from switch to light, or around from a second switch on the other side of the room to the first switch. Carter System switching satisfied that wish, too.

The logic of the design—how the switches work the lights

Carter System switching allowed electricians to do these things using two conductors rather than three, but there was a price.

To understand the price, look at how Carter System switching functions, when it is working. Sometimes the light is on and receiving power just the way it would as you wire it nowadays: the center contact is connected to a hot conductor, and the shell is connected to a grounded conductor. At other times, with Carter System wiring the light lights because it is fed backwards: the shell is connected to a hot conductor, and the center contact is connected to a grounded conductor.

Assuming that the lamp and fixture are good, when the lamp is dark, sometimes it is for the reason you would expect with correct modern wiring: no power is going to the lamp. Sometimes, however, the lamp does not light because hot conductors are connected to both the shell and the center contact! In other words, it has power, but not a complete circuit.

In short, one price the system always exacts is that lampholder (or receptacle, should you encounter a Carter System-wired receptacle) polarity goes out the window.

How the wires are run for multipoint switching

Describing the operation is not the same as explaining how raceways were wired, or cables run. Shortly, this discussion will get down to the nitty-gritty.

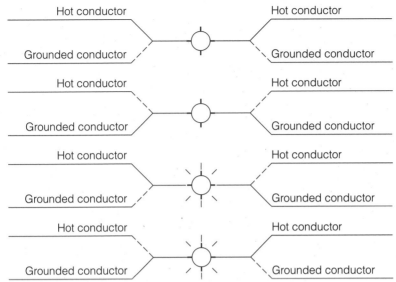

■ **4-1** *How lamps go on and off under Carter System switching.*

Standard multipoint switching

For contrast, here is how multipoint switching is now arranged. Normally, the hot wire coming from upstream feeds power into the COMMON terminal of a three-way switch. The switch directs the power to one of two TRAVELER terminals. The second switch is fed by conductors coming from those TRAVELERs. At any moment, one of the two TRAVELERs is connected through the COMMON terminal of that second switch to the nonidentified conductor or the black or coppery terminal at the switched outlet. The grounded conductor proceeds to the identified conductor lead or the white or silvery terminal of that switched outlet, uninterrupted.

Carter System lash-ups

Carter System wiring is different. One TRAVELER terminal at each switch is fed by the hot, the other by the grounded conductor. Thus, depending on handle position at any moment, each switch can connect its COMMON terminal to a hot or to a grounded conductor.

How does that translate into what was described at the lampholder? At the switched outlet, the conductor from one switch's COMMON terminal is connected to the lampholder's shell contact, and the conductor coming from the other switch's COMMON

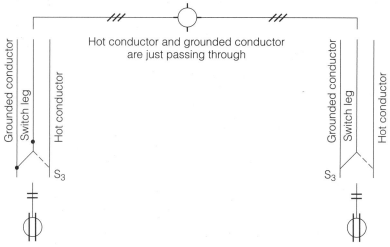

Hot conductor and grounded conductor
are just passing through

Grounded conductor | Switch leg | Hot conductor

S_3

Grounded conductor | Switch leg | Hot conductor

S_3

■ **4-2** *How lampholders are fed in a Carter System hookup.*

terminal is connected to the lampholder's center contact. If the switches are used to control a receptacle (it is fairly common to find receptacles as part of wall sconces), the polarity of the blade slots, and any conductor color coding, are meaningless and indeed misleading.

Fancier Carter System setups

A Carter System installation that employs more than two switches to control an outlet is rare, but you may come across one. In that case, you will find a hot and a grounded conductor fed into one four-way switch, two wires from that switch feeding any other four-way switches, and then the last four-way switch feeding the TRAVELER terminals of one three-way switch. The other three-way switch will be fed by a hot and a grounded conductor, either directly from a source of power, or, similarly, through one or more four-way switches. The three-way switches then connect to the fixture as described above.

Recognizing Carter System switching

An old troubleshooting rule says that it is almost always wise to start the investigation from the source of power. With switching systems, you initially will not know whether that source (the furthest upstream point of the circuit) is located at the lighting outlet or at the switches themselves. As mentioned in Chapter 2, "Troubleshooting," it often makes more sense to start your search at the switches. There are two reasons. First, since they have moving parts, switches can be expected to wear out. Second, they are usually easier to access—no ladder is needed. (There are arguments in the other direction, which will be mentioned anon.)

Looking inside the switch boxes

Carter System switching was most commonly employed where two or three multipoint switches were ganged together. The reason is twofold. First, the installer gained greatly in efficiency by reducing the number of TRAVELERs that otherwise would be needed. Second, one hot and one grounded conductor could feed all of them.

It is rarely hard to identify the presence of Carter System switching at such an enclosure. You will notice that several three-way switches have two of their three terminals wired together. You will find one of two things. Either there will be two splices, each of which contains one wire coming from each of the three-way switches, or wires will be daisy-chained from two terminals of one to two terminals of the next. If you are able to easily examine a switch, look for markings. If the terminal that is not connected to the other switches is marked, COMMON, that proves the case.

In standard three-way switching, in contrast, you might find one, and only one, terminal connected switch to switch: the COMMON. (Grounding terminals, the hexagonal green screws used for bonding to the box, of course have nothing to do with circuit logic.)

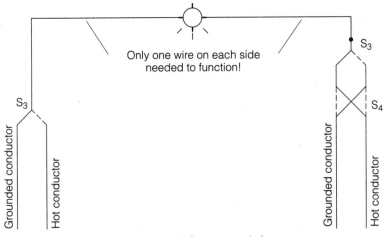

■ **4-3** *Carter System switching with four-way switches.*

Sometimes it is not clear where all the wires go, because the boxes are so crowded, the terminals are taped, or the wires are very short and hard to get at. It also can be quite hard to read the markings on switches, either because they are marked with faint embossing, or because they are not accessible without pulling the switches all the way out. Suppose you are sure at least one terminal on each switch is connected to the others but cannot readily see from markings on the switches which terminal is which. Then you lack a definitive indicator. The joined terminals could be the COMMON terminals, all connected to a hot conductor. You need to either pull out a switch or a splice, or test with your voltmeter, with power on.

Here is another ambivalent case. Suppose you find one three-way switch, in a single gang enclosure. It is fed by at least two cables or conduits, so you know that it contains more than just a switch leg. The conductors connected to two of the switch terminals come

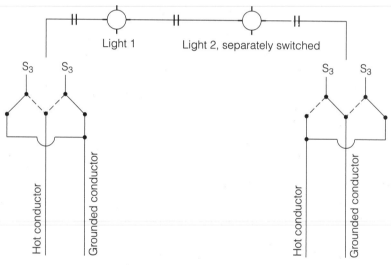

If there is one conductor (other than a grounding conductor) in a switch box, looped — or spliced and pigtailed — from one three-way switch to another to another, it could be the hot wire legitimately feeding the COMMON terminals of all of them. However, if two conductors follow such a pattern, it is quite a safe bet to assume that they are the hot and the grounded conductors of a Carter System setup. Identifying the terminals to which they are connected as TRAVELERs is merely confirmation.

■ **4-4** *Ganged Carter System switches. The markings on the terminals of these ganged switches usually are not very accessible, but you can identify the setup without reading them.*

from splices that contain more than one incoming wire and a pig-tail. In this case as well, you should strongly suspect presence of Carter System switching. Here is how to check. Presuming that you can safely test with power on, remove loads (for example, un-screw the lamp from the lampholder) to avoid backfeeds. Now confirm that one of the splices at the switch, or one of its terminals fed from a splice, is hot and the other is at ground potential; con-firm that the positions of the switch handles, both there and at the switch at the other end of the stairs or across the room, do not af-fect those readings. You have identified Carter System wiring. You can confirm its presence up at the light.

When you are troubleshooting, positive identification of the wiring system employed may not be quite that certain because your meter may mislead you. The hot or ground potential reading that should be present at one of the terminals could be missing because of a bad connection somewhere. Fortunately, when lighting systems go bad, switches are the most common culprit. Therefore, in most cases, you will find hot and grounded conductors at the switches.

If you are uncertain about a switch, you can always kill power, dis-connect the leads, and test the switch in isolation, using your conti-nuity tester or ohmmeter. At least, then, you know whether the switch is the source of the functional problem. Sure, on relatively rare occasions you will trace the problem to a bad splice or a broken conductor. Those problems have been touched on in earlier chapters.

The fact that the switch is usually the culprit makes the problem easy to cure, in theory. You find a Carter System-wired three-way switch with one apparent TRAVELER terminal hot, and the other at approximately ground potential. If the third, COMMON terminal does not alternate between those two polarities as you flip the handle, you have a defective switch. Do not replace it, though, un-til you have read through the discussion of fixes, a little further down. There are additional considerations.

If one switch tests out okay, you may need to try its partner. Just as with normally wired three-way switches, there are several failure modes. If one switch continues to conduct but stops switching, the other will be constructively turned into a single-pole, single-throw switch—possibly upside down, but otherwise working fine. As there will be reason to point out again a little later in the section on fixes, people more often than not rely on only one switch of a pair. Therefore, it is likely that, if the defective or dying switch were the unused or less-used one, the problem would not be noticed—not until the second switch also failed.

The other failure modes are simpler. The other common failure mode of a three-way switch is to stop conducting altogether through at least one of its contacts. If the inoperative contact is the one the switch is attempting to connect, the light will stay off until the switch is replaced. The third type of failure is one that few encounter: there is a short between the switch's two TRAV-ELER terminals. With Carter System wiring, that failure is a bolted fault, sure to trip the circuit's overcurrent device.

Troubleshooting at the switched outlet

If both switches test out okay, you have to investigate the light. There can be other reasons to start there. Perhaps you needed to get up on your ladder anyway, to check the lamp. Perhaps you are earnestly trying to avoid disturbing the expensive wallpaper abutting the switch plates. Perhaps the light fixture is a chandelier, with wiring that is visibly ratty. How can you identify Carter System switching from the lighting outlet?

There are several ways to identify Carter System wiring at the outlet; you may be able to do so even without taking it apart. Suppose that you find the lamp did not light because it was not making good contact—corrosion is clearly visible. Your tester scrapes through it, informing you that power is making its way up to the light. Without taking anything apart, if with the lamp removed you find both shell and center contact hot, you have fairly strong evidence of Carter System switching. Other cases are less clear. You have weaker evidence if their polarity is reversed, with shell hot and center contact at ground potential; and none if it is normal. (The latter case does not, however, prove the absence of Carter System wiring.) Flipping the switches and retesting at the light will provide further information.

There are other possible explanations for each of those findings. Reversed polarity can be just reversed polarity—the result of someone's ignorance, or haste, or of someone misreading the color coding of very faded insulation and not checking with a voltmeter. It is not rare for the white conductor to be hot and the black to be grounded. If both contacts are hot, it could be that the grounded conductor is broken somewhere upstream along the circuit, and another load is backfeeding to the light. (The reverse could be true if both test out at ground potential.)

Going into the enclosure behind the switched outlet will not automatically tell you more about whether you have Carter System switching. Sure, if the local source of power does not come up through the switches, but enters the room or area right there at the

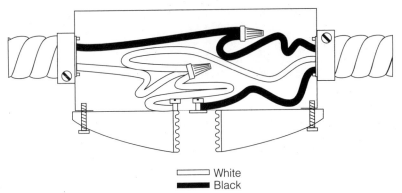

White
Black

■ **4-5** *A splice in the grounded conductor, going directly to the lampholder; it is proof that you do not have Carter System switching.*

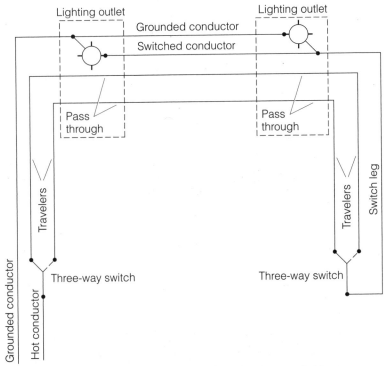

■ **4-6** *Paralleled lights controlled by modern three-way switching.*

lighting outlet box, that box may contain a grounded conductor splice from which a pigtail feeds the fixture shell. Finding that is proof that what you are investigating is definitely not Carter System switching. Even if that splice feeds the center contact, you simply may have a case of reversed polarity. That is still a different, and simpler-to-fix, violation than Carter system switching. In most other cases, especially when the hot or the grounded conductor is interrupted, the lighting outlet will not answer your question about the presence of Carter System switching.

One setup that might cause some confusion because the lighting outlets look the same whether there is standard or Carter System switching is where the paired switches control multiple lighting outlets simultaneously. In that case, splices may feed both lamp-holder terminals. From those splices, conductors continue on to other lighting outlets that turn off and on together with the one on which you are working.

There are several cases where the system used may not be apparent. If conduits or cables that contain several conductors enter the outlet box, and the lampholder is not fed from splices, you may not gain further information as to which switching system was used. In that case, you may have to use your tester at the switches.

A closer look at the Carter System's drawbacks

Before considering fixes for malfunctioning Carter System installations, consider the down sides of that system. There are several drawbacks to weigh, in case you are tempted to try for a grandfathering exemption that would leave Carter System switching logic in place. How deadly are those differences from modern wiring?

Polarity loss

The loss of polarization is a significant drawback to Carter System switching. It is nice to know that, when you have switched off a light, power is no longer going to it. That protection makes lamp-changing safer. In modern wiring, if a light is on, as soon as you start to unscrew the lamp, you disengage it from the source of power as it withdraws from the lampholder's center contact. That protection is part of what is lost when polarization goes out the window. Furthermore, without proper polarity future repair persons can get hurt if they rely upon conductors' color coding.

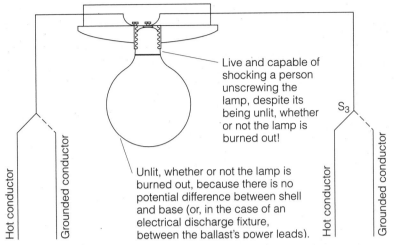

Live and capable of shocking a person unscrewing the lamp, despite its being unlit, whether or not the lamp is burned out!

Unlit, whether or not the lamp is burned out, because there is no potential difference between shell and base (or, in the case of an electrical discharge fixture, between the ballast's power leads).

Hot conductor

Grounded conductor

S₃

Hot conductor

Grounded conductor

■ **4-7** *A shell that can shock.*

Induced current

There is another cost besides the loss of polarization and color coding. Section 300-20(a) of the NEC requires that conductors be grouped together so as to avoid even local induced currents. Carter System wiring normally violates this provision. Theoretically, each switch could send conduit up to the fixture containing a single conductor.

More to the point, despite the fact that each cable or conduit contains several conductors, with Carter System wiring only one may carry power. Two real problems could result, and one possible. The first of the real problems is that current could be induced, and possibly flow across resistance so as to generate heat, whether in cable armor, in the walls of a metallic electrical box, or between armor, cable connector or clamp, and box. The second is that by inducing current in the surrounding metal, this configuration could result in a choke effect on the power wiring, causing an undervoltage.

Cancer risk?

The one problem identified as "possible" hinges on the validity of current concerns about bioelectromagnetic effects. It seems likely for several reasons that the occasional associations drawn between normal electric field exposures and bad health outcomes are spurious—that the electromagnetic exposure which seems to be associated with illness is a surrogate for some real

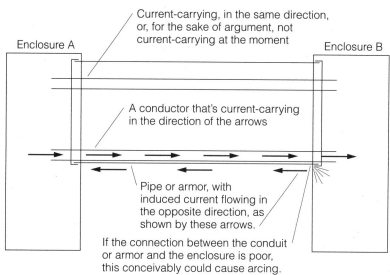

Enclosure A

Current-carrying, in the same direction, or, for the sake of argument, not current-carrying at the moment

Enclosure B

A conductor that's current-carrying in the direction of the arrows

Pipe or armor, with induced current flowing in the opposite direction, as shown by these arrows.

If the connection between the conduit or armor and the enclosure is poor, this conceivably could cause arcing.

■ **4-8** *Induced current.*

hazard. Still, if the 500-and-first study finally fixes blame on the power lines, any design that causes power to generate excessive exposure, for instance to travel to a light fixture by going up inside one wall and returning in the opposite wall, is pretty bad. It bathes the room's occupants in an electromagnetic field, by putting them inside the inductor.

Whether or not that danger is real, some customers believe in it and are willing to spend a little more to address their concerns. Even if you are not dealing with that worry on a particular job, these installations are clearly in violation of the NEC. It is time to consider fixes.

Even if it is broke, must you fix it?

Before you look at some creative solutions to the dangers posed by Carter System switching, it is worthwhile asking whether you really are required to rectify the situation or whether it is grandfathered. Some would say it is enough to restore the lights to service as you would any other problem caused by a broken switch, damaged conductor, loose termination or splice, or failed fixture (whatever generated the service call). That decision is between your conscience, your customer, and the local inspection authority. Lots of authorities having jurisdiction will say, "Gee, I'm not going to make you rewire the whole room just because a switch has gone bad." Others will quote Fernandel: "The [current] law is the law."

How they enforce the law can vary a great deal, as is discussed in more detail in Chapter 11, "Inspection Issues." The same inspector who will let you replace a switch will not necessarily let you replace a fixture. The circumstances are different.

There are arguments on each side. By one argument, it is okay to replace the switches, but not the fixture. You are making zero changes in the system's components when you replace a switch with an identical-to-all-purposes switch. Anybody working on the associated fixture will see that the fixture is old, runs this argument, and, accordingly, will be careful with assumptions about how it is wired.

By this first type of reasoning, fixture replacement is treated differently. (In a few jurisdictions, all fixture replacements call for permits, but device replacements do not.) Contrary to the situation with switches, if someone notes a new fixture, and knows or assumes that a professional electrician was the installer, he or she will be more inclined to assume that polarization is normal, and that unscrewing a lamp cannot result in being zapped by the live shell. That rationale is one way of evaluating Carter System repairs.

Some see things the other way around. It also can be argued that it is okay to replace the fixture, because there is nothing wrong with the fixture's wiring *per se*. The switches, on the other hand,

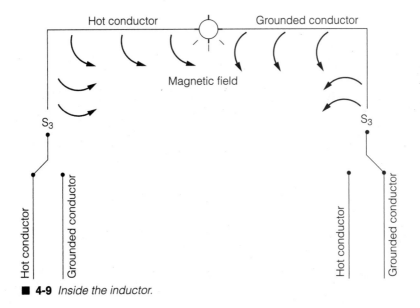

■ **4-9** *Inside the inductor.*

are the source of a problem and in fact are in violation. Therefore, if you replace one switch, you need to correct the whole setup.

Which argument is right? This is a case where there is no right or wrong. Experts say, "This is enforcement, not Code." When that is the case, it is helpful to step back and look at the bigger picture ("I don't care if the glass is half-anything; it's the wrong size glass."). The bigger picture is that Carter System wiring is old (presuming that it was installed back when such a design was legal), and relatively unsafe. That is all you should need to know in order to choose to upgrade Carter System switching to something more modern. Still, no one can lay down the law, saying you absolutely have to either replace it or walk off the job.

Fix-it options

Now it is time to explore how you might be able to modify the system. There is an assortment of options.

Elimination

Here is a simple, quick fix: use one single-pole, single-throw switch. Users of most rooms with multipoint switching do not really call on both switches, or, with four-way switching, all of the switches. If you eliminate all but one switch, you will be left with a switch box from which two conductors go up to a lighting outlet. You can deal with that, if Code permits—meaning that the second switch is not mandatory.

Here is how you can do it. Consider this layout. Suppose the source of power is at one switch. From there, a three-wire cable brings a hot conductor, a grounded conductor, and a switched conductor up to the light. From the light, hot and grounded conductors continue on to various other outlets along the circuit. The other switch receives a three-wire cable, which feeds it hot and grounded conductors from the lighting outlet (conductors that the second switch sends along downstream) and allows it to send a third, switched, conductor back to the light.

You want to change this from Carter System switching to single-location control. Eliminating one switch means that, from whichever one the customer decides to keep, three conductors will continue to the lighting outlet, with the switched one going to the lampholder's center contact. The shell will now be connected directly to the grounded conductor. The switch location that the customer decides to dispense with has its switched conductor disconnected and capped at both ends, its switch removed. Its cover is replaced with a blank. Alternately, its switch can be replaced with a receptacle, if legal on that circuit at that location.

It is possible to visualize exceptions, where this option may not be considered. If your paired switches are at either end of a staircase, NEC Section 210-70 (a) does not allow you to simply eliminate one. There are other cases where it will be similarly unsafe, or otherwise unacceptable to the customer. In some such cases, you indeed will have to run cable.

Remotes

Readily-available equipment offers you a few more solutions, though. Maybe one will appeal to your customer as being preferable to rewiring.

X-10

The first is low-voltage or powerline carrier switching, covered in Chapter 9, "Relatively Rare Situations." Electronic switching may be familiar to you from the controls that come with some paddle fans, controls that allow you to independently operate fan and light using two-wire switching, even from multiple locations. There are other systems similar to X-10.

Sound activation

Another easy fix is what some are inclined to misname a "snap" switch, because you can operate it by snapping your fingers. Eliminate all but one of the switches and replace that one with a sound-activated switch, and the light can be operated from anywhere in the room. Or, unfortunately, from the next. Depending on a customer's habits, those sound-activated switches have been known to operate inadvertently, in response to noises that were not directed at them.

Other sensors

Lutron's high-end "Spacer Personal Space Light Control™" pairs a stand-alone dimmer with a hand-held wireless remote control. No doubt there is a corollary to Murphy's Law saying that hand-held controls such as television remotes are always put down in the least convenient spot. Still, this system, occupancy sensors, and wireless versions of powerline carrier systems all provide alternatives to fishing, tearing up walls, or surface wiring. Occupancy sensors, if placed properly, certainly can be acceptable to occupants and authorities having jurisdiction as alternatives to switches at each landing.

Accept, Adapt, or Uproot?

THIS CHAPTER TALKS ABOUT DISTINGUISHING WHAT YOU can change from what you must change. It introduces two problems: being asked by a customer to make changes that will bring a system into violation; and dealing with crowded enclosures. It begins by looking at a challenge specific to the installation of fans and heavy chandeliers: retrofitting support required by the NEC. Chapters 6, 7, and 8 explore these issues further, from different angles.

The following discussion focuses on ceiling fans, but much of it applies to other utilization equipment and outlets as well. Suppose that your customer says, "I'd like you to hang this paddle fan in my old building." On the face of it, this is a very reasonable request. Paddle fans, located appropriately, are efficient air movers and can add greatly to comfort. Using very little electricity, they pose minimal strain on circuits and services. Since the late 1980s, though, doctrine has changed; fans now demand considerably more support than do most lights, as the electrical industry—and the NEC—have recognized that fans were dropping from the ceiling unacceptably often.

Until the Code changed, there was a rather convincing test that some would perform whenever installing a paddle fan. The electricians would perform pull-ups on the fan or fan support, chinning themselves as best they could, and make sure that nothing gave way. The Veterans' Administration used to rely on a specification that fan supports be sufficient to hold up a 150-pound static load. A 250-pound electrician utilizing the rough-and-ready test just described would make an even greater demand on the support.

Boxes

The old approaches no longer are necessary. The Code did change, and to simply replace an old light, suspended from two 8-32 screws attached to box ears, with a fan—or a heavy chandelier—is illegal.

Now paddle fans are to be supported only from boxes Listed for the purpose. Okay, electricians and inspectors also know that it is perfectly all right to rely on a Listed fan support that is secured above the box. Likewise, although it is not explicitly stated in fan installation instructions, nobody competent will give an installer grief for mounting fans directly to structural elements. That approach is indeed best of all.

In one small way, you may be better off in an old house than a newer one when it comes to mounting paddle fans. It is far more common in older structures than in newer buildings for a ceiling outlet to be mounted directly under a joist. On the other hand, it is far, far more common in older designs for that ceiling outlet to use an undersized enclosure, one that you should consider replacing.

Crowded boxes

Here a whole new issue is broached: volume. Pancake boxes, shallow round boxes, came in two sizes: small or tiny—$\frac{5}{8}$" deep by 4" or $3\frac{1}{4}$" in diameter. This means the volume often only allows two #14 conductors. There can be no devices, no hickeys, no ground wire that did not originate in the box. Apparently, they were strictly for ceiling-hung fixtures, though they have been encountered elsewhere. Certainly, conscientious electricians only used them with canopied fixtures, even before volume markings appeared on those canopies intended to be counted in box fill calculations.

Does this mean that you must never replace a light mounted over a pancake box without replacing the box? That decision lies between you and the local authority having jurisdiction. It is true that many pancake boxes, because they contain very old rubber-insulated conductors, are worth pulling out just so that you can get to good insulation. The thermoplastic insulation used in most installations during the latter part of the twentieth century is less commonly brittle or cracked. It is also true that, unless you damage something in the process, replacing an undersized box with a modern, deep one makes a better job.

You can be a wee bit more generous when you forget about boxes sandwiched between the bottom of the joist and the ceiling surface. Take a ceiling box, 3" or $3\frac{1}{4}$" round or octagonal by $1\frac{1}{2}$" deep. It is no longer even listed in NEC Table 370-16(a). Very commonly it will have been used as a junction box. Say it has one cable bringing power in, one carrying power onwards, one going down to a switch, cable clamps, and a fixture hickey (the hickey is attached to the hanger bar supporting the box between ceiling joists). Is the box illegally overcrowded? Probably, but do not be too hasty to judge. It depends.

Assume that the conductors are #14, and the cables are armored cable: there are no grounding conductors—or bonding either, for that matter—in the cables. Every item—each of the 6 wires, the pair of clamps (combined), and the hickey—demands two cubic inches. This comes to 16 cubic inches all told. The maximum volume the box can possibly offer is about 14.5 cubic inches. Still, if a fixture canopy were marked as adding 5 or 6 cubic inches, the total volume might suffice. Such a marking is quite unlikely to be present. Nonetheless, if the volume added by an unmarked canopy makes the job acceptable to the local inspector, that is almost as good. Replacing a woefully undersized box with a close-to-adequate box will improve matters greatly, and may be acceptable to the authority having jurisdiction, given the fudge factor created by the canopy.

Unless you are able to confer with the inspector ahead of time, though, you should not count on passing inspection with such a crowded enclosure. Now comes the question of how to add volume to a ceiling box. (You may be able to apply this to some wall outlets as well.) If the box is sufficiently, illegally, recessed that you need to add a box extension to bring it out to flush with the room surface, the situation impels its own solution. When this option does not present itself, use one of two methods: either move the box back up into the ceiling so you can add a box extension, or else replace the box. The latter is a far better job, but a little riskier to undertake and certainly more expensive.

Replacement: the benefits

With some boxes, replacement is the only reasonable choice. Even when the case is not quite that extreme, replacement is often worth considering. Here is the great advantage of replacing a box. Normally, you are replacing the overcrowded enclosure with a wider box, not merely a deeper one of the same diameter. (Oh, if what was there was a miniature pancake—a box 3" round by $\frac{5}{8}$" deep—indeed you almost certainly will be going to at least standard $1\frac{1}{2}$" depth; but the diameter will also grow). The increased diameter means that even if the wires are way too short, even if they do not look too good, meaning that the insulation is shot, you have given yourself some leeway to remedy those problems. You can cut back the cables at least half an inch to an inch to gain access to more wire and to get to insulation that has not suffered the same stress. You can free up another half inch at least per conductor by changing over from internal clamps to connectors and locknuts. Even better, having pulled the old box out of the ceiling, you very frequently can loosen a staple or two so that with care you can safely haul a couple more inches of cable slack into the box.

Replacing the box with a bigger one can make quite a difference. Doing so can mean changing from a box containing crowded splices of rotten conductors two inches long to a roomy, easy-to-work-with box, manufactured with a tapped 10-32 grounding hole: a box that now contains several inches of reasonably good conductor. Chapter 2 talks about reviving conductors. If there just is not that much conductor to be had, at least the new box gives you room to pigtail lengths of healthy conductor onto the tired old wires. If you need to heal the conductors, cutting back cable sheath can expose fairly good insulation for the end of your tape or shrink-tubing to grab.

If the box is fed by conduit, you may be somewhat less inclined to replace it. One consideration is that it may be harder to do—conduits do not push around as easy as do cables. If there are multiple conduits,

(a)

■ **5-1a–f** *Removing an old box, pulling out a little more cable in order to cut back the armor so as to reach viable insulation, taping the conductors, replacing the box with a modern one so as to be able to install a modern device ring of the correct rise, and then completing the installation.*

(b)

(c)

■ **5-1a–f** *(Continued)*

(d)

(e)

■ **5-1a–f** *(Continued)*

More Specialized Issues and Situations

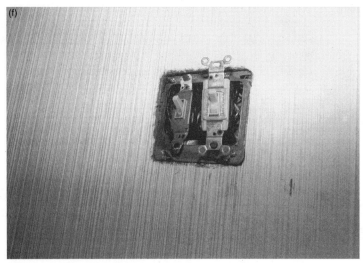
(f)

■ **5-1a–f** *(Continued)*

and they come into the box from opposite sides, that is even more true. Also, if the wiring method is conduit, you need not worry about cutting back cable armor to get to good conductor—if the wires are too short or too beat, you can pull them out and pull in shiny new color-coded THWN, with 6″ or more of free conductor.

Replacement: the problems. Chopping into joists

Suppose you do want to replace a pancake box, mounted right under a joist, with a larger enclosure. What you will be installing almost certainly is at least 1½″ (3.8 cm.) deep. This may mean you have to cut into the joist. It may not. Obviously, if the existing box is deeply recessed due to bad workmanship or to someone adding a layer or two of ceiling under it, you have that much less box depth to worry about concealing. It is also true that most fixture canopies are not flat. Their curvature may accommodate some of the depth that protrudes from the ceiling. Finally, some paddle fans explicitly instruct the installer to leave a small gap between canopy and ceiling. It is not clear whether this is to prevent heat build-up in the box, or, as one might presume, to prevent canopy vibration and movement from rocking it against the ceiling finish. If the former interpretation is correct, snugging the canopy against the edge of the box or the mud ring may be unwise.

While you may need to chop a bit, there are restrictions. The Code warns that you shall not weaken a building's structure in the course of installing electrical equipment. This rule would seem to be especially problematic when it comes to cutting into the underside of a joist. The reason is that the lowest part of the joist serves as the primary tensile member, just as the top takes more compressive loads. Hence the design of hollow I-beams.

Having said all that, and never forgetting that each installation must be evaluated individually, cutting even an inch out of the bottom of an old 2×8 or 2×10 does not normally lead to trouble. Circumstances would be entirely different if you were to cut into the top or bottom of a manufactured joist or wooden I-beam constructed of 2×2s or 2×3s top and bottom, bridged by a facing of plywood. Small likelihood: the only possible way such a joist would be found in an old building would be if the structure had been rebuilt. Although engineered wood products were first used in the 1950s, most of the early products were solid, glue-laminated boards. Incidentally, the latter generally are NOT to be drilled or notched at all. Wooden I-beams, in contrast, often have knockouts through the center, for use by plumbers and electricians. Even in repair or rebuilding, though, the use of manufactured joists would be highly unlikely. For instance, if, as does happen, a joist were partly rotten or eaten away by termites or carpenter ants, it is likely that the bad part would be cut away and

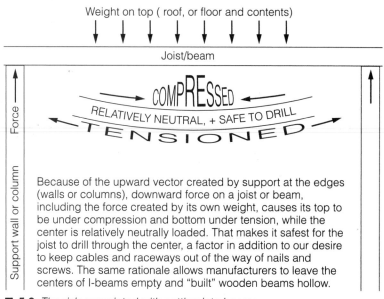

■ **5-2** *The risk associated with cutting into beams.*

the replacement joist section would be bolted against the non-damaged part. To do this, the carpenter would pretty much have to use a solid replacement.

There is one exception to the rule that you will be replacing an undersized pancake box with a box that is at least $1\frac{1}{2}$" deep. If your wiring system is nonmetallic sheathed cable, it may be that you can satisfy box fill requirements by using a box, Listed for fan support, that straddles the joist.

Replacement: other issues

If you are replacing an undersized ceiling box, perhaps for the reasons discussed in Chapter 2 related to deteriorated conductor insulation, you have a number of options. You need not be restricted to an 8B, a 4" diameter by $1\frac{1}{2}$" deep octagonal box. While the 4" round or octagon design is about the only standard type used for boxes explicitly Listed for paddle fans, you do have the option of supporting your fan in other acceptable ways. If more room would be better, or if you need to extend your enclosure further to the sides in order to let you shorten cable sheaths, you could employ bigger standard boxes and rely on alternate means of fixture support. Deep 4" octagon boxes ($2\frac{1}{8}$" as opposed to $1\frac{1}{2}$" deep) are less likely to serve your purpose than are 1900s or 11Bs (which are 4" and $4\frac{11}{16}$" square, respectively).

Those square boxes have their own disadvantages. With even a "flush" mud ring, a square box extends at least a hair deeper than $1\frac{1}{2}$". Also, a little spackling or plastering will be required around the lip of the mud ring if you are to do a tidy job, especially with a $4\frac{11}{16}$" square box. Still, this replacement will result in a better installation than you would get by trying to make do with an existing, crowded enclosure.

Here is a problem to watch for when attaching a new box, or any other equipment, to old structural elements. Old wood is brittle compared to newer wood. This is true of wood lath, as well as joists. Therefore, you are far better off sawing than whacking, chiseling, and gouging. Yes, electricians have worried away the occasional $\frac{1}{2}$" of joist that interfered with their installations using groove-joint pliers such as Channellocks™, and cleaned off the odd remaining wood fibers with a big "persuader" screwdriver, so as to mount a box far enough back for its front edge or mud ring to sit flush.

On the other hand, they have also to their discomfiture chiseled off a much bigger piece of joist than intended, hammering at that same screwdriver or wood chisel. And that has been true even when they first slit the joist on either side with a reciprocating saw such as a Sawzall™! Brittleness can be a real problem. You might even consider drilling pilot holes for screws that are to penetrate dried-out wood, lest they split it.

For further solutions to the problems of crowded boxes, see the section of Chapter 3, "Design Changes," called "Adding room."

Less-than-ideal locations On to the next problem. Suppose that you want to install a ceiling fan and the existing outlet is not under a joist.You may or may not have the option of replacing the box with one Listed for fan support. There are a number of reasons.

☐ Most such boxes are intended to be secured under or against a joist. If your chosen location is somewhere else, you need to go to some other plan.

☐ The boxes you carry in your truck, even the choice of boxes that you can buy at your local distributor, may not be Listed for use with the type of cable or conduit with which you are dealing. For instance, many fan support boxes are only suitable for use with nonmetallic sheathed cable, and maybe, just maybe, with loom.

☐ Finally, the boxes in stock may not have enough volume to accommodate the combination of conductors you find. It would take an odd inspector to refuse to grandfather an existing, undersized box, but at the same time to okay your installing a new, undersized box on the grounds that the (unmarked) canopy makes up for its volume deficiency. Incidentally, as of late 1997, there was no sign of a Listed fan support box that allows you to install a box extension and maintain the Listing. There always will be cases where you have to warn the customer that the ceiling needs to be chopped open and replastered, because you have to get up there to put in a big junction box separate from the fan location, and leave a blank cover visible against the ceiling.

If your location is right between joists, and you are not faced with the need to accommodate more than one or two cables, there are products designed to serve exactly your need. Inserted through the ceiling opening before the boxes, they bite into the joists, and there have been no reports of problems supporting fans from them. Admittedly, by using such an after-market kit, you may be deviating slightly from the installation instructions provided by the fan manufacturer. If you want to thoroughly protect yourself from liability, you can seek a written okay from the manufacturer before doing so. That would be going to an extreme; the products are investigated by an NRTL for this use.

If your opening is right near a spreader or joist brace, that obstructive material can be a nuisance when it comes to installing those fan supports. The usual solution has been to simply savage any wood that is in the way to get it out of the way. Electricians use a reciprocating saw when that will work; other times, they rely on a pry bar or other "persuader." The same goes for dispensing with an old sheet metal strap support that may have been adequate to secure the box holding the old fixture, but is only in the way of the fan braces.

Here is a sneaky trick that some have found quite satisfactory for supporting a paddle fan when the opening is within a few inches of a joist. Build up alongside the joist. With judicious preparation, a few chunks of 2×4 can be quite solidly nailed into the side of the joist, one against the other. Using 12-penny or 14-penny nails, or 3" screws, you can extend the "building structure" to which you are permitted to secure your fan box, widening it. Build it up right above where the fan outlet will go, and set it back from the surface to just where it needs to be in order for you to set the box flush.

There are three main problems with this solution, plus one secondary concern.

☐ It is tedious and time-consuming.

☐ It can mean scraped wrists and knuckles.

☐ As with most other mounting systems, sloppy work will not provide secure mounting—and this one involves more pieces, and more judgment, than do ready-made supports.

☐ Finally, it is nonstandard, which can make some inspectors uncomfortable.

This is a good place to examine the ideas of solid support and solid contact. It took the NEC many years to get around to saying that one 6-32 screw in the center of a receptacle does not provide an adequate path for bonding to a surface-mounted box. At the same time, they did come up with a specification. One 6-32 is not adequate to quickly and reliably pass the fault current needed to operate an overcurrent device; two such screws do suffice.

In most other cases involving hardware, you lack specifications. No studies have been reported in the trade press to help installers determine what size wood screws are needed to hold up a paddle fan. Sometimes one can fall back on the analysis implicit in a manufacturer's inclusion of wood screws with a mounting kit or their identification in installation instructions. Regardless of this wide-open status, many are uncomfortable using anything less than, minimally, #8 or #10 screws, penetrating at least $1\frac{1}{2}$" into solid wood. There is also the matter of head size and mounting hole size. If you have any doubts about retention, use a washer under the head of a mounting screw—especially if you use flat-head or panhead screws.

Old boxes and accessories

In dealing with older boxes, two more issues that have not yet been properly addressed frequently come up. One issue is the need to accommodate equipment not designed to mate with currently-used products. The second, which has been touched upon a little already and will be addressed throughout this book, is the tension between two options. One is merely doing what will meet Code, and thus minimizing present mess and expense; the other is going further to do a better, perhaps even an optimal, job.

Boxes lacking integral provision for covers

The first, and more-contained, issue is dealing with antiquated enclosures. One piece of equipment that mates with nothing modern is a box without screw holes at the perimeter, or the ends, or the corners. It has no ears to which to attach devices, rings, extensions, or covers. At the time when fixtures were mounted to hickeys rather than straps, some pancake boxes, as well as some deeper round boxes, were designed that way. Occasionally, you will have to replace an old fixture that was mounted to one of those. Without replacing the box itself, how are you to attach fixture canopies or straps?

There are a few less-than-wonderful ways of dealing with this type of old design. One is to use the hickey to securely support a strap that incorporates 8-32 holes. This, however, does not provide very reliable bonding between the one-hopes-grounded box and whatever equipment is being supported. A better choice is to drill and tap the box itself and to use mounting screws long enough to reach to those holes in the back of the box. Your holes, and thus the equipment you support from them, may be a little off-center; still, this is the best of the kludges.

■ **5-3** *A pancake box that lacks mounting ears. Note the damaged, loosely taped conductors. If your visual perception is very good, you may notice that insulation is missing from one of the conductors just where it emerges from the connector. Power should not be restored until this is corrected, at the least by cutting back the cable sheath and further taping.*

By far the poorest choice, especially in terms of safety grounding, is to take wood screws directly into the building structure in back of the box. What is particularly poor about that choice, for the moment putting aside issues of grounding and bonding, is the following. If you use wood screws to support, say, a canopy, each time you need to get at the box you will be asking wood screws to grip securely into wood from which they have been removed and then into which they have been reinserted. That is pushing their design, and the characteristics of wood, beyond reason. Presumably, you could try using hanger bolts and cap nuts. A far better job is accomplished by either replacing the box or drilling and tapping it.

Very old 4" square boxes

Another misfit piece of equipment is the very old 4" square box, whose tapped ears are just not quite where they need to be to line up with a modern device ring, plaster ring, or extension box. There are a number of reasons you might have to mate modern equipment with such an old box.

☐ You could be adding or removing a device, and thus need to go from single-gang to two-gang or vice versa.

☐ You could be adding a cable or raceway, and thus require the additional volume that would be afforded by adding a box extension.

☐ You could notice that the box and ring are recessed beyond the legally permissible depth and want to set it right. That problem is ever so common.

☐ Finally, you may have an old box with an old mud ring, one without tapped ears, and want to replace the existing fixture with one that does require a strap or ring with tapped holes.

If you have been doing old work for a while, you may have accumulated a collection of old device rings that will solve some of these problems. If not, most inspectors will see no harm at all in your drilling a modern device ring so as to place its mounting holes or slots holes where you need them—or very close to there. A $\frac{1}{8}$" mismatch should not concern you—but a $\frac{1}{4}$" miss probably would not line up with reasonable "persuasion." Technically, such modification might be a Listing violation, but reasonable accommodations are accepted by reasonable inspectors.

The problem may not even be real. You need not assume on first glance, when you notice that the old mud ring has its screws located away from the corners, that the box is going to be a royal nuisance of a misfit. There was a considerable period when 4" square boxes had ears tapped in two locations—the old and the modern. So remove the device ring and see whether you have lucked out.

Do not consider securing a modern ring or extension to an old box catty-corner. Should you do so, the 8-32 screws will barely grab a sliver of metal at the edge of each hole or slot. The result of that approach is very poor overlap and lineup of the old box and the item you are installing over it. The consequences are unreliable bonding between the two items, incomplete enclosure, and unreliable mechanical support. You would create a similar problem if you tried to install an octagonal extension box over a round box.

Another type of box that does not match what is used nowadays is the 13A, a round or octagonal box that measures $3\frac{1}{4}$" in diameter. It is a bit of overstatement to say such a box does not match anything used nowadays, in that modern mud rings still have a $3\frac{1}{4}$" opening. Some fixture canopies do, as well. What that old box, whose volume is almost always inadequate, does not match is any extension box.

■ **5-4** *The progression in the design of 4" square boxes: a cover designed for the very old version versus a modern one, with a transition box between the two covers.*

Extension boxes

This is a good place to discuss extension boxes in greater detail. Extensions certainly do add wiring volume, and in those rare cases when a box is recessed $1\frac{1}{2}$" or more behind the surface, they can do a nice job of bringing the front of the enclosure forward. On the other hand, using extension boxes does not yield as good a job as does installing boxes of the right size at the right locations. The reason is twofold. First, assuming ground is present, installing an extension box inserts into the grounding path an additional junction, whose conductance depends on two screws, plus whatever contact pressure they create. This attenuates the path on which you depend for fault clearing, compared to the solid path provided by the continuous metal of a unitary box of adequate depth (or, more than likely, even the path offered by a sectional or a tack-welded box). Second, using extension boxes makes it more difficult to gain access to conductors and splices.

Limiting access is a big problem when using one extension box on top of another extension box. Many inspectors will reject such an installation, especially one involving relatively narrow boxes, such as utility boxes (2" wide by 4" tall). Additional concerns revolve around the questions of how tightly the boxes will mate and how mechanically secure the assembly will be.

Mating fixtures to boxes and protecting combustibles

Another problem with a small wiring enclosure such as a 3" round box is that a lighting fixture's wiring compartment, which includes everything under the canopy, extends to the sides, beyond the box. Therefore some ceiling surface (or wall, in the case of sconces) is part of the enclosure. In some cases, that surface will be combustible—wood or wallpaper or fiberboard.

This problem is easily surmounted. It is a problem that comes up quite frequently with modern fixtures, too. The answer is to interpose something noncombustible, such as metal flashing, between the wiring compartment and building surface to, as it were, complete the enclosure. (Do be respectful of the materials you employ, considering issues such as the danger of cut fingers and the risk of nicking wires' insulation and shorting them.)

New fixtures meet old wiring

Ceiling fans are far from the only problem children when it comes to replacements for existing fixtures. Whenever a customer calls and says, "I have some new, Listed fixtures I'd like you to put in to replace what I've got," you would do well to respond with a number of questions before making an appointment. Just as with a paddle fan, the first is, "Is it a wet or damp location?" Explain that screened porches normally are not considered dry locations. Another, critical, question is this: "Look at the instructions that came with your fixture. Does it say anything like, 'Use only with supply conductors rated for 90° C or 75° C'? If so, how old is your house's wiring?" A house that was completed before the mid-1980s probably did not use nylon-jacketed conductor insulation, and has 60° C wire. Nobody will be happy if you show up to do the job and find that without rewiring you cannot use the fixtures the homeowner planned on. See Chapter 14, "Resources," for material that may help you deal with the problem of too-low a conductor temperature rating.

Respect for the elderly—old does not mean bad

Boxes are only one example of old designs that are difficult to match. Old does not automatically mean bad, even though antiquated designs can cause problems. Occasionally, old is even better than new. How many buildings today are wired in rigid conduit? Yet some old single-family residences get their power through conductors run in black iron pipe—which is just as good as galvanized rigid pipe, except for its inadequate rust-resistance in damp locations. (Plumbing pipe is quite a different matter. Particularly when coupled and connected with plumbing fittings, it is dangerous to use as a raceway and has been illegal for electrical work since quite early days.)

Similarly, small services are not automatically bad. Many large, modern, all-electric houses with heat pumps and electric cooking draw no more than 40–50 amperes maximum. And certainly there is nothing wrong with fuseboxes as circuit protection. Chapter 6, "Safety Surveys," will look more closely at some of these holdovers, in the context of advising customers as to what really has to go and what need not.

Outdated designs that have to go

Unfortunately, there are some old designs that would be perfectly okay except for the fatal fact that they are out of date. See Chapter 9, "Relatively Rare Situations," for further discussion beyond the suggestions that follow. If parts were available, this equipment could be repaired, rebuilt, and sustained indefinitely. Chapter 14, "Resources," contains a list of sources for exotic parts. Some pieces, however, just are not worth keeping alive unless your customer loves electrical antiques. See Chapter 10, "Historic Buildings," for further exploration of that special situation.

The following question is almost always worth putting to your customer: "Do you want me to fix this or to replace it, given that it may take as much in time and materials either way?" If replacement may well be the less expensive option, it is especially important that you ask, before proceeding with the repair you were called in to perform.

Superannuated devices

Here are a few designs that, while fine in their time, eventually ought to be replaced.

Despard devices

Despard, or interchangeable, devices, while still made, no longer are readily available. The closest equivalent that is easily, readily, available today is the rectangular, modular look, such as "Decora™" devices. It is closest only in that any device, single or duplex, takes the same cover plate as any other. The Despard system went further, but in the sense had the reverse kind of interchangeability. It enabled you to mix and match one to three devices on one yoke. You could choose from the menu: various configurations of single receptacle, single-pole or three-way switches, pilot lights, and nothing. You installed or replaced them by snapping them into the yoke's three openings.

The only pieces not fully interchangeable were the cover plates. A Despard cover for, say, a two-gang box containing one yoke with two devices and one yoke with three will fit only that combination. Worse, a Despard duplex cover will not fit a modern duplex receptacle. The latter should be no great surprise. Although some Despard receptacles accommodate modern, three-prong plugs, Despards are necessarily built more compactly than modern receptacles, or it would be impossible to fit three

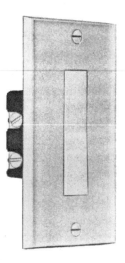

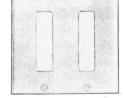

Single   Two-gang

■ **5-5a–d** *Some forerunners of the modern Decora™-type device and cover.*

on one yoke. A Despard duplex receptacle cover therefore has a blank space in the middle.

This incompatibility is disappointing when you do have to replace a Despard device. It is a bit more of a disappointment when one Despard device fails and you therefore have to replace every device in its multigang box, in order to fit a modern cover over the modern replacement device. It is far more than a bit of a disappointment when there is no room in the box to accommodate the modern equivalent. An example would be a single-gang box containing three switches. There are exotic switches available that do independently control three devices from one yoke, but are such pricey exotica worth keeping in stock?

Decora devices, and their forerunners, offer neither the "roll your own" flexibility of Despard-type devices, nor their repairability. On the other hand, they did and do offer the convenience of "one-size-fits-all" covers.

Crowfoot receptacles

Crowfoot receptacles also have to go. Those offered the choice of 120 and 240 volts at 10 to 20 amperes, almost exactly like a junior version of NEMA 50R range receptacles. (The latter are termed "range" receptacles, but their use is not restricted to serving rangetops; they legally can serve electric stoves, ranges, and ovens generally.) A second way that crowfoot receptacles are similar to range receptacles is that they provide no means of grounding except to a grounded (as opposed to a grounding) conductor. And even more than is now the case with range receptacles, due to NEC changes, crowfoot receptacles no longer match customers' equipment.

Other odd outlets

There are other oddly shaped receptacles that will not accommodate any modern cord connectors. Others will accept two-prong connectors but will not accommodate ground prongs. Like crowfoot receptacles, it may be possible to replace them with modern receptacles, once you check wiring polarities. Like crowfoots, though, they may well indicate the presence of old, deteriorated wiring.

Mercury switches

Another old item that is mighty uncommon is the mercury switch. A capsule history of switches will explain its background. The first switches were dangerous—plain knife switches. You still can find

(a)

(b)

■ **5-6a–b** *Other odd nongrounding receptacle shapes. Mostly used for 120 volts, they could alternately be wired for 240 volts.*

some in antiquated loadcenters, though we hope nowhere else. The next development was knife switches, still live-handle, with spring loading so as to minimize arcing. Then switches had their live parts enclosed. With consequently smaller contacts, it was necessary to develop them into "snap switches," switches making

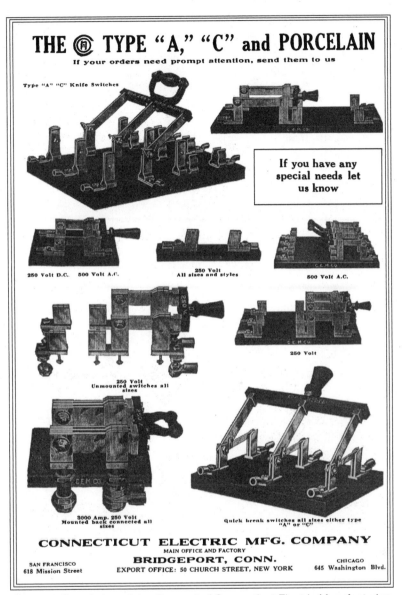

■ **5-7** *Knife switches: Advertisement of Connecticut Electric Manufacturing Company in* Electrical Contractor-Dealer, 18(1), *November, 1918.*

and breaking connections quickly enough that their contacts would not pit to uselessness in a matter of months.

They did "snap," though. For a quiet switch, an installer used one which made and broke contact at the boundary of a small pool of mercury. These had no springs to fatigue, no air-exposed contacts whose limited contact surfaces would burn up.

Mercury switches reportedly are still manufactured, but you may well be unable to find one for replacement. When a mercury switch does fail, therefore, it alas must be replaced with a regular switch which, while taking up less room in its enclosure, will not last nearly as long.

Incidentally, if someone ignorant has monkeyed with a mercury switch, it may malfunction even though it is not defective. Here is why. If it was reinstalled upside down, or even significantly off plumb, the pool of mercury may flow away from the contacts. Straighten that out, and you may well have "repaired" the switch.

Antique dimmers

High-low dimmer switches offered only two levels of light. They have no serious advantage over modern full-range ones, so you should have no qualms about replacing them.

Rheostat-type dimmers were far more rugged than modern semiconductor units. They were not destroyed by power surges or sags such as can be caused by a circuit tripping. They could be used with limited impunity to control motor loads. On the down side, they were also woefully inefficient, throwing away unneeded power as heat.

Rebuildables

Finally, you may run across rebuildable devices. It takes no genius to recognize that certain parts fatigue in normal use. In the early days, receptacles were screwed together so that you could renew them by replacing those parts that wore. Unfortunately, or in one colleague's view, fortunately, nobody is making the replacement parts any more. Besides, the time required to rebuild those small devices would prevent doing so from being cost-effective. All that this means is that if you see a worn-out non-grounding receptacle that is held together by screws rather than rivets, you still need to toss it out and install a new one.

Iffy wiring methods

That finishes the section on the old and quaint. The following are more serious issues. Installations wired in black iron pipe were described as being superior to most modern ones. The trustworthiness of the flexible counterparts of that black iron is less clear. Both armored and nonmetallic sheathed cables were used from the fairly early days of power wiring. Unfortunately, while it is often easy enough to pull new conductors through old pipe, you lack that option with cable or loom. (And flexible metal conduit can be a bear to rewire.) Yet degraded equipment cannot be tolerated, beyond a certain point. It needs to be replaced. Other installations should be replaced simply because they represent practices that are so far behind current standards of safety.

Grounding and bonding replacements

The evaluation and patching of deteriorated insulation are discussed in Chapter 2, "Troubleshooting." Deteriorated insulation is far from your only concern. The next biggest issue is inadequate grounding. (Bonding and grounding are also addressed at the end of Chapter 2.) Early nonmetallic sheathed cable had no ground wire; in fact, as of the 1996 edition, NEC Section 336-2 still did not require NM cable to contain anything more than two insulated conductors. There is no skirting that fact. If you have an installation with ungrounded nonmetallic cable, safety is greatly diminished compared to modern standards, unless you replace the cable. (Some ungrounded NM was installed with an exterior ground wire.) Section 250-50 Exception does permit you to augment the cable after the fact. Nonmetallic sheathed cable with a reduced grounding conductor was legal at one time. The black-and-white wires you find might be #14 AWG, and the bare one, #16. You need not worry about such an installation. So long as its splices and terminations are adequate, the grounding wire should maintain sufficiently low impedance to trip properly sized overload and overcurrent devices.

Early armored cable, and perhaps all BX *qua* BX, was manufactured without a bonding strip, which means that it has high inductive impedance. (That may be less true of the briefly manufactured double-wrap armored cable, which contained both an inner concave and an overlapping convex spiral.) Fortunately, unassisted cable armor has proven to successfully operate overcurrent devices by carrying faults. On the other hand, old-fashioned armored cable would not have been supplanted by AC cable had not that slender bonding strip significantly enhanced safety (calling the strip a "wire" has misled people into misapplication; see the discussion of language and safety at the beginning of Chapter 2, "Troubleshooting").

A sometimes-overlooked reason for bonding is found at the beginning of the NEC, in Section 110-3 (b). When manufacturers say to attach fixture straps to the ground wires of cable feeding lighting outlets, you realistically need not do exactly as they say. No inspector will blame you for bonding the fixture straps to the outlet boxes instead, where the boxes are suitably grounded. If you do not ground a fixture at all, though, you are straying further from a defensible position. In fact, even without explicit written instructions, if a fixture comes with a grounding lead, or a strap with a marked grounding hole, installing it without grounding may put you in the same untenable position even where the fixture is protected by elevation.

The isolation option

What about bonding switches? If a switch has a grounding screw, it could be argued that, if you install it ungrounded, you are in the same hard-to-defend position just mentioned. Even when you install a non-grounding switch, Code authorities generally consider its yoke something that may become live. In consequence, if you do not ground, you should at least use a nonmetallic cover plate. Some would argue that you should also use nonmetallic screws to mount that plate. Nylon 6-32 screws are cheap enough and easy enough to come by at specialty industrial supply distributors, if not at your local electrical wholesaler.

If you boast a more conservative and conscientious perspective, you will go beyond merely ensuring that the switch is safe as you leave the job. Recognizing that laypersons, maintenance workers, and home repairers often replace switch covers, you will not rely on using a nonmetallic cover, or even a nonmetallic cover plus nonmetallic mounting screws. Instead, you will see to it that the switch itself is bonded. (If there is no ground available at the switch box, though, bonding offers no benefit; isolate.)

At least as of the 1996 version, NEC Section 210-7 (d)(3)(a) permits worn-out ungrounded receptacles simply to be replaced with new ungrounded receptacles, except where GFCI protection is required. Doing so is not a terribly good idea. The problem is that customers' ignorance can endanger them. Unless you affix some kind of permanent—not to mention unsightly—warning, one that is impossible to overlook and difficult to remove or hide behind painting or papering, customers, or perhaps future occupants, are very likely to plug in three-prong devices, using outlet adapters. Such adapters are cheaters, in this case, not only by slang usage but by virtue of the fact that using one of them will turn on its head a manufacturer's effort to make equipment safe by using a grounding cord and plug.

If you do choose the Code-sanctioned option of replacing a worn, ungrounded receptacle with a new, similarly ungrounded, receptacle, consider replacing the cover plate and cover plate screw with nonmetallic ones. That both will reduce the shock hazard in the event of a short to the yoke, and just possibly may signal users that the cover plate screw is not to be used to ground an adapter. Similarly, you can replace an ungrounded light fixture with a porcelain or other nonmetallic lampholder. Doing so will offer very little hazard to users so long as the wiring itself, while ungrounded, is not ailing. Even if the wires short to the outlet, there is very little shock hazard so long as the porcelain does not have a metal pull chain. A brass lampholder, especially pendant, poses far greater shock hazard.

GFCIs

What if replacing a nongrounded receptacle with another two-prong receptacle will not do? The NEC offers two options besides rewiring. One is the use of a GFCI. The NEC is generally mum on the subject of grandfathering but breaks this pattern in the case of ungrounded receptacles. The installation of a three-prong receptacle, ungrounded, as replacement for a two-prong receptacle is explicitly prohibited. As an exception, the NEC explicitly authorizes the use of GFCI protection as a substitute for that grounding conductor. Or is it indeed a substitute? Not really. The NEC authorizes the "substitution" of GFCI protection on ungrounded circuits only when you mark outlets "No Equipment Ground" and "GFCI-protected." If GFCI protection and grounding were equivalent, that marking would be unnecessary.

Whenever some practice calls for a warning label, you can safely figure that the permitted practice is a compromise, embodying some inherent or potential hazard. There is no agreement as to what constitutes permanent marking, and if such a marking is indeed needful, there is little point in applying a marking that will be only temporary. As always, what you are permitted to do, in terms of lesser alternatives to a top-notch job, hinges upon the judgment of the local authority having jurisdiction.

Here is an example concerning the issue of affixing warning labels. In this case, the electrician was dealing with equipment that relied on two sources of energy. He was hired to implement recommendations stemming from an independent Listing laboratory's (NRTL's) field investigation. Among other things, the laboratory's engineers decided that, before they would affix their label to the equipment, permanent markings had to be attached, warning of the presence of two power sources. What kind of markings? The electrician was required to get plastic plaques engraved by a professional shop specializing in awards and signs such as are used on bureaucrats' desks. They were not cheap. That amount of trouble is never taken to mark ungrounded receptacles, but the permanence of the plaques made the NRTL's insistence on them reasonable.

A top-notch job provides a ground. The NEC continues to require grounding in more and more places. Section 380-9 of the 1999 NEC probably will require that not only snap switches but also dimmers be grounded—either by a jumper or through the mounting screws—even when the box presently has a nonmetallic cover. While there will be an exception for existing installations where no ground is available, grounding is clearly preferable.

Short of tearing out what the customer has in place and installing new cables, what are your options? (A separate earth ground is NOT an acceptable choice.) It used to be that you could simply run an equipment grounding conductor to the nearest cold water pipe. No more: that approach relied on intact metal plumbing pipe, continuous to the earth, and electrically secured to the main bonding jumper, to make the electrical system safe. Nowadays you are far better off running a line to a grounded part of the electrical system itself.

This leaves the questions of where and how to run a grounding conductor.

☐ You could run a conductor, sized in accordance with the circuit you are grounding, to any part of the grounding electrode system. This normally includes the first five feet of metal cold water pipe coming into the house; as of the 1999 NEC, it may include any "effectively grounded" cold water pipe.

☐ You could run to the service itself, as of the 1999 NEC.

☐ You could run to a subpanel fed off that service, if that is where the circuit originates, as of the 1999 NEC.

☐ With an accommodating inspector, you could possibly go to

any grounded circuit—although that option is not explicitly permitted by the NEC.

This last option points to a caution that you need to observe whenever you disconnect or tear out wires. When you kill part of a circuit, if you have not actually torn out the cable or conduit you must leave it accessible; and you never, never should disconnect its grounding. When you do tear out wiring, there are several points to keep in mind.

☐ Make sure you know where everything leads and get both ends of all cables.

☐ When you remove connectors, nipples, or conduit, use knockout seals.

☐ When you remove cables from cable clamps, tighten down the clamps so that they block the openings.

☐ When you remove a cable from a nonmetallic box that has no clamp, you may need to replace the box—there may be no listed system for closing the opening.

☐ Finally, when you eliminate wiring, whether because it is decrepit or because a wall is being taken out, keep on the lookout for extra wires such as may have been installed to ground another circuit—or even to ground a telephone system. These too need to be traced and dealt with before you can complete any demolition.

Back to the problem of ungrounded equipment. What is involved in running a separate grounding conductor? The Code is clear that all wiring must be suitably protected. Section 250-92 (c) (2) talks explicitly about individual grounding conductors. One easily could argue that an individual conductor, such as a grounding conductor, should be run precisely as open wiring was always run—with knobs, tubes, and loom to support and to some degree protect it. Inspectors do not require that. On the other hand, neither can a grounding conductor, which is relied upon for safety, be treated like low-voltage cable that can be harmed without causing any danger, merely inconvenience. (Even with low-voltage cable, this can be a deadly assumption. Consider cable associated with an invalid's hot line.)

This means it is unacceptable to stuff a ground wire behind baseboard or tack it like phone wire. It needs to be treated with at least as much respect as nonmetallic sheathed cable: secured, run through bored holes towards the centers of joists and studs, and protected with running boards and guard strips. And if it is exposed to damage, by golly you will have to enclose it. This could mean running $\frac{3}{8}$" flexible metal conduit (Greenfield or flex), electrical nonmetallic tubing (Smurf™ tube), or something more substantial.

Safety Surveys:
Looking at the Big Picture

THIS CHAPTER INTRODUCES THE COMPREHENSIVE SAFETY survey. It also tackles the challenge of educating customers concerning the difference between "shiny" and "safe." In addition, it begins to address the issues of boundaries and responsibilities. These challenges and issues always need to be considered, whether you are on a job to consult, inspect, install, or repair. Chapter 7, "Setting Limits and Avoiding Snares," goes further into the issues of boundaries, and Chapter 15, "Communication," offers detailed guidance about overcoming customers' misperceptions.

There are two conceptually separate pieces to this chapter. One is technical and has three parts: first, deciding how safe a system is; second, how Code-compliant; and, third, for an optional added fillip, how old. The other piece is communicating your conclusions to your customer. This piece involves a number of aspects as well. You have to communicate what you find. You also have to communicate what your findings signify, in terms of the seriousness of any problems, and in terms of the trade-offs associated with various solutions. In the course of all this, you have to overcome any misunderstandings your customer brings to the dialogue.

Not all aspects of safety surveys can be addressed here. To avoid breaking the flow of this chapter, issues associated with aluminum wiring are relegated to Chapter 9, "Relatively Rare Situations."

When does a customer ask for a safety survey?

How might the request for a "comprehensive safety survey" come about? Perhaps someone is buying a building, and the engineer or home inspector says, "You had better get an electrician to look it over," or perhaps "Have an electrician check out items 2 and 7 on my punch list." This may be a rush job, especially if contract settlement awaits your findings.

Perhaps a customer says something like this: "My wiring is kind of old, although I did put in three-hole plugs a while back. My neighbor had a "heavy-up," which wasn't all that expensive. Could you bring my system up-to-date by replacing my fuses, putting in switches?"

This second example starts you right off with some electrical myths and misnomers to correct. The customer has started out with at least one incorrect assumption: of itself, a heavy-up does not commonly make a system markedly safer. Appendix 3 recounts a goodly list of such myths and offers wording that has proven helpful in straightening out laypersons' misunderstandings. Ultimately, your customer wants to know what it will take to make the house safe. The only way you can answer that question with any specificity is if you sell him or her on having a survey.

Establishing a framework

There are many kinds and levels of surveys you can undertake. In order to begin any job intelligently, you need to be clear as to what you have been hired to do. Exhaustive investigations are almost always prohibitively expensive. Fortunately, comprehensiveness may be unnecessary. So the first thing to ask is what your customer needs to know.

"How likely is this store to have major electrical problems over the three-year period of my lease?" will spur one type of investigation; "How dangerous is our home?" asks for a different one; "How out of date is this wiring?" will get a third; and "Can this building's system support a pottery studio?" warrants another one entirely.

Having said that, serving your customer is not the selfsame thing as doing no more than what your customer thinks he or she needs. You are the expert, not he or she.

A simple example will bring that home. Consider the people who ask, as is frequently the case, about upgrading their 60- or 80-year-old electrical panels. You probably should tell them that it is more important that you look at the wiring first, because its deficiencies are far more likely both to inconvenience them and to put them at risk of being burned out of their homes. Changing over to circuit breakers and heavier services normally will do nothing to avert those risks.

Another factor to consider in determining what you look at and what you leave unexamined is that your customer is not the only person you serve as you deal with the wiring, despite the fact that he or she writes the check. Especially when you carry a professional license, you bear some responsibility to the community, to tenants, and to future purchasers of buildings you have touched. When you discover unsafe or illegal wiring, it is part of your fiduciary responsibility—the trust associated with your role—to take note. Mention the danger to your customer, and note it in your records, on your invoice or report. More discussion of fiduciary responsibility appears in Chapter 7, "Setting Limits and Avoiding Snares."

Criteria to use in designing the survey

Please do not blow the discussion of higher purpose out of proportion. Your time is purchased for your customer's purposes, and it can be unethical to wander off without asking and devote your energy to poking your nose into the rest of the system. The framework created by your customer's needs should establish the major criteria for your investigation and analysis.

For instance, that pottery studio will require a kiln or kilns, meaning that the wiring will have to support major heating loads, plus ventilation and smaller motors. The store's major load almost certainly will be continuous lighting and HVAC. For some stores, the major load will be refrigeration. Accordingly, that capacity is what you will spend most of your time evaluating. If it has any very large rooms in which customers will gather, it may qualify as a place of assembly, a category that carries extra life-safety requirements.

Home purchasers probably will want the most precise, detailed analysis. Safety is most essential to them, as they and their children will be sleeping in the building you are checking out. Also, home purchasers often tie up a major part of their assets in the purchase that hinges on your findings. Other homeowners are surprisingly unworried or penny-pinching.

Besides trying to answer any specific questions that have been posed, or that are evident from the nature of the structure and its intended use, you should apply a hierarchy of priorities in checking out electrical systems. The first thing to check for always is imminent danger. Immediately after that come "accidents waiting to happen." Following those come gross illegalities. After catching those urgent matters, look for devices and equipment that are deteriorated or antiquated and inconvenient but, nonetheless, may have some life left. Then, finally, there are minor violations and no-longer-legal designs.

Imminent danger

What constitutes an imminent danger? Electricity is dangerous because it carries the threat of shock and fire. Clearly, if a device or piece of equipment shocks people, that problem needs to be corrected, and corrected now. If something is smoldering, you cannot wait for later to kill power and extinguish it. Similarly, if something smells like burning or hot insulation, or feels unduly warm, or looks charred, that constitutes a call to action—at the least, it demands immediate investigation.

Accidents waiting to happen

There are a number of "heads-ups" that must be fixed, even though they may not require you to drop everything and deal with them immediately. If you notice a bare conductor, its insulation damaged or gone, something needs to be done about it without fail. (If you are certain that the bared wire is part of a grounding conductor, the urgency recedes.) If you notice a discolored ballast without thermal protection, that is likewise a call to action. The following items need to be mentioned, even though the remedies do not necessarily require an electrician:

- ☐ overlamping;
- ☐ cord connectors with third prongs removed;
- ☐ other means of bypassing grounding;
- ☐ cords run under rugs, equipment, or furniture or through windows or doorways;
- ☐ readily-accessible live terminals;
- ☐ broken insulators.
- ☐ Dying fluorescent lamps, and bearings that sound as though they are about to give out, may not constitute dangers, but where they have not been noticed you will do your customers a favor by mentioning them—and possibly will generate additional billable work.

You need not go out of your way to inventory them, if that is not what your customer wants. Still, as you come across them, take note.

Gross illegalities

Customers can sincerely be unaware of the most blatant Code violations. It matters not whether they genuinely are oblivious or merely are playing games: you need to point out those violations.

If the customers are oblivious, you are doing them a signal service. If they are testing you to see what they can get away with, again you are doing them a service—by setting limits. You also are doing yourself a service by not leading the customers to think you are willing to wink at the unsafe. You do not even necessarily need to blow the whistle to the authorities. Just beware of getting into a position where your reputation and your license are jeopardized by your customers' (or their predecessors') penny-pinching creativity.

Benighted ignorance Here is a fine example of customer oblivious-ness in a house where an electrician was called to repair a flicker-ing fluorescent. He discovered that the misbehaving fixture was a shop light hung above a translucent panel in the kitchen's sus-pended ceiling. The lamps were a month old, but the ballast was older. Before ever touching the ballast, he had to work to clean up the installation in terms of legality and workmanship. An old 3" round outlet box, about 1" deep with a rounded bottom, was located in the structural ceiling. It contained cable clamps and six #14 wires, run to it in armored cable. Presumably it had fed the original kitchen light. The ballast leads ran out through a knockout in the fluorescent fixture to join in midair with wires coming from the (coverless) outlet box. The grounded conductor was con-nected with a twist-on wire connector; the switched conductor was simply doubled over the black ballast lead, with nary a splicing device or insulator.

Clearly, this never had been touched by a professional, and never inspected. The electrician put it in order. In the course of his work, he noticed that the service equipment had no circuit labeling, no visible inspection sticker, no obvious grounding electrode conduc-tor, and nothing but 20-ampere circuit breakers. He told the cus-tomers that the panel looked illegal and asked whether it had been inspected. "Sure it was!" said the lady of the house. "And the in-spector didn't find anything wrong with it." This was certainly sur-prising. He wondered to himself, "Did somebody pocket a payoff?" He inquired further about the reported inspection. "Oh, the circuit breakers were here when we bought the house ten years ago, and our certified home inspector looked everything over."

In this case, there clearly had been no electrical inspection, but the customer had no intention of deceiving the electrician. It was simply a matter of ignorance.

Violations are certainly not unique to old buildings. Illegalities are, however, more likely to be found in structures that have been in place for many years, because, somewhere along the line, some-one tried to improve on inadequate wiring. Furthermore, because old buildings may be less fire-safe than modern ones, it is more important to be alert for electrical hazards. Nonelectrical ways in which they may present higher risks include everything from lack of sprinklers to narrow exits to tinder-dry wood.

The patent, gross illegalities frequently betray themselves by lousy workmanship. These are not simply junction boxes whose contents exceed permitted box fill by one conductor. Five conductors, maybe. And no connector, just a cable shoved through a knockout without its sheath stripped, without any ground connected.

This is not fussing about a box missing a cover screw, but about one with no cover and maybe a cable coming out from the front of the box rather than through a knockout. Other, similar, illegalities include splices without boxes (not, in these cases, in legitimate old-fashioned open wiring) and material such as zip cord used for permanent wiring.

There are any number of surface clues that can lead you to uncover such illegalities. One clue to amateur and thus potentially hazardous wiring is finding something alien, such as a drywall screw substituted for a machine screw as an electrical fastener. Another clue to amateur work is that well-intentioned amateurs sometimes confuse the very skinny internal bonding strip in modern AC cable with a grounding wire. In one case, where an engineer told an electrician that he had connected a fixture lead to the incoming red wire, it turned out that by "the red wire," he meant the bare copper grounding conductor.

Trivial illegalities

Sometimes a minor oversight or a technical violation can mislead a hasty inspector. Consider the case of ungrounded three-prong receptacles. When a patent receptacle tester says that there is no ground at a three-prong receptacle, you have found a violation. It may be a "so-what" violation. NEC Section 210-7(d)(3)(c) says that a grounding-type receptacle may be used there, provided that it is protected by a GFCI, and provided that it is appropriately marked. As violations go, the lack of a marking sticker does not stack up very high. To determine whether this is the case, you will have to either search for the GFCI or else bring a good ground to the receptacle, so you can try to trip the GFCI with, say, your Wiggy™ or other solenoidal voltmeter.

The point is that it is unprofessional and embarrassing to create a foofaraw over an observation that may signify nothing, when a couple of minutes' checking can put you on solid ground.

Deteriorated insulation and devices

It is taking a long time to get to worn-out wiring: that is because deteriorated insulation is not the first thing to look for in old buildings. Certainly any equipment that operates intermittently bears a close look. While it may just be bedeviled by a loosened termination, it is equally likely that a part is failing. Chapter 2, "Troubleshooting," addresses that issue at great length. If there is any suspicion of intermittency, challenge the equipment: jiggle those switches; try those receptacles repeatedly; start and stop that motor, and repeat shortly.

Water over time equals danger Water is a major cause of deterioration. An IEEE monograph authored by Dan Friedman, and referenced in Chapter 14, "Resources," reviewed reports from over a thousand home inspections in which service panels were examined. Corrosion resulting from the entry of moisture was present in 12 percent of loadcenters. The problem was worse than just rusty sheet metal; in 9 percent of the cases, circuit breaker terminals also showed signs of corrosion.

Compared to moisture damage, the inspections uncovered far smaller numbers of other problems, including burned connections, conductors damaged by the substitution of sheet metal screws for machine screws, and cracked components. Rust never sleeps, and moisture damage can progress rapidly. Moisture strongly accelerates the deterioration caused by various chemicals and the mutual damage resulting from contact between certain dissimilar metals.

Most commonly, the water came in along the service cable's penetration through the outside wall—sometimes inside the cable jacket, sometimes around it. Water entry most commonly stemmed from failure to use duct sealant or caulk, or from the use of interior connectors outdoors, or from the failure to use standoffs in damp locations. Keep on the lookout for these practices. Such shoddy workmanship yields installations that may start out problem-free but suffer premature failure—albeit long after the installer is gone.

Even where you do not see serious deterioration, it behooves you to take note of openings to the outdoors and report them. Should you be inclined to dismiss such potential problems, think of the future. Consider what happened to the grounding continuity of the armored cable or flexible conduit pictured in Figs. 6-1A and 6-1B, as the steel rusted through.

105

a

b

■ **6-1a–b** *The closeup (b) shows how little armor was left, after rust had its way. The larger photo (a) shows beyond any doubt that the problem was neither age nor current, but damp; the steel stood the years perfectly well, above the section exposed to moisture. It shows the worst rust where a strap held it against the wall and trapped drips or condensation. You can see only remains of the strap, its base, which is still screwed to the wall, and a darkened section of armor of the same width.*

More Specialized Issues and Situations

Plan of attack: priorities and sequence

Now it is time to walk through a real safety survey. It is unnecessary to explain minor violations and no-longer-legal designs to customers, unless they are especially curious and interested. At any rate, such problems are the last items to look for. A comprehensive safety survey is something you will just about never perform. It is too expensive a luxury to open all outlets, look at every splice, and trace each circuit throughout a building, unless you or your loved ones will live there. That usually unacceptable expense is why you must set priorities, as mentioned earlier. Those priorities are not given the form of a standard checklist because surveys vary so greatly depending on customers' needs. In case you find checklists reassuring, at least one such list has been published by a very knowledgeable authority, Joseph Tedesco. It is referenced in Chapter 14, "Resources."

To set priorities, you have to "out-think the opposition." If there is deterioration, where is it most likely to be found? If there has been dangerous, illegal work, where is that most likely to have taken place? What bits should you check out even though they are unlikely to have major defects, simply because they are essential to system function, or to safety? What will provide the best clues as to the quality of previous workmanship and inspection?

Deterioration

The first question is a gimme. Other factors being equal, if the wiring is of multiple vintages, deterioration is most likely to show up in the oldest part of the system. Even more important than age is stress; that which has experienced most stress will show the most wear. Usually the most stressed equipment is whatever has been used most and subjected to the most heat. Anything likely to have been abused gets top billing.

The setting determines what gets used most. In a home, check the light in the center of the kitchen ceiling. Kitchens are hot, due to cooking and refrigeration equipment, lights generate heat, and kitchens generally get lots of use. In a factory or workshop, check where the heavy equipment is located.

Sometimes the sources of stress are idiosyncratic. In one old tavern, electrical equipment in the basement suffered worst, not because of heat or loading but because moisture—sewage, to be blunt—leaked down onto and into it. The second most deteriorated wiring at that pub was not in the kitchen, where one might expect it, but adjacent to the bird's nest of wires at the disk jockey's station.

Illegalities

That tavern rewarded the pursuit of a more standard clue as to where to look for kludged wiring. The best place to look for bad wiring is where a change has been made, especially one with the earmarks of not having been a professional effort complete with permits. In the establishment just described, the cigarette machine had been moved when the bar was extended. Its outlet suffered a near meltdown due to the haphazard splicing the owners had relied upon to allow them to change the machine's location.

In a home, a prime candidate for such checking is a homeowner-finished basement, attic, or porch. Sure, if you find that the installation of wood paneling has resulted in a switch box being recessed in violation of NEC Section 370-20, adjustable box extenders are an easy fix. That is not the end of the matter. Finding such a violation in the initial sampling cues you to check further—someone ignorant of the NEC has been at work. If time does not permit this, your report should say that the installation bears further investigation.

Essential equipment

After sampling for deterioration and illegalities, next check the essential components of the system. The essentials are the service, the grounding electrode system, and all loadcenters. In an industrial establishment, you could very well include exit and emergency system, motor connections, and lighting contactors. Whatever the occupancy, smoke detectors and GFCIs deserve at least to be sampled, if no one else has tested them.

Miswired GFCIs

If a GFCI is miswired, its test button may trip it, but an actual ground fault—including one that you simulate with your tester—may not.

If there is no ground present at the GFCI outlet, you will get the same symptoms. Furthermore, if there is no ground present at the outlet, but you inadvertently touch a grounded object as you use your GFCI tester, 120 volts may momentarily take a path to ground through your body. If the GFCI is miswired, it may take that path more than momentarily. As you test, beware of that risk.

In most cases you should spend considerable time on the part of the system that is the more expensive to correct—the minutiae. Unlike some of the other aspects of the system, the essentials are a relatively clear-cut chunk both to examine and to repair. Therefore, there is an advantage to saving them till your watch tells you that you have only so much time left to complete your survey. Admittedly, one could as easily argue for checking them first as for checking them last. This discussion really might have covered the essentials first, as they are the one category you will tend to survey exhaustively. However, in many cases they have already been at least checked by a home inspector.

Surveying the essentials will give you some idea of your predecessors' workmanship. The service equipment, if not the grounding electrode system, tends to be a showpiece. If this is a mess, it bodes ill for the rest of the electrical system. Contrariwise, wiring in out-of-the-way corners such as crawlspaces tends to show installers at their worst—and most dangerous.

Overfusing

Unclosed openings and missing covers in a service panel suggest indifference to safety. The odd missing trim or screw is less of a concern. Overfusing, though, is a more insidious danger than leaving live parts exposed.

In early 1987, a major, reputable manufacturer inadvertently sold some two-pole 50- and 60-ampere circuit breakers in merchandising packages marked for 30-ampere breakers. One would think that an installer, even a homeowner or an appliance store service technician, would check the marking on the breaker's handle after taking it out of the package, but that assumption is unwarranted. There have been many cases where, even without any deceptive packaging, breakers were selected and installed in simple ignorance of the requirements for ampacity matching. With fuses, of course, installing the wrong item often is far easier.

System grounding

The grounding electrode system does deserve special attention. It is not uncommon for old wiring to lack a belt-and-suspenders type grounding electrode system. For decades, grounding to the cold water pipe near its entrance to the building was considered adequate. Even when the water pipe was supplemented with a ground rod, the conductor diameter and the clamp design were not necessarily anything now considered adequate or suitable.

Such inadequacies are not necessarily a bad enough problem for you to raise a hullabaloo. Doctrine changes. Worse though, there have been any number of cases where telecom installers loosened grounding electrode clamps to stick their grounding wires in alongside the grounding electrode conductors. Such loosening often can result in a high-impedance connection, making this a far worse problem than the simple lack of a double-threat grounding electrode. Even more troubling, in a few cases you will find the grounding electrode system identical with the lightning protection system ground connection. (Note that, as of the 1999 NEC, a code provision is likely to be added requiring that the electrical system and lightning protection system grounding electrodes be bonded together.) Finally, there are the following assorted miseries. Keep an eye out for all these problems:

- [ ] broken grounding electrode clamps;
- [ ] "short-cut" (literally, cut down to an easier-to-drive length) ground rods;
- [ ] grounding electrode conductors removed and not replaced when water pipes used for system grounding (as well as for the bonding of interior pipes) were replaced;
- [ ] water services redone, so the incoming water pipes no longer offer ten feet or more of metal underground.

Other big questions

Your customers are liable to have other big questions about their systems, especially when the customers are home buyers. If they want to know how old the wiring is, there are some clues in Chapter 13, "Dating and History." Often they will want to know whether they can get three-prong receptacles installed, if those are not already present throughout. If you do not find ground readily, that question may require you to determine what kind of cables or raceways are in place. Many knowledgeable purchasers will urgently want to know one thing about their wiring: is it aluminum? Potential problems with aluminum branch circuit wiring, and possible fixes, are logically discussed in Chapter 9, "Relatively Rare Situations," but identifying it rightfully belongs in this chapter. Customers do want to know if it is present, but, to avoid breaking the flow, identification of aluminum conductors is discussed in Chapter 9 along with their problems and remediation. Parenthetically, in most areas of the country few safety surveys will come across aluminum branch circuit wiring.

The report

When you have completed your survey, the next step may be to make your report. Ideally, though, there is an intermediate stage: thinking over what you have found. You would be well-advised to allot some on-the-meter time to ordering your thoughts. If you are comfortable with it, there is nothing wrong with having customers tag along as you do your surveys, if you are comfortable doing so, unless tagging along will expose them to hazards they are not trained or equipped to handle. Some enjoy the companionship and interaction. You might even encourage them to take notes. But you cannot be shy about needing time and space to review and reflect on your raw impressions and notes.

If there is some simple question that brought your customer to hire you, it is wise to address that first in your report. If it has a simple answer, give it: "Your wiring is in good shape;" or "This store would take a crew weeks just to rip out the violations." If there is no simple answer, say so right off the bat. Then, when you have done with your discussion of the complexities, get back to the questions that inspired your customer to bring you onto the premises. Make sure your customer has not lost track of your answers, in the course of hearing your catalog of ills.

A responsible electrician has had customers say, "Is this safe to use?" and (1) told them that he had no simple, certain answer; (2) explained that the item seemed intact; (3) touched on how it was not up to modern standards; and (4) balanced that by saying that failure is nevertheless pretty infrequent.

After all that, with them nodding their understanding all along, customers have asked, "So is it safe to use?" The customers were not being dense or cute. He had lost them in the undergrowth, and needed to summarize the mass of information he had presented.

Offering the customer some perspective

It is unwise to guarantee that something is safe, especially when you are not the installer. "It's safe as they make 'em" is something you might say about a new installation you have completed. This is different from a flat "It's safe." Undoubtedly you guarantee (and trust) your work, but also carry insurance—just in case. An unkind aphorism says, "Nothing's foolproof; fools are too ingenious."

What you might do for a customer who wants a simple, seat-of-the pants answer is to give them a subjective opinion and label it as such. For instance, you might say, "I wouldn't worry about it if this were my home." Or, "I wouldn't leave an open knife switch in my factory, whether or not it had been there for 60 years. Sure, they're still in use. You could conceivably even run into, to use an appropriately painful expression, open knife switches in a residential high-rise. As late as the 1950s, apartment house panels in New York City used them in panels. But I hate 'em."

You can help confused customers by offering further perspective. For instance, you could tell them that you would put up with ungrounded receptacles in a relatively short-term rental; but you would upgrade, unquestionably, if you owned the building rather than renting.

Here is how an electrician offered two kinds of perspective in responding to a phone call from a woman about to buy a house valued in the six-figure range. She was concerned not about deteriorating wiring but about radiation from the nearby power line. The example nevertheless applies.

Without going into elaborate detail, he told her what he knew about the bioelectromagnetism controversy, and then asked her whether she and her daughter use a hair dryer. "Yes." An electric blanket? "Yes." Then, he explained, the transmission line would subject them to a field that is orders of magnitude smaller than that to which they already voluntarily exposed themselves. He did not go into the scientific subtleties, but still gave her the assurance that she sought, and that he felt was warranted. That was one type of perspective.

Would he live in such a home? she asked. Yes, he told her. He did not stop at that: he pointed out that some electrical workers seem to be at heightened risk for cancers, similar to some kids growing up near power lines—and yet he chooses to remain an electrician. For further perspective on his decision-making processes and his level of safety consciousness, he also mentioned that he has driven motorcycles for decades.

Making recommendations

The reason for giving an example of how an electrician talked to a customer about her concerns is that a list of findings is not sufficient to constitute a report. What makes a coherent report so important is that you are up against two related challenges: the prevalence of phony fixes and ignorant myth.

What good are quick fixes, and what harm can they do?

Phony fixes are the bane of the conscientious electrician. Phony fixes are solutions that, while often simple or lucrative or both, deceive the customer because they fail to address serious or core problems. The phone call described at the beginning of this chapter mentioned two often-phony fixes: service upgrades and upgrading outlets with three-prong receptacles. Both jobs are performed unnecessarily, and, worse, done wrong, by people selling cosmetic "modernizations." Done badly, these jobs expose buildings' occupants to far greater risk than was posed by the existing systems. Even when performed properly, even when in fact past due, these are phony fixes when applied to the problems of inadequate and deteriorated circuits and cables.

A two-prong receptacle can be simply replaced with a three-prong receptacle where—and only where—it is in a grounded box, served by an appropriate circuit, utilizing conductors protected by healthy insulation. Replacing such a receptacle is a worthwhile. though not necessarily urgent, endeavor. On the other hand, fooling people into thinking their wiring has been "modernized" because their receptacles have been upgraded, perhaps even to decora-style "state-of-the-art," is a dirty trick. First, it may deceive them into not worrying about the fire hazard created by deteriorated insulation and overloaded circuits. Second, where done improperly it can trick them into using grounding-type appliances. These certainly are more dangerous to use ungrounded than are double-insulated versions. They may well be even more dangerous to use with their grounds defeated than are old, pregrounding two-wire appliances in good condition. Arguably, the worst oversight in the government's widely circulated electrical safety checklist is that it urges consumers to use well-secured three-prong adapters in two-prong outlets but does not caution them about the danger of doing so without confirming the presence of a ground. This practice potentially poses the same danger as inappropriate replacement.

Service upgrades raise similar issues. Replacing a fusebox whose plug fuseholders are pitted or damaged, or whose cartridge fuse clips have lost their spring, is a worthwhile endeavor. Replacing a service whose meter base or cable is deteriorated makes excellent sense. Contrariwise, replacing a fusebox because circuit breakers are "more modern" is a deceptive practice when it is used to assure the customer, or the prospective tenants or purchasers of a building, that the electrical system has been overhauled.

The worst aspect of a quick-fix upgrade to circuit breakers is that it feeds on a common myth. As with most myths, this is based on finding an answer to some question or a cure for some discomfort that is, in Mencken's words, "short, simple, and wrong." The source of the discomfort is this: power goes off, and the owners or tenants

■ **6-2** *A panel with switched fuses. The nice aspect of this design is that unscrewing fuses is not the normal means of interrupting circuits. The nasty aspect is that such panels are ancient, and when a switch heads for Valhalla, you may not find a suitable replacement.*

have to change a fuse. The myth is that the blown fuse is the problem, and that substituting something that is easier to deal with reduces this problem. This is related to ostrich-like "kill the messenger bearing bad tidings" thinking.

Sure, eventually that tripping breaker will force the customer to call a more conscientious electrician. Eventually the real problem will be addressed. By then, the sharp operators who sold them the circuit breaker panel as a fix will have taken their money and run. The new electrician called in to track down and correct the source of the overload or short may end up making a big mess, spending a lot of time, and thus charging lots of money, with very little visible product as a result. It is hard to make that effort look good, compared with one that gave the customer a shiny new loadcenter at a fixed price. And no, the situation is not fair. That is why customer education benefits the more conscientious contractor.

More troubling than the questionable integrity of making easy money with a phony fix, though, is the fact that the take-the-money-and-run electrician left the customer with an unsolved problem, at least for a time. And unsolved problems sometimes cause fires, rather than tripping breakers. Even when faults do trip breakers, if the faults are overloads rather than short circuits they can cause two additional problems. One is that they can decalibrate the breakers, changing their tripping curves in unpredictable ways. The other is that they can pyrolyze structural members. This means that as the cables pass through wooden studs and joists, the wood can be converted into charcoal, weakening it and making it far more flammable.

That is enough of phony fixes. For myths, see Appendix 3. Back to safety surveys. The most important part of the report, for the customer, usually is your recommendation. It works better to separate that part from the description of findings. Findings are the part that give your recommendation credibility.

Specific recommendations: service equipment changes

There will be cases where the electrical service definitely needs upgrading, in which case you almost always will be installing a circuit breaker panel. In other situations, the need for a service upgrade may be less pressing, and the idea of putting in a new, roomy, circuit breaker panel may be attractive but not compelling. The probable 1999 NEC change eliminating the option of 60-ampere services for very small single-family dwellings will not, of course, affect existing services.

Fuses versus breakers There is plenty besides prejudice to go on in making protective device recommendations. In case you have not thought it over, circuit breakers have an inherent safety disadvantage compared to fuses. Fuses have only one failure mode—they can blow prematurely. This may be due to factory error, to having sustained repeated borderline overloads, to high ambient temperature, to corrosion, or to other causes. A circuit breaker can fail for similar reasons. To its disadvantage, it has more than one failure mode. It too can fail by opening prematurely—or it can open too late, or not at all. Sure, many prefer to deal with circuit breakers; it is generally easier to tell when one is tripped, and to restore it. But you cannot recommend them as inherently safer than fuses.

While there is some reason to be sanguine about the continued use of old fuseboxes that are in good condition, the same is not always true of all old circuit breakers. Some electricians shake their heads when they come across certain brands of panelboard. The reason is that they have encountered samples of various circuit breaker brands that "just wouldn't blow."

You will hear this accusation leveled at a number of brands. Among some electricians Federal Pacific (FPE)'s cheap thermal-only breakers, and, likewise, old Zinsco breakers had the very bad reputation of not tripping. Square D's XO series purportedly was rushed onto the market to compete with them. In the 1930s and 1940s, according to one source, Cutler-Hammer sold XO-type breakers as well. The Pushmatics®, manufactured by Bulldog, and later by ITE, at one point had a second-class reputation. Finally, there was a brief period when Westinghouse's and ITE's bolt-in breakers of the Quick-lag® type, in response to harmonic currents, would get much hotter than people were comfortable with, without tripping; apparently the designs were quickly changed, but at least some of the breakers that were in use probably are still out in the field.

This accusation, by the by, has not been leveled at the more recently manufactured GTE-Sylvania breakers of the Zinsco design.

No one can tell you what to do with this hearsay, in the context of a safety survey. The issue is what some describe as a can of worms. Rather than making an explicit recommendation, some will pass the hearsay on to customers and let them decide whether they want their existing panels torn out, just in case.

There are a couple of very simple tests that you can perform to determine whether breakers and panelboards, reputable or ill-reputed, warrant replacement. First, with a multipole breaker, especially, some advise, with a Pushmatic®, operate the handle and test with your voltmeter to make sure that both (or all three) poles come on and off together.

Second, with any old plug-in breaker, some would say especially with a Stab-lok®, make sure that the breaker makes tight contact with the busbar stab.

Some jiggle the wires, which also checks for inadequately torqued terminations; some tug on the circuit breakers themselves. UL tests new breakers to make sure the spring metal that carries current into the breaker, as well as holding the breaker in place, grips the busbar stab firmly. They do not, however, have any standard or test to guard against the possibility that the heating and cooling associated with years of carrying current right through that spring metal will cause it to lose its temper, and thus its grip, and then, under heavier loads its cool, and eventually its calibration. When you do find a defective termination or breaker, it is wise to remove it and look and sniff for damage to its underside, and to the busbar where it was attached, and even the adjacent positions. When the busbar shows damage, or many circuit breakers appear damaged, it is judicious to replace the panelboard.

Another concern is that circuit breakers have changed over the years in a number of ways. Interrupting ratings are one obvious way—from 5 kAIC to 10 kAIC to much higher. Also, in the mid-1960s, the ambient temperature standard changed from 25 degrees C to 40 degrees C. This offers a reason to recommend replacing old breakers when installing a new loadcenter, in some settings even one apparently compatible with the old breakers.

Of course, fuseholders do go bad, and some fuses get out of date. If a customer has more than a 10,000 kA available fault current, or needs to feed motor loads, Class H renewable fuses will not do. In those circumstances, even though the replacement links are still available for "renewing" the fuses, once you have removed those fuses you cannot stick them back in. The application no longer matches the fuses' listing.

Aside from overfusing, the most common problem with fuses is that fuseholders get loose as their spring tension relaxes. And the looser they get, the more the connection heats, furthering the metallurgical change and thus further marginalizing the contact. It would be nice to somehow eliminate that concern. At least one fuse manufacturer, Gould Shawmut, makes fuse clip clamps that perform that very function of enhancing contact between fuse clip and fuse. The permanency of that fix is unclear, and it, of course, does nothing to reverse corrosion damage.

Dealing with inadequately sized services This section of specific recommendations began by acknowledging that, in some cases, the existing service clearly needs upgrading. It is pretty rare to find an older single-family residence that cannot handle its loads on a 100-ampere service. Sure, there are still some 30- and 60-ampere services in use, even some two-wire 120-volt services (with either single- or two-pole disconnects). Generally, if additional circuits are needed, you will need to upgrade those minimal (under 100-amp) services even if the service equipment is in good condition.

A loadcenter can be inadequate to serve additional circuits—or even all those presently connected—despite adequately-sized service conductors and hefty busbars. In older, full-up panels, you frequently will find multiple conductors attached to one overcurrent device. This in itself may be a violation of the manufacturer's instructions, and thus of NEC Section 110-3 (b).

Such connections affect the terminal's heat dissipation. Also, the terminal may not be able to make reliable contact with more than one conductor. If those were the only problems, the fix would be simple: pigtail. There are at least two other factors that could complicate your recommendation. First, when circuits are doubled up, it is far more likely that the panel directory is inadequate, in violation of NEC Sections 110-22 and 384-13. Second, the circuit may be overloaded.

Worse, there is always the chance that one of the conductors serves a small appliance or bathroom circuit, which should not be shared with the other conductor's load. Before you can recommend any fix with certainty, you have to check out such possibilities. If a small appliance circuit is doubled up in violation of NEC Section 110-3(b), you may have to double up two other circuits with a pigtail, instead, in order to free up a space in compliance with NEC Section 210-52 (b)(2).

Subpanels can be an appropriate answer So the existing panel may spell trouble in any number of ways. A service upgrade can be a great, straightforward solution. With plenty of room, and no deteriorated overcurrent devices or busbars, a pretty, clean, new panel also finds favor in the eyes of customers and home inspectors. However, service upgrades are not always a significantly superior solution, even putting aside issues of cost. Subpanels are a very useful, less-expensive option from which to run new circuits, when the service itself is adequate and shows no deterioration. Thoughtful customers will appreciate your making a relatively conservative recommendation.

Subpanels: benefits By installing a subpanel, you can untangle that spaghetti. It is not terribly uncommon, among the smallest, oldest residential panels, to find six fuseholders, period. This was legal, in accordance with NEC Section 384-16 Exception 2. It is common in old residential panels to find MAIN sections with two to six pairs of fuseholders, all lacking overcurrent backup. Moreover, it is all too common to find single-pole 15-ampere branch circuits illegally fed from that section. The latter usage probably was illegal whenever it was installed.

Suppose you do add a subpanel. Now you can remove all those MAIN section fuses except for the fuseholder containing two 60s that protect the BRANCH section busbars. Remove the cables from those other MAIN section fuseholders, feeding loads requiring less than 60 ampere protection, and instead, run those cables to the fuses or breakers in the subpanel or subpanels. (The NEC does not limit the number of subpanels.) Now use a nipple, or a cable, containing three suitably sized insulated conductors plus grounding means, to feed the subpanel (or several of the same to multiple subpanels), from the 60-ampere, two-pole fuseholders you freed in the MAIN section.

Here is what you will have created in such a subpanel. First, you now have a main lug fuse or circuit breaker loadcenter, with room for plenty of circuits, so long as their loading as calculated under NEC Article 220 does not exceed 14,400 volt-amperes (60 amps × 240 volts), or 12,640 (60 amps × 208 volts). It has new busbars, whose rating is probably at least 100 amperes; and plenty of wire-bending room. If you want to use a GFCI circuit breaker, it can be accommodated. If you want to use an arc-detecting circuit breaker (AFCI), a protective device discussed at the end of this chapter, that too is possible. Even better, instead of backup (coordinated) protection of 200 amperes for your #14 and #12 conductors, as you might have provided in a heavy-up, you have provided much closer backup at 60 amperes. While you may not have formally coordinated tripping curves found in a single main breaker panel, knowledgeable experts have suggested you probably will not be far off.

Subpanels: shortcomings Despite all those good points, the subpanel option is not worth proposing where incoming service is undersized, or where the service equipment shows signs of deterioration. A subpanel will not do much good if the main overcurrent device blows or trips regularly. And of course, in itself a subpanel does nothing for the worn-out insulation on existing conductors. But subpanels are an option worth keeping in mind to help implement partial solutions to that problem, as described below, in addition to solving certain other problems.

Specific recommendations: addressing decrepit wiring

Now it is time to further explore the choices you can offer customers whose systems show serious signs of age, such that isolated repairs will not bring everything up to speed.

Just do it: replacement In the words of the quite-mortal Dave Barry, some systems should be taken out and shot. Immediately. These are uncommon, especially if you restrict the category to systems that have died from age. Yes, if the wiring is hanging together by the kindness of the gods, say so and advise your customer to break open the piggy bank. Other systems should be rewired now simply because the customers are game to have the job done, and system condition indicates that rewiring should be done sooner rather than later. Chapter 8, "Tearing It Out and Rewiring," talks about the process of major rewiring. How much of those details you discuss with a customer who has you in for a safety survey depends on your findings and the customer.

Hold off: living with bad wiring There are less radical fixes. Thoughtful electricians do not simply dismiss unfortunate customers who own buildings that need nothing less than rewiring when told that that big a job cannot or will not be authorized. On the other hand, neither do judicious electricians proceed to do minor electrical repairs and ignore the big picture. Problems that threaten imminent danger need to be fixed. If a customer will not go along with you on that, let someone else have that customer's work. Liability insurance is there to cover unforeseen eventualities. If you let imminent dangers slide without corrective action, it is all too easy to foresee a fire or accident in your customer's future, with a hungry liability lawyer chasing at its heels.

If the wiring is bad enough, you should not touch it. You do not want to work on the contents of an enclosure unless you can stuff away splices and restore the cover with a reasonable expectation that the contents will not short out. In those cases where performing minor repairs would not constitute a genuine service, and major ones are not a viable option due to customer constraints, you might consider offering the customer advice instead of repairs. First on the list of recommendations is the installation and maintenance of battery-operated or dual-mode (hard-wired with battery backup) smoke detectors. Second is anything else to enhance fire safety. If the customer is interested, you can mention issues such as planning evacuation routes, safe storage of combustibles, and, as a distant third, purchase of ABC-rated extinguishers. Do caution customers about overlamping, overfusing, and circuit overloading, and encourage them to distribute loads as evenly and gently as possible.

Preventive measures: retesting If the wiring does not demand instant replacement, yet is beyond its design life, perhaps it deserves periodic reinspection. Since no authority questioned has been willing to specify the expected longevity of wire insulations, it is not surprising that there are no official standards for evaluating wire. Yes, the Consumer Product Safety Commission urges homeowners to have old wiring inspected somewhere between every one and four decades, but their recommendations are lacking in specifics. Therefore, what you offer customers is merely going to be your personal sense of what is reasonable.

Certainly, one would evaluate wiring when purchasing or moving into a building. If a spot check finds any deterioration, or any old rubber-insulated conductors, recommend an exhaustive, complete survey. After that, every time a device or fixture needs replacement check the wiring in its enclosure. The rationale behind this suggestion is that the frequency with which equipment needs replacement is bound to be roughly proportional to the amount of use it gets. Even the number of operations which switches are subjected to correlates somewhat with the amount of energy handled; wear on receptacles and lights correlates much more closely with the amount of electricity used and heat generated. Current and heat over time are also the factors that cause wire insulation to deteriorate, and those two factors plus gross movement cause terminations to loosen.

Preventive measures: arc detection Here is an odd new technology that may be of great benefit for older buildings whose wiring presently cannot be ripped out and replaced. Arc-sensing circuit breakers trip not only when a circuit draws too much voltage or current, but also when the wave form of the voltage or current being drawn suggests a fault rather than a legitimate load. If you are concerned about the condition of the conductor insulation in a building, this technology, along with smoke detectors, will provide great protection—presuming that the new technology stands the test of time. Before recommending anything that is new on the market, you have to evaluate the pros and cons of promoting something that has not been broadly tested in the marketplace—but that could well save lives and buildings.

Several brands of arcing-fault-interrupting circuit breakers should be available on the market by the time this is in print. At least two different systems were prototyped in 1997. Square D's purportedly is more sensitive to the precise waveform; Cutler-Hammer's purportedly is less likely to be fooled by intervening equipment, such as power conditioners, that modify the waveform of the arcing wire or device. ITE/Siemens's lineups, including Murray and Challenger, also should offer such devices, and additional brands will get on the bandwagon if the products are well received. Some should be Classified to work with many brands of panel. A couple of bold NEC proposals would have required that most branch circuits in new residences be protected by AFCIs as of the turn of the century. The electrical industry, however, does not trust new, expensive products to be glitch-free. Should you recommend AFCIs to protect old wiring? They are relatively unproven, as were the early GFCIs. Nevertheless, they could prevent some fires. They certainly do not promise enhanced protection from glowing, as opposed to arcing, faults.

Another option you can suggest to slender-pursed customers is piecemeal replacement. Instead of tearing everything out, run new lines to heavily used outlets. This is where the empty slots in a new panel can come in handy. New lines can enhance safety significantly for customers who simply cannot see their way to major rewiring. Unfortunately, some of the most burdened outlets are lighting outlets located smack dab in the hard-to-get-to center of the ceiling. Still, adding a few new circuits will enhance safety far more than many a shiny service upgrade performed without replacing rotten branch circuit wiring.

Setting Limits and Avoiding Snares

THIS CHAPTER IS ABOUT SETTING BOUNDARIES THAT WILL protect your licenses, your cash flow, and your customers. Two related issues have been discussed already: system safety and communicating with customers. (The latter is more extensively addressed in Chapter 15.) The present focus is more legalistic. Recognize that this book does not pretend to be a substitute for a lawyer, and only your own local attorney can tell you how to protect yourself in the jurisdictions where you work.

Taking on somebody else's problems

One thing you are guaranteed to encounter when taking on old work is wiring that someone else has installed improperly. Assuming responsibility for such installations as though they were your own is a sucker's game. Yet it is not reasonable to insist that anything that bears the taint of not having been done professionally, and covered under a permit, always be ripped out entirely.

The last person in can be left holding the bag

Inherited installations call for judgment on the part of both contractors and inspectors. Differentiating between contractors who rescue messes and devil-may-care sorts who rent out their licenses is a necessary part of the inspector's job. Sometimes, jurisdictional authorities make it easy for the conscientious contractor. Right on the permit application are boxes to indicate, "This job was/was not begun by someone else." And just as inspectors may well perform interim inspections when a job is taken over from a contractor or sub who was fired, they may be willing to walk through some old hodgepodges and say, "This we'll let slide; that one you've got to yank out and do right."

Different inspection departments work in different ways. If a simple phone call or office visit fails to yield clear guidance, there are

**PRINCE GEORGE'S COUNTY, MARYLAND
DEPARTMENT OF ENVIRONMENTAL RESOURCES
CONSTRUCTION STANDARDS DIVISION
COUNTY ADMINISTRATION BUILDING
UPPER MARLBORO, MD 20772
952-4456**

Permit Processed By:

_____
Permit Specialist

_____ _____
Signature of Permit Specialist   Date

APPLICATION FOR ELECTRICAL PERMIT AND INSPECTION
(Please Print or Type)

PERMIT NUMBER

Date _____

Has this work been started by another? _____ Yes _____ No

Permit No. _____   Tax Map # _____

Work Location _____
                    Street No.   Street   Town or City   Inspection Area

Lot _____ Block _____ Subdivision _____   Land Property Tax # _____
                                                                              Election Dist.   Account No.

Building – New ☐ Old ☐   Work – New ☐ Additional ☐   Building Permit No. _____

Owner _____   Owner's Address _____ Apt. # _____

Occupant's Name: _____

Phone (Res.): _____ Work: _____   Property Use: ☐ Commercial   ☐ Residential

EQUIPMENT AND WIRING

NUMBER OF ROUGH WIRING OUTLETS			NUMBER OF FIXTURES				MOTORS AND GENERATORS	APPARATUS– MISCELLANEOUS
Switch	Lighting	Recepts.	Medium Base	Mogul Base	Fluor-escent	Mercury Vapor	No. and H.P. or K.W. of each	

NOTE: DO NOT REQUEST INSPECTIONS ON DAY OF PERMIT ISSUANCE.

Type of Work: Open ☐   Concealed ☐   Armored Cable ☐   Non Metallic Cable ☐   Conduit-Moulding ☐   Knob & Tube ☐

Current Supplied by: PEPCO ☐   BALTO. G&E CO. ☐   SO. MD. CO-OP ☐

To expedite all inspection requests, call the Automated Inspection System by touch-tone phone 925-5390 from 7 am to 11 pm daily Mon. through Fri. If no touch-tone phone is available, contact the Dispatcher at 952-4914 7:30 to 4:00 Mon. through Fri.

Homeowner Permit ☐   Contractor/Master ☐   License No. _____

_____   _____   _____
Company Name/Applicant   Master Signature/Homeowner   Phone

_____
Address   Street   City or Town   Zip Code

Date	Type Insp.	Inspector

PERMIT APPROVED: (Application becomes permit when approved by Chief Electrical Inspector.)

CUT IN CERTIFICATES	
Temporary	
Final	

TOTAL ELECTRICAL PERMIT FEE: TREASURY RECEIPT STAMP:

$ _____

P.G.C. FORM #159 (Rev. 6/91)

PERMIT NUMBER

TREASURY RECEIPT STAMP:

ACCOUNT NUMBER	FEE COLLECTED
RESIDENTIAL GF01420134541102	$

ACCOUNT NUMBER	FEE COLLECTED
COMMERCIAL GF01420136541102	$

■ **7-1** *This jurisdiction recognizes the history an installation can carry, complicating issues of responsibility.*

More Specialized Issues and Situations

other ways to learn what should be acceptable. Your local contractors' association is one avenue. The local chapter of a national group is another—the National Electrical Contractors' Association, if you are union; Independent Electrical Contractors, Inc. or Associated Builders and Contractors, if you are not. The International Association of Electrical Inspectors, IAEI, is a group devoted to consistent application of the NEC far more than it is a professional guild. It has easily twice as many associate members, such as contractors, as inspector members. Meetings of its local chapters are excellent places to ask, "How do inspectors think this should be handled?"

These are the main national and international electricians' groups that have local chapters:

National Electrical Contractors' Association (exclusively union) 301-657-3110.

International Brotherhood of Electrical Workers (the union) 202-833-7000.

Independent Electrical Contractors Association (nonunion) 703-549-7351.

International Association of Electrical Inspectors (noninspectors are associate members) 214-235-1455.

What if none of those groups are in your area, and there are no inspectors to query-or you are the inspector? If you can spare the time, you can turn to Code authorities at the major electrical magazines. Sometimes you can seek guidance from a manufacturer, a product's original Listing laboratory, or the National Fire Protection Association. Unfortunately, these latter resources are of a whole lot more help when you seek to understand the application of new products than when you want to know how the Code should be applied with regard to existing violations.

When you do take over a job that someone else botched, especially when the nature of the work is such that your jurisdiction requires a permit, there are two intelligent choices. One—and it may be necessary—is to rip out all that has come before. The other is to take on the job of correcting and completing what was started but to notify everyone who might have a direct interest. Interested parties include the inspection department and also at least arguably the owner, if you are hired by a lessee.

THE PRINCE GEORGE'S COUNTY GOVERNMENT

DEPARTMENT OF ENVIRONMENTAL RESOURCES
Construction Standards Division

December 20, 1989

David E. Shapiro, M.S., M.Ed.
Safety First Electrical Contracting,
Consulting and Safety Education
3419 Forty-First Avenue
Colmar Manor, Maryland 20722-1904

Re: 809 Montrose Avenue
Laurel, Maryland

Dear Mr. Shapiro:

As I indicated in our telecom yesterday morning, I had the opportunity of inspecting this house and must agree that much of the electrical work is in violation.

Inasmuch as the work is of a type that does not require a permit, I would suggest you develop a plan with the owners to make the repairs, starting with the most urgent work first. If the owner feels that she wants it inspected by us then of course a permit is necessary and the work must be completed within one year, i.e. before the permit expires. The description of the work to be performed on the permit can probably be best identified by the words "To make repairs and corrections to electrical work performed by others working without electrical permit". The cost will be $20.00.

Thanks for bringing these shortcomings to the owners and our attention.

Respectfully,

Arthur W. Hesse
Chief Electrical Inspector

County Administration Building — Upper Marlboro, Maryland 20772

■ **7-2** *A stalwart inspector respectfully acknowledging a contractor for consulting him on encountering a mess of violations.*

Such work is still fraught with risk. Unless you can examine every inch of the wiring, you cannot be certain of its condition. But that is true of any existing wiring, newer or worn-out, installed by amateurs or professionals. When problems emerge, it always is easy to blame the last contractor to have touched the job. A policy of "Let the buyer beware" on the part of a skilled tradesperson working on a home does not impress a judge. That precedent apparently was established way back in 1957, in the Ohio case,

Vanderschrier versus Aaron. (The reference is in Chapter 14, Resources.) It simply means that, as you well know, your expertise
saddles you with responsibilities that persist after you finish a job.
Legal concepts of liability are discussed in some detail later in this
chapter.

Amateur assistance

Almost every electrician has opportunities to perform "old
work," where someone else has been previously, and thus faces
this problem. Another set of problems and decisions is associated with a different option, one proposed more frequently in
cases where modest or poorer residences require major rewiring.
That option is sweat equity, a term that can mean different
things. You may not even want to evaluate the possibilities and
perils associated with it. Even a contractor who will consider taking on "work-with" jobs will not touch some other varieties of
what is called sweat equity. If you are sure that you would never,
never consider letting a customer help, and never pick up work
that an amateur has touched, you may want to skip ahead to the
sections on "Consulting" and "The long and short of liability"
near the end of this chapter.

The unworkable: do-it-yourselfers

Here is an opportunity that could mean trouble. A customer calls
and says, "Can you look over the wiring I did? My Home Wiring
book said to leave the last part to a professional. I don't want to
go inside the breaker box to hook things up " This is a sucker
pitch. A suitable answer is, "I'd be glad to look it over. However,
I'm not about to put the final touches on what you started;
I could risk my license. That said, here's what I charge for consultation."

A home repair book that tells readers to start a job and expect a
professional to complete it performs a notable disservice. When
someone asks you to "finish my job," there are likely to be hurt
feelings if you take on the assignment and treat it with the caution
it deserves. You would be well-advised to go over every inch of the
installation, ripping out and redoing anything that looks questionable. You also would need to address permit issues. Many homeowners neglect to pull permits to cover their work. Cocky
do-it-yourselfers frequently object to the expense of permits and
to the nuisance and intrusiveness of inspections. Even when they
have taken that trouble, once you take over the jobs you are

responsible to pull new permits in your name, or to transfer the existing ones.

The potentially workable: "work-with"

There is a place for sweat equity. Consider this situation. You look over an old house and give your customer the bad news about its wiring needs. Your customer, a thoughtful type, says, "This sounds like a big job. Way beyond what we can afford. I know we don't have a guarantee from heaven saying we have the right to own a home, but we've got this house and we plan to keep on living in it. I realize from what you've told me that the wiring is pretty dangerous. I wonder if there's some way we can both come out okay. I'd like to give you my business. Can I maybe do some parts of the job, to save money—under your supervision of course?"

This sounds reasonable, but there are pluses and minuses to the proposal. Before agreeing to any such arrangement, you have to make sure that liability issues can be handled. This is not something trivial, something you simply can assume will be okay. Conceivably there could be liability associated with accepting cooperation as minor as the customer performing nonelectrical parts of the job such as patching and cleanup.

The arrangement also has to fit how you want to do business. If your business plan revolves around keeping crews of journeymen and apprentices busy, sweat equity may have no place in it. If you choose a different view of what you are about, sweat equity may make sense if it is a good use of time. The risks will be addressed in greater detail after a number of the advantages have been described.

The pluses of "work-with"

First, when customers work next to you, they know without any doubt that you have given value for money billed. They have sweated, and they have seen that you sweat. Furthermore, unless they are accustomed to the exercise, the work is probably a lot harder for them than for you—making your accomplishments look that much more impressive. It is nice to move away from distrust and adversarial relations to a sense of being on the same team—even the same crew.

Second, it introduces a fudge factor, one that a reasonable customer will readily recognize and accept. Whenever you deal with old wiring, indeed with old buildings, there are many small uncertainties that make it hard to come up with a precise grand total estimate. You have to gamble on what you will encounter, so you need to figure some slack into your estimate to account for unknowns. If the labor to be factored in includes some amount of your customer's efforts, that unknowable quantity patently opens up the estimate.

A third advantage is that when your customers are working with you, it becomes easier for you to help them understand problems. They are right there to make decisions when the need arises.

Example: having the customer handy The following example illustrates the benefit, to you and to your customer, of having the customer at hand versus the cost of not having him or her available.

An electrician was called in by a general contractor, John, to troubleshoot a malfunctioning ceiling light fixture. He discovered that the wiring to it was severely deteriorated, damaged to the point where cutting back conductors would not get to reliable, healthy insulation. He told John that the fixture needed rewiring. John gave him the go-ahead to start pulling fresh wire in through the conduit. (John had given his customer an estimate before calling in the electrician; the electrical portion, appropriately, incorporated a fair amount of slack.)

Replacement was not as straightforward a job as the electrician had hoped, due to a very roundabout wiring layout. John was right there with him as he started replacing the wires going out the first of four conduits feeding the fixture box. In fact, John gave him a hand on the pull. So John knew the need to revise the original labor estimate was on the up-and-up. By the time he had completed that first pull and respliced at the other end, they were already an hour-plus over John's initial guesstimate.

Unfortunately, John did not feel as though he had the power to authorize replacement of the rest of the wires, even though he could see that they too were shot—albeit not quite as badly; the electrician had wisely started by attacking that which looked worst. With the customer out of town, he was stuck. (Your options in such situations will be discussed a little later.) Had the customer been at hand, rather than his GC, the electrician would have been able to proceed. As it was, he had to patch the insulation well enough to button things up. (Had that been impossible, he would have had to leave power to that part of the wiring turned off.) This way, they had to have him back later, which entailed the extra expense of a second trip.

A fourth advantage to sweat equity is that working alongside makes it easier to teach—and drive home—lessons about electrical safety. "You see this light's rotten wiring? It's no older than the wiring at your stairwell light, but look how much worse it is. That's what I am concerned about when I warn you about overlamping." Besides underscoring what you have taught your customers about safety, such demonstrations of your expertise reduces their inclination to dismiss or second-guess your advice and decisions.

A fifth advantage is economic leverage. Permitting customers to work alongside you is a special service. This is true both in the sense that such a service may be unique in your area and in the sense that it makes unusual demands on you. For both reasons, you reasonably can charge a premium.

A sixth advantage is that by entertaining the work-with option, you can put safe wiring in homes that might otherwise burn. People in every profession and trade contribute services to community charitable projects or to the destitute. The "Habitat for Humanity" project is in fact a sweat equity program. It is to your credit if you choose to participate in that sort of noble enterprise. Sweat equity jobs allow you to offer a similar benefit to a slightly different population—with no sacrifice.

There is a level of financial health intermediate between rich, or even comfortably well-off, and poor. Homeowners surviving at that level, while not penniless, can feel caught in the "working poor" or the "house poor" trap. They cannot afford to hire someone to rip it all out and do it right. Yet their wiring is bad enough that a radical redo really is what they need. This is where you come in. By offering the sweat equity option, you can enable them to greatly multiply the value they derive from your services and make safety affordable.

Some people who have earned the right to be taken seriously are afraid that you can create a monster this way—a layperson, quite ignorant, who believes that he or she can do electrical work without knowledge of the NEC, supervision, or sufficient training (not to mention a license or permit). Such monsters abound. But consider—the presence of fussy master electricians looking over laypersons' shoulders makes it less likely that they will develop into such monsters than if they are left to rely on how-to books and home center seminars.

The risks and drawbacks associated with "work-with"

Even if you decide to specialize in old wiring, there are excellent reasons to reject the sweat equity option. First, you need to find out whether you can keep your liability coverage untainted. Is anybody going to demand that you cover your customers under Worker's Compensation? Are you going to be considered liable for any injury they incur? Are you going to be considered liable for any damage they do? Only your insurance agent or attorney can answer these questions.

A second basic issue is whether you (or your foremen) have the time and patience to supervise ignorant and possibly wilful novices. Do you have the ability to set and keep limits? Even charging premium rates, does it make enough sense, given the advantages listed above, to put in one-on-one supervisory time? If there is enough work around, and you have enough reliable journeymen and helpers, you could employ that same supervisory effort to run a crew, earning profit off every man-hour on the job worked by every crew member.

Third, another question you must answer, on a case-by-case basis, is how well you trust your customers. Is this family going to blame you if they hurt themselves or break something, even though you have made it clear that they have to be completely self-responsible? Are these persons going to accept your dictates in the role of job supervisor—boss—even though they are signing the checks?

One nice thing about sweat equity is that you never have the disappointment of paying, or the hassle of firing, somebody who turns out to be a goof-off. Furthermore, somebody working on his or her own house is likely to be quality-conscious. But the latter is not always true. This person you have agreed to supervise is equivalent to the rawest apprentice, with one difference. At home he or she is used to being in charge. This can be an explosive combination. One unpleasant situation is where the customers have second thoughts about putting in all that labor—but still magically expect the installation to cost as little and proceed as fast as it would if they were doing their part.

The final question coming out of that has to do with your fall-back in the event that they change their minds. You need to be prepared for the possibility that your customers will decide that the work is too much for them to handle, and you then have to take on the whole thing yourself. Suppose they decide that they just cannot afford the work after all, given the effort it requires of them or the hours you need to spend? Suppose they decide part way through that they want to hire the neighbor kid to work with you in their place? Or, alternately, suppose that they think they see what needs to be done and want to take over, terminating your contract? Contingency plans are important. In cases where people are committing themselves to new and unfamiliar behavior, it often is well worth your while to review the implications of contract terms with your customers.

Example: Wilful customers Customer disobedience may not affect life safety. It still can cause problems. Each of you must decide just what tasks you will and will not allow a customer to undertake. Surely patching drywall or plaster is a job that most electricians are willing to leave to another—sweat equity or no sweat equity. It is a simple enough task, and frankly not one of earth-shattering consequence. Yet that patching is your responsibility, not the painter's or carpenter's or customer's. There is no need to call on the "neat and workmanlike" clause of Section 110-12. If that part of the job is not performed, your installation is in violation of NEC Section 370-21.

Where customers have forgotten or blown off that assignment, easygoing contractors (who failed to check) have nearly gotten nailed on that clause by inspectors. In some cases, the contractor has happened by and said, "Whoa! You were going to patch those openings!" In others, the inspector has missed the oversight. In yet others, the inspector has generously said, "I'll pass this one, but I expect you—not the customer, and not some handyman—to fill that hole."

Contractors who offer sweat equity explain to their customers that it is necessary to complete all their tasks before scheduling the inspection, but these oversights occur despite that precaution. Sure, if the work were red-tagged and the contractors had to make extra trips, they could bill the customers for those call-backs. Doing so would not make them look any better to the inspectors. Certainly it would be better to avert the problem.

This type of experience underscores the need for vigilant supervision when working with amateurs. You would be well-advised not to let customers pull wire or cable, at least not unless you spend as much time teaching them the appropriate care and the type of force needed as you would spend teaching a green helper.

Policies and procedures for "work-with"

Should you decide that some sort of sweat equity arrangement looks workable, you have to decide how to work it. One piece of advice is to make no assumptions. Just because a customer has "worked construction" does not mean that he or she knows wiring, or how to follow instructions, or even how to drill a straight hole through a joist. Being an engineer, an electronics whiz, or even an electrical engineer or electrical salesperson does not mean someone has the appreciation to treat 120 volts with suitable respect. Read Chapter 15, "Communication."

Consultation

That is enough about sweat equity, an option that only the few and the hardy will take on. In dismissing the sucker job, the previous discussion touched on a position carrying an intermediate level of risk—consulting work. There is a real and valuable place for consultation, and it is laudable when amateurs get professionals to double-check what they have done or what a home inspector has reported. When a responsible homeowner has pulled a permit and done his or her own wiring, the jurisdictional inspector tends to spend more time reviewing that job than he or she would spend on a professional installation. Even so, it certainly does not hurt when the homeowner hires a consultant to give the installation a first look-over and to answer questions, before the inspector is called. As a consultant, you are not doing any of the work yourself and you are not covering the installation under your license. You do need to make these limits on your involvement clear, in addition to making sure the customer understands that your consultation does not substitute for a jurisdictional inspection or for the specialized services of an architect or P.E.

If you become a niche specialist, consulting opportunities eventually will come your way. Whether you are called upon by lawyers looking for expert witnesses or by customers left in the lurch by irresponsible contractors, you will have something valuable to provide as an old-wiring expert. It still is your decision to take offers or leave them.

Example: Just say no (thanks)

Saying "No thanks" is part of professionalism. An electrician who advertises his consulting at least as widely as his contracting received a telephone call asking whether he could inspect a duct or plenum. There were odors, the caller complained, and he suspected that gases were leaking. Could the consultant come out and check?

His answer was, "That is outside my area of competence." His going over and sniffing would not have been worth the $125–150 or more that the consultation would have cost. Sure, he could have poked around, albeit not as knowledgeably as an HVAC contractor. If action had seemed called-for, action more sophisticated than, say, removal of an odiferously dead rodent, his recommendation would have been uncomfortably like guesswork. If a report or deposition had been called for, he would not have been a particularly credible witness. So his choice was to refer the customer elsewhere.

The pluses to consulting

The pluses to consulting are simple. You can charge more for this work than you can for straightforward contracting, and receive more respect. You can do work that faces less competition. You can spend less time sweating at repetitive "grunt" tasks, or supervising them. You can add variety to your workload, doing something out of the ordinary. Finally, you can make a name for yourself; doing so can feed your other business.

The minuses to consulting

Consulting is not an unmixed blessing. Some people cringe at the thought of lawyerly dealings. Who wants to be cross-examined, to face an opposition lawyer's challenges to his or her competence and honesty? Who has a conscience so clear that he or she can talk about how another contractor has screwed up without thinking that at another time the tables might be turned?

Then there is the issue of "What do you mean, 'consulting' fee?" This is not likely to be a problem when you are called to be an expert witness. Nonlawyers, people who are not used to dealing with consultants, may need to be educated. If you have been in business any length of time, you will have run into people who think—who indeed expect—to receive design or troubleshooting service gratis if they call contractors and ask for free estimates. The way the electrical industry has developed, this is not a particularly off-the-wall expectation, at least when it comes to new work.

Finally, there is the theory that the more you expose yourself, the more roles you take on, the more vulnerable you become. Consider how construction contractors get sued. According to Steven M. Siegfried's 1987 book, *Introduction to Construction Law*, there are four common theories of liability used to protect the purchasers of new homes from bad builders:

☐ One is violation of an implied warranty of habitability or fitness. (Flick the switch, and a light should go on, written guarantee or no written guarantee.)

☐ A second is violation of an express warranty. (Your contract said you would replace the heat pump if it blew up in its first year. It did; you did nothing to repair or replace it.)

- [ ] A third is negligence, where you fail to exercise a reasonable standard of care, diligence, skill, and ability.
- [ ] The fourth is the rarely applied theory of "strict liability," meaning that the house just is unreasonably dangerous.

As a consultant, it is conceivable that you could be held to the same legal standard as a design professional. On top of the standard ways a construction contractor is likely to be sued, this adds a number of additional ways to get into trouble:

- [ ] One is breach of a duty to disclose known defects.
- [ ] Another involves a third-party beneficiary contract, where someone else besides you and your customer expects to benefit from your work, and is disappointed in that expectation.
- [ ] If your plans, or your assertions about what was likely to come up, failed to foresee serious problems, you potentially can be charged with negligent misrepresentation.
- [ ] Another odd charge is that of impropriety as an arbitrator. To be exposed to that liability, you do have to arbitrate over some disagreement. (Arbitration, by the way, is a profession; perform the work of any profession without suitable training, and by golly you expose yourself to liability.)
- [ ] Finally, *respondeat superior* is the charge when employees or subcontractors cause any of these problems, and you are held responsible for the consequences of their actions.

Example: poor boundary maintenance—being too easygoing can mean trouble

Here is a consulting assignment that went far from perfectly; you can learn from the experience, as did the electrician in the story. Very frequently, "work-with" and "consulting" are taken as coded language for the illegal practice of "renting out your license." In this example, the electrician hired on only after demanding and receiving assurances that he would have the authority that goes with taking responsibility for an installation. He believes that his trust was betrayed—possibly not because the customer was trying to cheat anyone but merely because the customer was impatient and was less concerned about the proprieties than was the electrician. Judge for yourself.

A nonlocal contractor was hired to install a telecom installation— no power wiring at all. However, under the local electrical code, it required a permit and inspection. In order to punch down their 22-gauge pairs into 66 and 110 blocks in racks for voice and data, all the telecom installers had to apply for electrical apprentice licenses and work under the supervision of a local master electrician. (Incidentally, the installation was still covered under the installer's insurances.)

In order to avoid creating unnecessary billable work, the master decided to supervise only that part of the job which he considered warranted the attention of a master electrician: the interface with the electrical system proper. While he did require that the telecom folks familiarize him and keep him up-to-date with the rest of what they were doing, he focused on system bonding. Once bonding had been handled, he did not hover around the job. Instead, he waited for them to call him in as they entered each new phase. Then there was an interval of about a month during which he did not hear from them.

He called in and learned that they had finished, settled with the inspector, and left. He should have been part of that loop. He had failed in his responsibility to authorize, and possibly participate in, the closing inspection, because he had trusted the installers to stand by their promise to keep the electrician informed. In this case no harm was done . Still, his not wanting to take advantage had exposed him to an awkward situation, wherein they took advantage of his license.

The long and short of liability

"Last person in gets the blame" is not a formal legal doctrine. Whether you are held liable for fire, accident, or follow-up repairs depends not only on what you did and did not do but also on your connections, your luck, and your lawyer. Still when you put yourself in the position of being "the last person leaving the scene of the crime," you had best cover yourself by carefully documenting what you did and did not do—especially if you, or your insurance company, may be viewed by a rapacious lawyer as having "deep pockets."

Law books tell you that negligence takes four elements:

☐ You have a responsibility to someone.

☐ You violate that responsibility.

☐ They sustain damage.

☐ The damage is not merely subsequent to but actually the result of that behavior or lack of behavior on your part.

Despite general agreement on these elements, there is still plenty of room for argument about whether an electrician was negligent. Did his or her judgment error result in work that can be characterized as showing less than "normal care by the standard of a reasonable person"? Did it fail the "highest standard of care of a prudent person reasonably skilled in his occupation"? Different judges have used these different ways of looking at the four elements.

Was his or her contribution a "proximate cause" or "immediate cause" or "direct cause"? Different judges define these differently. Some have said that only if you knew of a danger did you have (and fail to utilize) the "last clear chance" to warn and thus protect your customer; others assume that if the danger was "reasonably obvious" you knew of it.

Take it from the chief

Alan Nadon, a very experienced chief inspector, and a guy who is not shy about carrying a screwdriver and a tester, summed it up from his perspective. (Not being a lawyer, he is not offering legal counsel.) He said,

> The electrician's responsibility starts as soon as he answers the phone, or looks at a job, or turns a screw, depending on the quality of [his] legal counsel. A permit is usually the cheapest insurance an electrician can get. If a building has an electrical fire the day after you finished working on it, and you had it inspected with no violations noted, your lawyer will be smiling. The cause had to be a "hidden defect" that was not evident to a person with "superior knowledge," i.e., you and the inspector...In my opinion a permit will not save you if it is obvious that what you did was done in a way that would cause loss or injury. It may be a help in proving that you intended to do right. Alternately, the lack of a permit starts with the presumption that you intended to do wrong...[With regard to the question] about whom you warn if you see a hazardous situation: The basic rule is that the property owner is responsible for his property. The property owner is assumed to lack the knowledge that the electrician possesses. Your responsibility is to explain the problem to the owner. The owner's failure to take action can be used against him at a later date if there is injury or loss.

Incidentally, to a lawyer there is ambiguity even in the language of this plain-talking chief. If the inspector misses something, Nadon presumes it can be considered a hidden defect. Most electricians certainly would not argue the point. Can we rely on that interpretation? One lawyer tried arguing that a pendant double lampholder with a lamp in only one socket constituted a "hidden defect," because the open socket presented a shock hazard. Happily, the judge slapped him down.

This is not to say there was no legitimate safety consideration in the lawyer's argument, even though fixed wiring is not involved. The owner of a rental property in a college town (College Park, MD) was observed busily inserting lamps in unused ceiling lampholders. "Why?" asked his electrician. Because, he explained, the following week he expected the annual visit from the town's safety inspector—and empty sockets get safety citations, in that burg where most tenants are college kids!

Example: good boundary maintenance—agreement, documentation, remediation

This chapter will close with the story of a minor triumph. It is not about sweat equity or about getting in trouble. Instead, it concerns a job that started out as consultation. It points out a new aspect of taking on jobs where predecessors may have ignored the Code. Sometimes you have to fire up your creativity and do substantial research in order to leave your customer feeling better off for having called you in.

The job started as a safety survey for an elderly—but not dull— widow. For several reasons, she was concerned about the age of her wiring. First, now that her husband was dead, she wanted some reassurance that as she got older and perhaps feeble she would not be caught in an electrical fire. Second, if the house were sold, she did not want to be surprised by the prospective purchaser's inspector. Third, a neighbor whose house was of similar vintage had suffered an electrical fire.

It required no tools to recognize one part of her wiring as a clear, blatant, violation; correcting it in an acceptable manner was, however, a challenge. The light illuminating the front steps had a cable sneaking out from under its canopy to feed an illegal security light that had been added later at the carport. NEC sections violated include 110-3(b), 370-17(b), and, as could be seen once the fixture was removed, 370-20, and 370-25 (b). (Interestingly, Section 370-21 does not apply to building finishes other than plaster and plasterboard, so leaving as large a hole around a box as one wishes is quite legal, unless an inspector wishes to invoke the dread general duty clause of NEC Section 110-12.)

a

b

c

d

■ **7-3a–d** *There is no question that when installed, these were clear violations— and could never have passed a competent inspection.*

e

■ **7-3e** *Previous figures (7-3a–d) showed the illegal way power was obtained at the entrance to the house. This is where it went: an open splice to flexible cord feeding a fixture on the underside of the carport roof, without an enclosure.*

What had happened is that, more than 20 years earlier, the customer had a carport added. A security light was included as part of that installation. The customer clearly thought that it all was covered by a permit and all passed inspection. She and her husband knew no better. Now, in response to the electrician's survey report, she went through the papers more carefully, and to her dismay found otherwise.

a

b

■ **7-4** *These show the new, improved arrangement.*

More Specialized Issues and Situations

He told the customer that he saw two straightforward ways to make things legal and safe. One would be to remove the cable that snuck out from under the fixture canopy. Instead of the previous lash-up, he would feed a legal carport fixture from a separate switch run from an outlet located on the wall nearest the carport. That option was unacceptable. For security, she wanted to control both lights from the same wall timer/switch. (Running cable indoors from the existing timer over to the other wall was out of the question.)

The other straightforward option was to install an extension box over the gem box recessed in the wall, using the existing UF cable to feed a legal carport fixture, but properly securing the UF to the extension box through a weatherproof connector. Unfortunately, local supply house personnel all told him that their outdoor fixtures "could be made to work" mounted over a gem box extension. In other words, they were not Listed for the purpose, but the salespersons were alas not knowledgeable or honest enough to say so. This suggested that the only two ways to replace the light by the front door were either to use a weatherproof lampholder, plain or fancy, mounted through a threaded knockout on the weatherproof box extension or its cover plate, or else to blank off the extension and run from its end knockout to a nearby weatherproof box, surface-mounted, that could support a slightly more attractive fixture.

141

While these options were preferable to chopping into the wall, they were quite unsatisfactory to the customer. Fortunately, research revealed a happy alternative. One mail-order outfit, Ruud Lighting, sells a surface-mount weatherproof box onto which a number of their reasonably attractive weatherproof fixtures can be mounted. (Stonco, which sells through electrical distributors, also has products that would have done the job.) The nicest aspect of the design is that this box has a large rear knockout, rather like those found in the backs of many modern surface fluorescent fixtures, allowing him to install it as an extension, mounted over a flush (in this case, actually recessed) box. Like other weatherproof extension boxes, it also has threaded side knockouts that he was able to use to install a weatherproof connector and run the UF cable to the new security light—as well as to install a short EMT 90 to a new, GFCI-protected, outdoor receptacle. The lack of such an outlet was not a wiring violation, given the age of the house; but its lack certainly was a safety hazard, grandfathered or not.

Tearing It Out and Rewiring 8

THIS CHAPTER FOCUSES ON MAJOR REWIRING. IT REVIEWS how to decide that it is time to recommend rewiring (an issue previously touched on in Chapter 6, "Safety Surveys") and discusses the options for doing so as elegantly and painlessly as possible.

Reasons to choose—or avoid—major rewiring

There are several rather selfish reasons to put off rewiring, but not only are none of them especially worthy, they do not pay off in the long term. Why might you avoid tearing out the existing electrical system? For one thing, while rewiring is a big, potentially lucrative job, it is likely to eliminate further need for your services in the foreseeable future. Furthermore, rewiring, especially as a "gut" job, demands far less skill than does a conscientious job of patching and extending the life of frail fixtures. This means that you face low-end competition. Lastly, most customers will hate the idea of having major rewiring done, because of the mess and the cost. For some it will constitute a genuine hardship.

Contrariwise, here are two subjective, selfish reasons for pushing the total-tear-out option. First, it eliminates the uncertainty that otherwise would linger after you walked away from the job. Second, while the work faces more competition, because it is easier than troubleshooting and piecemeal remediation, it also may enable you to make efficient use of your less-skilled employees.

All of these can be persuasive arguments, but they do not address the most important point. Sometimes, tearing it out is the only right answer—the alternatives are not even halfway safe or suitable. When that is your clear judgment about an electrical system, service calls that consist of diddling around with the old wiring are as ethically questionable as are inspection visits spent chewing the fat instead of examining the wiring.

When is rewiring critical?

The time to recommend rewiring is when you discover how bad the present system is. Here are two examples.

An electrician was hired to repair a flickering, slow-starting fluorescent fixture in a kitchen ceiling. Its ballast needed replacing, and, because the fixture's original installation had been very far from legal, he spent an hour correcting its wiring. In the course of redoing the connections, he cut back on cables containing rotten insulation.

Did the house need total rewiring? He could not know. He did warn the customer that other ceiling fixtures could well be in equally bad shape, but he had no way to know without checking them.

A few weeks later, he was called back because wires were shorting at a switch box. He discovered that even there, at an enclosure not located near a heat source, the insulation was so bad that an incoming conductor had shorted to the cable connector. Thus he knew beyond reasonable doubt that the system was shot. He did tape the wires to get them working temporarily, but he also told the customer that she needed her house rewired. He knew she could not afford rewiring, but it was still his obligation to warn her.

The second example is not all that different, except in nuance. An electrician was called in to troubleshoot a receptacle that had gone dead. He traced the circuit up to a light fixture, where he found rotten insulation—but nothing broken. All conductors were continuous. He mentioned the condition of the insulation to the customer and kept hunting. The second light fixture that he examined was just as bad. One of the solderless connectors had burned off from a splice. At the third lighting outlet that he examined (and that again disclosed wires with rotten insulation), he located the bad splice responsible for the service call (the connector was completely gone) and told the customer that the wires needed to be torn out. He offered to put the four outlets back together, but warned that the best he could do in restoring them without replacement would not make them safe.

He presented the customer with another option, one at which some might look askance. It would be far safer, he said, if he capped the wires and left them hanging until he was ready to rewire. Yes, that would be leaving the job unfinished. Yes, the wires then would not be barricaded within enclosures that would hold in any sparks. The reason for this recommendation is that temporarily leaving them dangling seemed far safer than further torturing the brittle wires by stuffing them back into their overcrowded boxes.

Options for rewiring

They briefly discussed options for rewiring. The customer had been considering making some architectural changes in his old house anyway. Did the electrician have any idea what it would cost to fish in all new wiring? He gave the customer a ballpark number that curled his eyebrows. "Out of the question," was the response. That did not end the conversation.

Then the electrician suggested an option that made more sense for the customer. Given the advanced age of the house, there probably was enough decrepitude in other systems to make it worthwhile to open the walls and ceilings for general upgrading rather than have the electrician fish. This would make rewiring far faster, hence far cheaper.

Mimicking new construction: the gut or semigut job

Once the walls and ceilings have been opened, what do you have? Suddenly the job looks more familiar and straightforward. You are faced with a bit of demolition plus what looks a whole lot like new work. The advantages of this approach are manyfold. You get to completely rationalize the system, in the sense of creating a logical, straightforward arrangement.

☐ One circuit per room? Fine.

☐ Dedicated circuits for special loads? Certainly.

☐ Properly-supported paddle fans? Easy.

☐ Intercommunicating smoke detectors throughout? Sure, for a relatively small increase in the fee.

☐ Telecom wiring in the walls? Why not?

Unfortunately, not everyone takes the long view and thinks rationally about how to maximize return on expenditure. In this case, the customer had already done major work drywalling over the old, cracked ceilings. (That was obvious, from the way pancake boxes were recessed about $\frac{3}{4}$".) While deciding on this basis may not be terribly logical, such factors can be important enough to your customer to determine the rewiring method.

Cooperation and coordination with other trades

Opening up walls and ceilings is useful but costly. It is well worth your while to find out if it is likely that another trade will need to cut a chase or opening. In some cases, that is of no help. Sometimes there will be no room for wiring, for instance when ductwork will completely fill the space between the ceiling and the floor above. Other times, it is simply unwise to run the wiring in the same area—for instance in the vicinity of a hot water pipe. In many other cases, it may not be "cheaper by the dozen" but two trades can be accommodated almost as easily as one.

Either job discussed so far would have presented an excellent opportunity for coordination between the trades. Given what was visible of the houses, the plumbing was undoubtedly rusty iron pipe rather than copper, brass, or plastic.

Even duct, which is unlikely to rust out, can benefit from attention. In early 1996, the Electric Power Research Institute reported a study they performed that involved sealing leaky ducts and plenums in the Pacific Northwest. In the houses they studied, most of the ducts were in uninsulated spaces. They achieved a 43.5% decrease in electricity usage by the HVAC system. This should impress even customers whose ducts are within interior partitions.

Lucky breaks: when the existing system makes your job easy

The obvious should not be ignored: if there is a raceway system in good condition, your customer may be in luck. Even if its existing enclosures are undersized by present-day standards, there are solutions. You may be able to attach extension boxes; even the use of surface metal raceway extension boxes may be acceptable to your customer, if it saves enough mess and money. Also, doughty electricians have had considerable success yanking out old boxes and muscling in new, larger ones, even when the old boxes had rigid metal conduit attached to opposite sides.

Such luck normally will carry you only so far. A major drawback of reusing the old raceway system is that, if you do not extend it, you are stuck with the old wiring layout. Furthermore, most such systems combine pipe and cable. Therefore, this option, while very attractive, rarely eliminates the need for other types of effort.

Toughing it out—more difficult approaches

The other approaches to rewiring fall into two categories. To paint with a broad brush, they can be characterized as "lots of little holes" and "ugly." The first category, little holes, implies fishing. Incidentally, fishing does not always require holes that will need to be patched; there are shortcuts whereby a clever or painstaking worker sometimes can finesse access.

Holes are hard on both the customer and the electrician. First, they mean mess. Second, there is a rough equation between how few and small the holes are, on the one hand, and how long the fishing takes and how scraped up your hands and forearms get, on the other. Special tools make some difference, but they do not upset that equation. Therefore, even when fishing, it is worth your while to consider options such as pulling away a baseboard, if it does not appear brittle, so you can easily drill studs concealed behind it. Diving down to an unfinished basement or up to an unfinished attic so you can inoffensively run exposed cable also makes eminent sense. Even if you need to drill joists in such spaces, you will be doing much less aesthetic damage that requires patching. Do beware of weakening joists by chewing them up too much. Section 2517 of the Uniform Building Code restricts hole diameters to $\frac{1}{3}$ the depth of a joist. The May–June 1997 issue of *IAEI News*, pp. 18–19, has an excellent article by Joe Andre, "Electrical Installation: boring holes in wood studs."

In older buildings, you are often not dealing with drywall. When you have to cut holes in the wall or ceiling to the side of each stud or joist, the patching job is major. Some find it dirt-simple to patch plaster or gypsum block walls using plaster of paris or patching plaster. Traditionally, one stuffs balled-up newspaper or other filler into the hole to give the plaster something to bear against. If the hole is too big for that, nailing or stapling some wire lath (an expanded metal mesh) in place is by far the best way to underpin your patch. The biggest problem with plaster ceilings is that they can be relatively brittle. If enough ceiling comes down, it may be necessary to replaster properly—with lath followed by sand mix followed by plaster—or to resurface the ceiling with drywall.

■ **8-1** *Legend: Stuffing and patching. This hole is awfully large for this system of patching.*

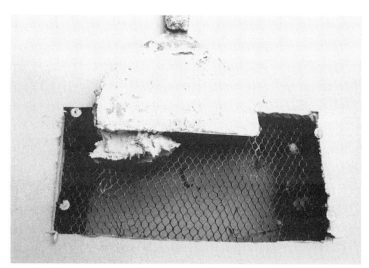

■ **8-2** *Wire lath and patching.*

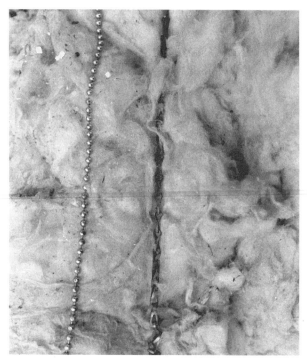

■ **8-3** *Sash chain versus ball chain.*

More Specialized Issues and Situations

Fishing varies in difficulty. Fishing down (or up, if you really must, even though doing so means fighting against gravity) can be very easy if a house has balloon construction. More often than not, though, you will run into firestops, braces that stiffen the studs but that at the same time prevent your cable from dropping straight down. There often are also problems with HVAC ducts, plumbing pipes, and other wires. Finally, there may be wads of newspaper that previous workers have stuffed into holes, *a la* Figure 8-1, to make them easier to plaster over.

There are many ways to get your cable through. A plumb bob can be used for fishing, to lead a pull line. Some use chain, either as a leader at the end of a fish line or, in greater lengths, to serve in itself as a pull line. Some people advocate the use of sash chain, but far preferable is ball chain, a larger version of the type that can be found on some pull-chain light fixtures. Sash chain hangs up on obstructions far more easily than does ball chain. "Fish line," incidentally, can mean different things. Actual nylon fishing line normally is used just to find a path. It then is used to drag in a heavier line that has the strength to pull in cables.

Sometimes cable itself has just the right weight and slipperiness to make its way where you want it to go. Electricians' snakes (fish tapes, to use the formal name, though even the new, non-metallic, ones have no relation to fishermen's nylon line) are less useful—they are designed for the inside of conduit, not for wall cavities. A 1" wide tape measure sometimes is useful in fishing, especially for sliding between ceiling joists. A folding foot-rule has come in handy any number of times for fishing in hollow spaces; on the other hand, any number of foot rules have been broken by being employed for that purpose. Proprietary products designed for fishing inside walls are widely sold. See the section called "Fishing Rods" in Chapter 14, "Resources."

Here is a warning about obstructions. It used to be that when electricians or plumbers came upon obstruction in walls, they broke past. Figuring that they were probably firestops, they would screw additional pipes to what they were sending down, and ram them on through. The warning may be rather obvious: if you do not know what is in your way, smashing through it is taking a chance. For the same reason, when using a long drill bit or bit extension, be very leery of blundering along. When your bit seems to be biting, you want to know that it is about where you

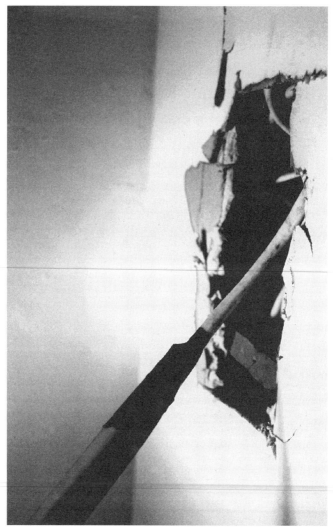

■ **8-4** *Tape measure used for fishing, here successfully pulling a cable (taped to it) out of a wall.*

expect wood—and not cherished hardwood floor. If you are unsure, you would be wise to back off. Even if all that you have encountered is fiber insulation, it will do you no good at all to wrap your auger bit into a great mass of it. Untangling your bit can be the devil of a job.

An example: going fishing

To bring this down to earth, consider a typical rewiring job performed under a contract that specifies fishing rather than demolition. Assume an occupied single-family residence, with two fully habitable stories, an unfinished attic, and a partly finished basement. The service equipment is in the basement, which is substantially below grade level.

With a building that is in use, the customer understandably does not welcome the disruption associated with major demolition. Nevertheless, you need to help the tenant or owner understand two things. First, even the most careful fishing job generates some dust, debris, and disorder. Second, minimizing mess by doing things the hard way to some degree trades off tidiness against efficiency. Less efficiency not only means that work costs more time and money, but also means less efficiency and rationality of design—hence, lower long-term convenience. As you will see, a fished electrical system will tend to end up less conceptually tidy and workable than it could be if the building were gutted.

Leaving the existing system alone

With instructions to keep mess to a minimum, you may want to blank off rather than remove old outlets. This choice represents a

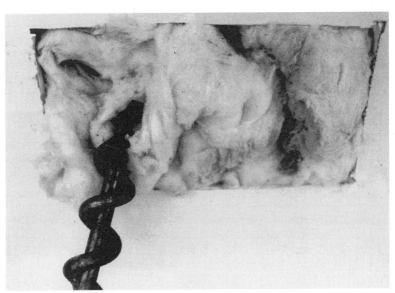

■ **8-5** *Bit caught in rockwool.*

real trade-off in terms of aesthetics, as well as in terms of potential confusion. On the other hand, it has the advantage that you need not disconnect the old wiring until your new system is in place and ready to be energized. The main exception is where a new outlet simply must be located in the identical place the old one occupied—the center of the ceiling in the case of a lighting outlet, or 4'0" AFF just between two adjacent door frames, in the case of a switch.

Only in rare cases does it make sense to reuse the old boxes with new cables. For one thing, doing so does require pulling out the old cables, stapled though they probably are. Otherwise, they occupy needed box volume and also present too great a potential for mix-ups. Furthermore, old boxes may not have tapped grounding holes, often are unnecessarily or illegally recessed, may be undersized for your needs even with the old cables removed, and frequently are damaged. New boxes can have the type of internal clamps you want plus all the volume you need. Abandoned, the old boxes can sit there peacefully, with blank covers, till the building falls down.

If someone does decide to bury abandoned boxes, at least they are no longer connected to the electrical system. So long as their wiring at no point travels to an enclosure containing live wires, it even is arguably legal to plaster over them. You would not want to have to dispute that one with an inspector. All it would take is one overlooked connection between the discontinued and the new wiring, and you would have a potential hazard—not to mention possibly a real puzzler when troubleshooting. Should such a potential hazard bear fruit, you would have the chance to argue the appropriateness of your choice not with an inspector but with a lawyer.

Further choices: cabling layout

Electricians sometimes do permit the old system to dictate the design of the new at least in part. If you are going to remove all the old wiring, you probably will have to open walls and ceilings to so great an extent that you might as well use those same routes for some of your new wiring. You might even be able to reuse some existing holes. You are faced with the trade-off of running the cables where it is easy versus running them in the way that will produce the most rational system. Two types of consideration must be weighed in making that decision. The first is materials cost: minimizing cable runs. The second is labor and mess: minimizing chopping and drilling.

Minimizing the number and length of cable runs is a relatively unimportant consideration. Doubling the amount of branch circuit cable used in order to cut your labor time by 10–20% is a net win. Doubling it in order to produce a rational system, at no great increase in labor costs, does the customer a substantial service. That is true despite the fact that longer runs mean more voltage drop, as well as more drilling, stapling, or strapping. Certainly, increasing the number of cable runs because you are increasing the number of circuits reduces the risk of overloading and gives the customer better control.

More detailed examples: the up-or-down dodge   The next section will make the ideas about maximizing fishing convenience more concrete. Consider the rewiring of a top-floor bedroom, a 10-foot-by-15-foot rectangle with one entrance. It initially has one receptacle on three of the walls, and a pull chain light in the middle of the ceiling. When you are finished, it is to have eight receptacles and two wall switches. One switch will control a fixture in the center of the ceiling; the other will control half of a split-wired receptacle. There also will be a dual-mode smoke detector, intercommunicating with the detectors elsewhere in the house.

You may benefit from the fact that this bedroom is on the top floor. If there are few firestops, you can fish without making additional wall openings beyond those required for the boxes you are installing. You can loop from one new outlet to the next through the attic, protecting the cable up there on running boards if needed, or using whatever other system of protection is convenient and legal.

There is no need to burrow horizontally from one receptacle to the next. There is little or no need to cut holes in the wall in order to drill studs. You can go up and down from receptacle to receptacle to switch to light, and then bring it all home on one circuit. If you prefer lighting to be on a separate circuit from receptacles, that too is easy to arrange. The smoke detector cable can simply travel up and down through the attic from one detector to the next.

If this bedroom were on the first floor, over the unfinished portion of the basement, you could accomplish the same thing by looping down and back up from receptacle to receptable to switch. (You probably would have to chop holes in order to fish cables to ceiling outlets.)

One thing to watch for in drilling down into the basement at a building's periphery is that you can damage your bit, or puzzlingly fail to find your hole from below, because you have drilled down over foundation masonry. Take this on faith: concrete sills are harder than auger bits. Another common danger when drilling down into a basement ceiling is drilling into ducts run just inside the basement ceiling. Sometimes these dangers seem unlikely, based upon your examination of the building. Then you may be comfortable cutting a hole in the wall for an outlet box and drilling down. When in doubt, you should not be too proud to drill a $\frac{1}{8}$" hole into the floor (between floorboards, sometimes, or through the edge of the baseboard). Then leave a light shining down, and go below to locate the light. Or stick a long wire through, and go downstairs to locate it. That way you can make sure that you are coming down where you want to. Similarly, when drilling up into an attic at the edge of a building, make sure that you do not go right on up through the roof.

One old-timer's trick recommended by Steve Jones, of Jones Electric, Texas, is to use a straight, fairly short, piece of wire hanger as your drill bit. This is said to be less likely to ravel carpet than is a twist bit. At the same time, the wire is hard enough to cut through hardwood, and it can be sharpened by trimming it back with side-cutting pliers.

TAKE WARNING: should you decide to attempt this, be aware that you are NOT using a tool in the manner for which it was designed. Mention here is not an endorsement of the approach. No one has evaluated that piece of metal for safe use in a drill motor, the way manufacturers evaluate drill bits they offer for sale. A specific warning handed on by Jones is to avoid leaning on your drill. Too much pressure could bend the piece of hanger, causing it to whip around dangerously.

A far-milder departure from the manufacturer's intended use is to employ a regular twist drill in this situation, but run the drill motor in reverse once the bit has penetrated the underside of flooring and entered the carpet (or, from above, run it in reverse until you have reached wood). This way, the flutes are less likely to grab and draw it into their spiral.

Warning: crawlspace hazards Where there is no unfinished basement or attic as such, sometimes there is a crawlspace. If you have not done much old work, take warning. Crawlspaces can be much nastier than mere unfinished rooms. Bugs, rodents, brickbats, and broken glass are a few of the unpleasant contents electricians run into or crawl across. Consider bringing down at least a drop cloth on which you can lie or crawl.

Also, especially in under-the-building crawlspaces, beware of shock hazard. Should you brush against an unenclosed splice whose connector has fallen off, you may be in a very bad way. You are likely to be well grounded, and you may be wedged against the wires so a flinch is restricted and ineffective. Since crawlspaces rarely if ever see an inspector, they are more likely than other locations to sustain such dangerous illegalities. If no one is keeping an eye on you, you could easily die from such an encounter.

More detailed examples: doing it the hard way

When you need to wire over and under finished spaces, access is more difficult. In many situations, there just is no unfinished space from which to work. This is when the trade-off between rational and simple really comes up. You cannot loop up and down, or down and up. Therefore, if fishing is specified, you have to engage in the tedious and messy work of drilling through every stud between one receptacle and the next. That is to say, you have to do so unless you are willing to create a not-very-rational wiring layout.

Here is an illustration of how logic may be compromised for convenience. Suppose that a first-floor room needing an outlet is right under a second-floor room, and that is right under the unfinished attic. The house is built on a slab. You drop cable from the attic down to a receptacle in that second-floor room. Normally, you would be looping back up, if daisy-chaining, or terminating at that outlet, if wiring radially from a large junction box in the attic. Instead of taking that rational route, you could drill through the floor plate under that second-floor receptacle, or, if the receptacle is in the outer wall in balloon construction, just drop down. If there are no obstructions, you can proceed down to a first floor outlet location, without any further drilling, and certainly without drilling horizontally through the half-dozen studs you might otherwise have had to deal with in order to wire that first-floor location from an adjacent outlet. Many, many old houses have circuiting much like this, where the odd receptacle was added by picking up power from an outlet upstairs or downstairs.

This approach is hell on future troubleshooting, not to mention on creating an intelligible panelboard directory. On the other hand, if the square footage and loads are such that it makes sense to have most of the habitable rooms on one circuit, it is not harmful, just confusing. In those circumstances, there is nothing in the NEC forbidding it, so long as you heed the restrictions in Articles 210 and 220.

New wiring does have to meet the Code's current circuiting requirements. If, in the old system, bathroom receptacles were on circuits that also served bedrooms, the provision in Section 2 10-52 (d) requiring separate bathroom receptacle circuits means that the old layout no longer will fly. Section 210-52 (b) speaks similarly with regard to small appliance circuits. If you were replacing a length of bad cable, most inspectors would allow you to retain the grandfathered layout. In the case of massive rewiring, most inspectors will require you to come up to date, except where there are strong grounds for a special pleading.

If you truly must do the least possible, check for local rules. In several jurisdictions (Prince George's County, Maryland is one such), inspectors permit additions to existing wiring to use the same wiring method that is in place, even if the system no longer is legal. As an example, those jurisdictions have permitted adding more NM to existing NM cable in places of assembly. Only when the cost of remodeling work is estimated at 50% or more of the value of the building is this generous grandfathering policy voided. This is a very liberal variation on the "25/50%" rule mentioned in Chapter 10, "Historic Buildings."

Further choices: chopping or disassembling

When you see no way to avoid lots of drilling, you will have to damage or pull apart some portion of the structure. What is the best procedure?

Temporary removal Sometimes you can pull away baseboard (or, in very rare cases, crown molding), demolish the plaster behind it, and drill merrily through the studs you have exposed using your right-angle drill. Some old-timers swear by the option of cutting through the tongues of tongue-and-groove wooden flooring, removing a plank, and then fishing and drilling to get wiring to wall outlets—or to ceiling outlets below. When attempting either of these approaches, be careful with the baseboard or floorboard you have removed. It would be a shame for that to suffer damage before you can put it back in place.

Cutting up walls When removing and replacing baseboard will not do the job, you can open up the wall between two studs or joists and drill each of them. Since studs and joists are not terribly far apart, with luck you can get away with opening the wall or ceiling only between every other pair of studs or joists. Drill a hole through each stud, stick your cable or fish tape or other implement of construction through, and then fiddle and feel till you can find the other hole from the blind side. Then grab your cable or fishing device with your hand or pliers and pull it through. When you have pulled enough slack, you can shove it through the second access hole, wiggling and jabbing and aiming it until it finds the next hole you have drilled. In Chapter 14, "Resources," you will find descriptions of custom products that might help in fishing.

This description of blind fishing underscores three points. One is the need to make your holes in the wall large enough that you can reach in to drill fairly straight and manipulate your cable or fishing device with reasonable comfort. The second is that the straighter (and larger) the holes you drill through the studs, the easier it is to aim through one hole, blind, and find the next. The third is the familiar refrain that fishing, at least through studs that are 16" on center, often is a lot more time-consuming and troublesome than opening things up, even accounting for the time spent patching in either case.

Fishing holes Here are some cautions for when you cannot simply drop a line, and therefore must open up walls or floors to drill studs or joists. First of all, if you are not using a right-angle drill or a Kett™ drill, you probably will be drilling at an angle. Therefore, especially if you are using spade bits rather than auger (self-pulling, "Selfeed™") bits, one common practice is to ream the hole some with the sides of the bit, if you can do so safely. Be very careful not to wrench your wrists by hanging up the bit. A far better approach to ensuring smooth pulls is to go back through the hole from the other side. (This often is not possible where you are fishing blind, trying to minimize disruption to the structure.) One old-timer swears by an alternating, zigzag pattern. He angles up through one joist, down through the next. He believes that the cable gets hung up less this way. Others' experience has been different. Many find it easiest to pull cable in a straight line or a smooth curve. You can easily find out what works best for you.

Assorted less-obtrusive options

There are a few more ways to reduce the tedious fishing required when a customer does not want walls and ceilings ravaged. With wood paneling, sometimes you can remove a whole sheet of paneling and, after doing what is equivalent to new work, restore it. In those rare cases where an old building has drywall, you often can cut a neat rectangle extending from one stud to another, overlapping both, and replace it afterwards, or cut a longer but narrow strip across the ceiling or wall that can be later replaced.

Sometimes, with drywall (but, generally, not with plaster) you can cut a flap that can be screwed back in place when you are done drilling and fishing. Leave the "hinge" side over hollow wall or ceiling, and the "latch" side of the flap over the stud or joist, so you can screw it back down when you are done. And be gentle with it, especially the hinge side, so the hinge does not tear off while you are drilling or when you are wrestling your hand into

157

the wall or ceiling to urge the cable where it needs to go. When you are ready to screw the flap back down, make sure no rubble gets caught up in the "hinge."

As a final option for concealment, consider using finished spaces that simply do not qualify as living areas. If you run wiring exposed in such spaces as an alternative to fishing it is usually not offensive. For a $100 saving in labor costs, many a customer has been very happy to authorize surface wiring runs along the back of a closet or pantry. You do need to consider the issue of adequate protection from mechanical damage in these installations, especially if you employ nonmetallic-sheathed cable on the surface of a wall.

Solid struggle

This exploration has not yet described the unpleasant job of running wiring through solid walls and ceilings. To chisel out a chase in a concrete ceiling, in order to embed EMT or AC cable, is always a nasty job. Most electricians experienced at old work have chopped and fished their way through cinderblock walls. Such work is much the same whether the building is old or new, so long as you are unfortunate enough to arrive on the job after the masons have finished. In cases where you may have to chop a channel, it is worth your while to talk to your customer about alternative options such as surface wiring, wireless switching, even the installation of a false ceiling. Some of those options are discussed below.

The ugly ones Bringing up surface wiring moves the discussion into a range of choices that many find unacceptably ugly. This is not your judgment to make for the customer. On the other hand, it is wise to take pains ahead of time to help your customers understand what these options look like, so they avoid disappointment that could result in anger and blame.

Surface Cable The first such option is, quite simply, running cable on the surface. True, that is almost never going to be authorized in a bedroom or family room. It is more acceptable in other settings. In many a residential or commercial workshop, in some commercial and even home offices, and in any number of "nonpublic" areas, frugal customers have been perfectly satisfied with even nonmetallic cable run along the wall. This option does offer a rock-bottom price. Even more then when running cable along the back of the closet, you need to make sure that it is not going to be subject to physical damage, in accordance with NEC Section 300-4.

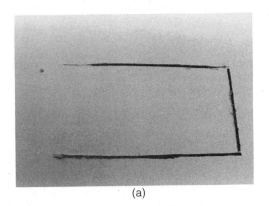

(a)

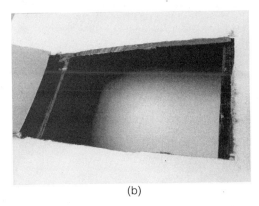

(b)

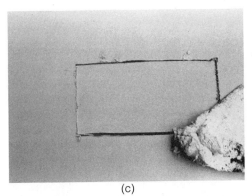

(c)

■ **8-6a–c** *A "window of opportunity." Cutting away a strip and replacing it after wiring behind. With luck, you can hinge it out from the surface, as in 8-6(a) rather than removing it completely.*

Armored cable is considerably tougher, although no prettier and considerably costlier. It still demands suitable protection. For a step up in terms of quality and appearance, EMT and NMRC are not a whole lot more expensive to install, and they hold up yet better. Any of these options is likely to be far less expensive than fishing.

Surface raceway Surface metal and nonmetallic raceways are options intermediate in price and appearance between these previous choices and concealed wiring. The type that is said to simulate baseboard, such as #2100 Wiremold™, is more expensive but to many people more attractive than standard surface raceway, such as #700 Wiremold™. Either is a dust catcher, and either can take only so much abuse. If you suspect some risk of mechanical injury, employ two-hole straps to secure standard sizes of surface raceway, rather than just back straps. Surface metal raceway is partly knocked off the wall way too often in homes, offices, and stores. Besides interrupting grounding continuity, that abuse causes the raceway to hang on the wires inside. This experience is not healthy for the insulation, nor for the splices or other terminations.

In and out The final system to consider is at the same time very ugly and in certain setups very discreet. Whether you are trying to get power from a basement to a higher story, or from receptacle to receptacle along an outside wall, sometime a fast and easy way to go is to run outside the building. The appearance of this on the front wall of a house could be terribly offensive, and perhaps even in violation of neighborhood covenants. What about buildings whose sides abut blind alleys? In these cases, no one may care if you have RNMC or UF cable neatly secured on the outside of the wall. (Note that in jurisdictions such as New York City, for instance, outdoor wiring may have to be entirely in a raceway.) In a factory, the wiring actually may be safer installed outdoors than it would be if run from ceiling to floor in NM-B or UF-B, with a minimal six inches of mechanical protection as it emerges from floors, as required by Section 336-6(b).

Factories facing on blind alleys are not the only settings where frugal customers have utilized this option. Financially strapped homeowners who would not consider fishing, nor exposed systems indoors, have authorized their contractors to secure cable discreetly alongside downspouts, following the exposed wood of timber-and-stucco houses, or above window frames. (Note that cable is considered exposed to injury when installed below windows that can be opened.)

Incidentally, the 1999 NEC will explicitly forbid installing cables on the outside of buildings where subject to physical damage. This suggests that, in the past, installers have not exercised due care in such installations.

Support Keep in mind that cable out in the open does not qualify as "fished." It matters not that you drilled out from, and dropped the cable from, the third story of the factory and brought it in again at basement level. An alley is treated differently than the inside of a balloon wall. You have to get on your extension ladder and secure the wiring. It needs to be tidied in accordance with both the specific requirements of, say, Article 339 for Type UF cable and the blanket injunctions of Section 110-12, "Mechanical Execution of Work."

How close must the supports for Type UF be placed? Inside buildings, Article 339 refers you to the rules for type NM cable—supports $4\frac{1}{2}$ feet apart, maximum. Outside, interestingly, it is left up to you. The article says nothing on that head. This question is discussed further in Chapter 12, "Inspection Issues," along with other concerns that come up in the course of rewiring.

One final matter. While working outside to secure a raceway or cable that penetrates walls, do caulk around its entrance and exit to bar moisture and vermin.

Advanced Topics

Relatively Rare Situations

THIS CHAPTER COVERS AN ASSORTMENT OF INSTALLATIONS that you will not encounter in many buildings. Some of these designs are still around, and some, while possibly unfamiliar, still are being installed.

The oddball—Loadcenters in unexpected locations

To start, here is something that is simply oddball, something you will find in very very few old buildings. The practice of mounting a fused cutout in a ceiling was once very common where uncluttered wall space was unavailable. Nowadays, no electrician would use this location for a loadcenter, nor would any inspector approve it. Nevertheless, only one requirement in the NEC could be used to outlaw the installations.

Even now, there are loopholes. Neither Article 110, 230, nor 384 applies to the examples you are likely to encounter, which contain nothing but plug fuses. Article 380 would apply to such a loadcenter if it contained circuit breakers—and the circuit breakers were used as switches; or to a fusebox that included one or more switch disconnects. Even in those cases, there are exceptions. In a low-ceilinged room, or if a working platform brought users to within 6'7" of the switches, the installation would be legal.

Why concern yourself with this odd layout? Like any other overcurrent devices, fuses in these oddly-located loadcenters blow. If you fail to consider the possibility that there might be a subpanel in a ceiling, you may face a frustrating and fruitless search elsewhere. You might even figure that the dead #14 black wire you found at an outlet is broken somewhere upstream along the cable, and unfairly assume that it was protected solely by the two-pole 40-amp or 60-amp fuse or circuit breaker at the service equipment that actually protects its feeder.

This warning about hidden panelboards applies beyond the specific example of loadcenters located in ceilings. If power is dead at several outlets, and all overcurrent devices in the main panel are conducting, and there are unmarked two-pole circuits, you should consider the possibility that an overcurrent device is tripped in a subpanel before delving into the wiring at too many outlets .

Here is how to proceed. First eliminate the obvious: the unmarked two-pole circuits could feed electric ranges or dryers, or unused outlets for second ranges or dryers. They could feed water heaters or electric furnaces or air conditioners or heat pumps. Except for electric furnaces, and heat pumps with resistance backup elements, the overcurrent devices for such loads generally will have lower ratings than those that feed subpanels.

If none of these possibilities bears out, consider the prospect that a subpanel hides where you would never place one. Look behind art, furniture, file cabinets, and other storage. Include places where clearly no electrical enclosure should lurk, and certainly no overcurrent devices.

☐ Could a subpanel be living in a bathroom? Yes: outside of dwellings and guest rooms, this usage has even been legal. Adoption of the 1999 NEC should change this.

☐ Behind a permanently installed bookcase? Certainly.

☐ In the back or side of a clothes closet or kitchen cabinet? Alas, yes.

How to deal with oddly located overcurrent devices

The most important thing to know about these oddball installations is that they exist. Keeping an eye out for them can save you ever so much time. Once you find a fusebox in an unexpected location, you can change the blown fuse. But is that enough? No.

There are two more things to do, at a minimum, when you discover one of these oddly located panels. One is to alert your customer as to its location and to educate your customer regarding any associated illegality or danger. The other is to create a record—something more than just your own notes, or even written notification on, say, your invoice. It is wise to create a record, in any event, whenever you observe a significant code violation. Here you would do well to go one extra step.

In accordance with Section 225-8(d) and 230-2(b), when multiple sources of power are brought into a building you generally have to affix a permanent sign at each service equipment location. It is intelligent and professional to do so even when not required. When you find subpanels at nonobvious locations, it is similarly appropriate, even though not required, to leave a permanent note at the service equipment locating those subpanels as well as identifying the over-current devices controlling them. True, subpanels do not present the shock hazard associated with multiple sources of power. That reason for putting a note at the service equipment does not apply. But where else could the note be placed? Besides, without it the feeder breaker or fuses will not have been adequately labeled. The service equipment is the logical place to post any warnings—this is where the circuit directory is located and thus where people look when they want to know which circuit controls a particular outlet.

Should at least some of these subpanel installations be changed, corrected, redone? Yes, even when they are not in violation of NEC Article 384 or 110. You certainly are not obliged to perform that work on the basis of having come upon such an installation. Leaving your imprint on the system would bring you more responsibility; do not add an additional circuit to a loadcenter located in the back of a clothes closet. Well...at the customer's insistence, you might go ahead if the local inspector says it is okay, but try to propose alternatives.

Odd locations aside, you are likely to encounter many examples of loadcenters and even service equipment with inadequate clearance. In many cases, these did not start out in violation. Subsequently, though, the owner or tenant installed equipment within the required clearance or set something in front of the panel. Storage basements are notorious for this. So are commercial or institutional establishments whose electrical closets start out as dedicated spaces and end up doing double duty holding cleaning supplies.

One example involved service equipment consisting of a fusebox mounted low on the front wall of a house, in the kitchen. Despite its odd placement, perhaps two feet AFF, the fusebox started out quite legal. Unfortunately, someone mounted a shelf above it and also extended a peninsula counter and cabinets out from the wall right next to it, in violation of NEC Section 110-16 (a).

Depending on the egregiousness of the violation, on how completely and permanently access to the panel is obstructed, there is no need to get in too much of a tizzy over this sort of setup. Yes, tell the customer he or she has a violation. Yes, insist on clearing impediments out of your way, rather than taking the risk of working on a panel while impeded by clutter. But usually, digging in your heels and fussing would be overreaction. If you temporarily abate violations created by storage, after you leave the customers are likely to put it back the way it was and there is nothing you can do. With luck, they will hold off until after the inspectors' visits. The odds are that if they do not hold off, the inspectors will at most have a word with them, rather than red-tagging your work.

A conversation on this very subject took place at the end of 1996 between Bill King of the U.S. Consumer Product Safety Commission and Phil Cox of the International Association of Electrical Inspectors at a board meeting of the National Electrical Safety Foundation. They had just seen a film on electrical safety in older homes, and the narrator had failed to cry foul over a clothes dryer shown in what should have been dedicated clear space in front of a panelboard. While those viewing commented on the omission, there was general agreement that this is an example of a battle one might not elect to fight.

Suppose you do want to correct one of the more extreme oddball or illegal loadcenter installations. Perhaps it appears on the punch list a home inspector provided a purchaser who insists on moving into a house without Code violations. Doing so may or may not be simple. When a subpanel hides in a ceiling or in the back of a closet, you may be able to fish multiple circuits from it to a new subpanel that you install at a more appropriate location. Then you can blank off the old subpanel's cabinet, converting it into a large pull box. Doing so is cumbersome but does not require structural changes.

When the loadcenter is illegal because access is largely blocked by a fixed part of the building structure, the solution is more difficult. One such installation involved a subpanel in a kitchen with two quibbling flaws and one real problem. First, it was located not two feet but a mere two inches off the floor. Why was it set down there? Probably to make it easy to fish cables up from the basement. There is nothing illegal about setting a loadcenter in that awkward location. (Yes, an inspector could characterize it as "not readily accessible," and red-tag it, but doing so would entail an idiosyncratic, nonstandard definition of "readily accessible," exceeding the intent of the NEC. In a flood zone, it might have violated FEMA rules.) Second, the cables emerged from the floor without protective sleeving. To object to this would also be absurd. With only two inches exposed, they were quite adequately protected from mechanical injury by the cabinet itself jutting out above them. Third, there was a clear illegality: a permanently installed bench, sort of a window seat, right in front of the loadcenter. Not only did the bench interfere with the clearance required under Section 110-16, it partly blocked the panel cover.

The best way to correct that situation would have been to relocate the window seat. Failing that, the fix described above—running cables to a new subpanel in, say, the basement—would have been a partial solution. A shortcoming of that solution is that NEC Section 370-29 requires access to wiring in junction boxes; that access still would have been missing. Still, many inspectors would go along with such a partial solution. As it happened, the customer would not authorize any work to correct that violation. Therefore, the contractor on the job simply found the dead fuse and replaced it with a properly sized no-tamper fuse and a fustat adapter.

He also marked the service equipment in order to remind him of the problem on his next visit, or else to alert the next electrician.

Multiwire mischief

The next odd situation found in older buildings is not nearly as unusual as one would wish. There are quite a few rules about multiwire circuits. Every one of them has been trampled upon. Multiwire circuits should consist of two or three hot conductors sharing one grounded conductor, which should be the neutral and should be identified in accordance with the requirements of NEC Section 200-6. This is one of the rare places in this book where the four terms, "return," "neutral," "identified," and "grounded conductor" can and therefore, for contrast, will be used interchangeably. Hot and return wires should run in the same cable or conduit. The grounded conductor, if it carries any current, should be no more heavily loaded than it would be in two-wire 120-volt circuits serving the same loads. The grounded conductor should be broken only in splices, rather than daisy-chaining via device terminals. The hot conductors should originate from different busbars in the same loadcenter. The circuit's overcurrent devices should be arranged so as to be disconnected simultaneously when they feed 208 or 240 volt loads, and, in dwelling units, when two or more overcurrent devices feed multiple receptacles on a single yoke. In other words, such multiwire loads should be controlled either by multipole circuit breakers, or by circuit breakers with handle ties, or by, say, multiple fuses in a single fuse block. By rights, meaning in accordance with the NEC, that is how it should be.

There are three differences between the modern design just described and the multiwire setups you are likely to find in older buildings. First, Section 210-4 did not always require simultaneous disconnection. You have to be alert to the possibility that in an old building a circuit you have turned off may be backfed through a shared grounded conductor. A preliminary (but only preliminary) indication that you are safe from the risk of multiwire backfeeds can be gained by checking inside the loadcenter. Trace the conductors of the circuit you are working on between their overcurrent devices and where the conductors exit. Those that leave the panel as two-wire circuits should not pose this danger, in the absence of violations.

Second, to ensure continuity Section 300-13(b) forbids daisy-chaining a multiwire circuit's neutral through device terminals; it too was not always in place. Therefore, in older buildings it is especially important to remain alert to the possibility that one loose connection could result in 240 volts being divided by whatever loads are connected; in a careless moment those loads could include your fingers. For what very little reassurance it offers, the voltage to ground remains 120 (or 277 in some commercial/industrial settings), a mere three or more times the voltage needed to electrocute a healthy adult male; it seems reasonable to presume that the greater the degree of overkill present in a hazard, the more likely it is that any single exposure to it will kill or cause serious injury. On the other hand, the reason that so many more people die from 120 volts than from higher voltages is only partly because there is vastly more exposure to 120 volts, especially among poorly-trained people. Another reason is that 120 volts is treated with insufficient caution. See Chapter 17, "Hazards and benefits," for related discussion.

Third, as with so many aspects of older buildings, the longer a system has been around, the more likely it is that something went bad and was kludged back into operation. The type of case that will be described below—and it can be very complex to troubleshoot—is where some wire or connection failed and the replacement was "borrowed" from a convenient nearby outlet, in one variant creating an *ad hoc* multiwire circuit. In old installations, you also often will find cross connections between grounded and grounding conductors. This means that any voltage imposed on the neutral of a multiwire circuit may be present—hazardously—on the exposed surfaces of electrical equipment.

Complex kludges you may encounter

An example or two will clarify this problem of "borrowed" lines. The description is quite complex, but it does represent setups you are likely to run across. A large enclosure—call it "A"—is serving, and fed from, a lighting outlet. "A" contains three two-wire cables and one three-wire cable. The three-wire cable contains, first, a switched conductor going to the lampholder or ballast; second, a grounded conductor going to a splice that feeds the lampholder or ballast and also one of the two-wire cables; and third, a hot wire that is spliced to the other conductor in that two-wire cable. The other two two-wire cables are spliced to each other, black to black and white to white. In the electrical panel, the cables come from circuit 23 and 35. Originally the first two had been connected to circuit 23 and the last two to circuit 35.

At some point, a so-called electrician was called in because power to part of the circuit had been lost. Power was still getting to the light; hot and return both still came through on circuit 23 to point "A." Power continued on along circuit 23 to point "B," the next outlet downstream on that circuit. The power that was missing was on circuit 35, the one represented by the two two-wire cables that were spliced to each other—merely using point "A" as a junction box. The loss was traced back from some other point "C," where power was off, to the lighting outlet or in this case the switch enclosure, point "A." However, neither for love nor money could the electrician find the upstream location of the fault.

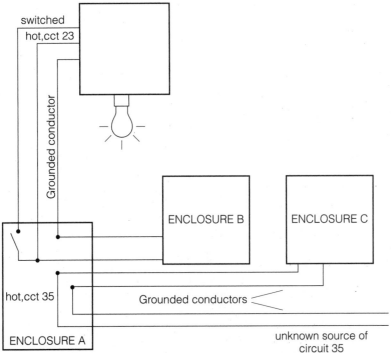

■ **9-1** *Schematic for an enclosure that was turned into "the splice box from hell."*

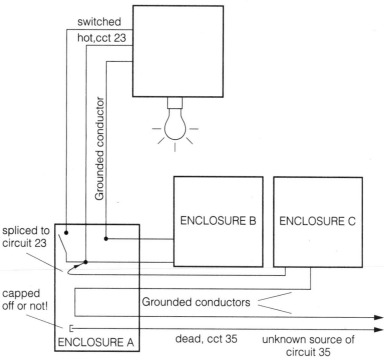

switched
hot,cct 23

Grounded conductor

ENCLOSURE B   ENCLOSURE C

spliced to
circuit 23

capped
off or not!

Grounded conductors

ENCLOSURE A   dead, cct 35   unknown source of
                                       circuit 35

■ **9-2** *Schematic for kludges associated with the "splice box from hell."*

Here is what the unthinking, "get it working" hazard-on-two-legs did. He or she disconnected the dead two-wire cables at point "A." Then this fix-it innovator determined which one continued on to "C," the dead outlet downstream. He or she then capped the up-stream (dead) hot wire from circuit 35 at "A," (one would cer-tainly hope so) and brought the downstream hot wire into the splice with circuit 23's hot wires.

The result was that the one hot wire from circuit 23 now fed into two, parallel, grounded conductors. While circuit 23 is live, you would have some chance of being shocked by interrupting and getting in series with either conductor. Because you would be in parallel with the other grounded conductor, that risk is not as se-vere as some. Of greater concern is the fact that the conductors now carrying the power of circuit 23 are following different paths. Inductive effects will be particularly severe where metal raceways or enclosures are used. Note that if both hot and grounded conductors in the dead two-wire cable going on down-stream to point "C" had been disconnected from the dead up-stream two-wire cable and changed to the other splice, there only would be a problem or violation if the circuit directory was not changed with respect to circuit 23 and circuit 35. (Of course, if this change added impermissible loads to circuit 23, for instance by overloading it, that too would be illegal and possibly dangerous.)

The problem created by borrowing power is more serious in cases where the grounded conductor had the break. In many such a case, the ignorant installer brought the grounded conductor that was heading downstream into the splice containing the other circuit's grounded conductors. Doing so turned the home run into a multi-wire circuit. Legalities aside, the one grounded wire from circuit 23 was now fed by two, parallel, hot conductors. You now risked being shocked by interrupting and getting in series with the grounded conductor when either circuit 23 or circuit 35 was live. Further-more, the installer did not install handle ties on the breakers feeding the two circuits (given their spacing, it would have been physically impossible) to ensure that when someone wanting to work on the circuit flips one breaker, the other is also turned off.

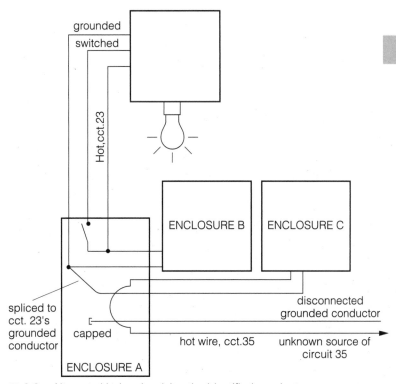

■ **9-3** *Alternate kludge, involving the identified conductor.*

When ignorant or unthinking problem-solvers create that sort of unofficial, hazardous, multiwire circuit, they also introduce another danger. When the two circuits originate from the same busbar, they can easily overload the grounded conductor, generating excessive heat that can potentially damage equipment and even cause fire. The numbers "23" and "35" were chosen for the examples because in most loadcenters those numbers would originate from the same busbar, though they are so far apart that their common origin might not be obvious.

The identical type of problem would have been created if, instead of stealing power from another circuit passing through the same box, the creative ignoramus borrowed power by running a cable over to a nearby outlet that was on a different circuit. In other words, if box "A" did not contain both circuits 23 and 35, he or she might have had to run a cable over to box "A," which contained circuit 23, from box "C," which was dead, having been fed by the somewhere-interrupted circuit 35.

One difference in the latter situation is that the worker, ignorant or not, might have been more inclined to do the job right, rather than creating a problem. Or maybe not. The reason for doing the job right is that when running a cable from a different box, one has two conductors, so there is some intuitive inclination to use them to run both hot and grounded conductors, rather than continuing to use one of the conductors from the defective circuit.

Doing it right would also have meant disconnecting and capping both upstream conductors and relabeling the circuit directory accordingly. Then everything could be safe and legal. At any rate, the installation could be safe provided that it was not overloading the circuit; that it was not causing dangerous crowding in the boxes; and that it was not admixing forbidden combinations, such as small appliance circuits and other loads such as lighting.

The other case where such "borrowing" occurs without creating a hazard is where the two circuits in an enclosure are a legitimate multiwire circuit. Instead of being controlled by breakers 23 and 35, they are far more likely to be on breakers 21 and 23; they may even be handle-tied. That certainly is not a given. This set of scenarios underscores two points. First, the previous repairperson may have made an assumption that created a hazard, without the slightest awareness. Second, if you do not watch out for such treacherous pre-historic gaffes, you could get injured.

A similar but rarer case of modifying a multiwire circuit and overloading the neutral is where a fuseholder went bad and the conductor attached to it was moved to a free fuseholder. In other instances, where all fuseholders were already occupied, the "orphaned" conductor was spliced to the wire that went to a good fuseholder, and the splice was pigtailed to the new location. What could be wrong with those practices? Suppose the wire that was moved was part of a multiwire circuit. Unless the installer was aware of that and took suitable care in choosing which fuse to join, the neutral may be carrying 30 amperes to the hot's 15.

Note that when circuits are improperly mixed, you have an increased chance of violating Section 300-20(a) and inducing currents in conduit and cable armor. The associated concerns with respect to heat, shock, and putative bioelectromagnetic effects are discussed elsewhere.

Photovoltaics

Here is a problem you may find with some old, but not terribly old, houses. It is especially likely to be present in homes that were at some time owned by jack-of-all-trades types—even more so if they were owned by survivalists or others strongly committed to self-sufficiency. If the owners converted to wind or solar power in order to dispense with reliance on a utility, they may very well have installed a DC system. Or they may have purchased a 120-volt inverter and converted the electrical panel to two-wire operation. In doing so, they may well have overloaded the "neutrals" of multiwire circuits, which since the conversion have suffered from double loading as above.

The 1999 NEC may add an exception permitting that usage, with restrictions designed to protect the "neutral." Still, be wary of that practice; where someone has misunderstood the permission, you may very well be faced with overloaded conductors.

Only with the 1996 edition was an article on photovoltaics finally introduced into the NEC. Some of the material was gathered from elsewhere in the Code, but much was brand-new as formal doctrine. Older photovoltaic installations that lacked its guidance may suffer from remarkably creative wiring. See also the discussion at the very end of this chapter.

Grounding to plumbing, or to the grounded conductor— returns to haunt you

The ignorant mistakes described thus far, including the multiwire miswiring, account for only a few of the circuits whose current returns along unexpected paths. Section 250-50 (exception) permits formerly ungrounded circuits to be grounded to any point on the grounding electrode conductor system. The NEC used to permit grounding to suitably bonded water pipe. As the result of various plumbing disruptions, some of those grounds added to originally ungrounded outlets during earlier upgrades may have grown less certain. This means that some circuits that ought to be grounded are not. The consequences, as demonstrated below, can be deadly.

Perhaps worse is the fact that some raceways have deteriorated to the point that the systems feeding them may be reasonably grounded, but parts that once were grounded—perhaps were grounded from day one—are no longer reliably grounded. For instance, after being installed in a concrete slab treated with the common additive calcium chloride, over 30 years even heavy-wall, rigid metal conduit has corroded through completely!

Such deterioration is more than an annoyance; it has resulted in electrocution. In one instance, described by Donald Barrett, former Chief Electrical Inspector for Prince George's County, Maryland, someone was killed by touching a defective air conditioner while standing outdoors on the slab through which the air conditioner circuit ran. Ironically and fatally, the makeup of that slab's concrete had destroyed the conduit it contained, so that circuit grounding was interrupted before the conduit reached the defective air conditioner. Had the ground not been interrupted, the circuit presumably would have blown as soon as the air conditioner developed a fault rather than shorting to the slab through the person bridging between air conditioner and concrete.

For the same reason that a three-prong tool with the third prong removed, or with a cheater employed on the cord, can be far more dangerous than even an old-style two-prong tool, three-prong equipment connected to a circuit that has lost its ground presents increased hazard. All accessible metal parts not intended to carry current are bonded together. The result is that any ground fault makes all those parts live. If grounding on the circuit, rather than just at the one tool or outlet, is defeated, one piece of shorted equipment will create a shock hazard at all three-prong equipment on that circuit—even at undamaged equipment connected to other outlets!

Certain other changes in how equipment is wired will not create any surprise for readers of this first edition, but are worth mentioning for the sake of future readers.

Ranges and dryers

The grounded and grounding conductors are supposed to be joined at one place only: at the service equipment (including that associated with a separately derived system) or at its equivalent at a separate building. The grounded conductor is expected to carry current, but the grounding conductor is never intended to be live except, momentarily, during a fault. There are, however, exceptions in NEC Section 250-60. And they used to be broader.

NEMA 30R and NEMA 50R receptacles, intended for, respectively, electric dryers and electric ranges or ovens, offer combined 250/125 volt outlets but no separate ground. The chassis of an appliance using a corresponding NEMA 30P or NEMA 50P cord connector was bonded to the connector's third prong. Type SEC service cable with a concentric neutral, covered but not insulated, was permitted as the feed, provided that it originated from a loadcenter that served as the service equipment. This design was introduced into the NEC during World War II to reduce copper use and thus save metal; it was not removed until the 1993 NEC. (An alternate theory is that the exception was accepted to make electric dryer installations more competitive with gas dryers.) In future years, these installations will grow more and more rare, but there probably still will be a few around by the year 2100.

Incidentally, should you change the system layout such that the loadcenter feeding a NEMA 30R or NEMA 50R receptacle is converted into a subpanel, perhaps to economize during a service upgrade, you will be required to make additional changes. That SEC cable no longer will be legal for the purpose of feeding an electric range or dryer, which now requires both an insulated grounded conductor and a separate means of equipment grounding. Since you will have created that violation by your changes, it will not normally be grandfathered.

Antiques

You may find some additional types of equipment that ground to the identified/grounded conductor or, worse, that substitute other grounding means, whether in the form of a raceway or a conductor, for an identified/grounded conductor. Some are easy to up-grade; others are a trial.

Crowfoot receptacles

First is the crowfoot receptacle, introduced in Chapter 5, "Accept, Adapt, or Uproot." The crowfoot is a standard-sized receptacle, single or duplex, whose openings look like, and are arranged exactly like, those of a NEMA 50R range receptacle, except for being smaller. Like a NEMA 50R, the crowfoot has no separate ground. If you find a crowfoot receptacle, it probably will not be in use—nothing made in recent years will plug into it. The last one observed in use held a standard two-prong plug whose prongs had been brutally twisted to make them go in.

If the wires feeding a crowfoot are in good condition, you can substitute a modern receptacle. If the enclosure is grounded, you are set. If it is not grounded, you can deal with the situation as you would the replacement of any ungrounded receptacle, with one difference. If it happens to be a dedicated outlet, you have a spare wire that most inspectors will permit you to use as a grounding conductor, painted or taped green at both ends.

Sheer ignorance

Crowfoots constitute the more promising case. Other arrangements, where someone used the ground as a return simply because doing so made the outlet work, can be beyond any reasonable repair. In many of those situations, the only answer is to rip it out and do it right. See Chapter 8, "Tearing It Out and Rewiring."

Using the ground as a return often results in hazardous conditions. In some of the worst of these situations, the danger is compounded because not only was the electrical ground used as the return, but that ground relied upon the plumbing system for continuity. Then some plumber came along and replaced a rotten water pipe with plastic, and the outlet went dead. If the light or appliance happened to have been removed or turned off, fine. Otherwise, the plumber probably got zapped as he interrupted the current's return path. If the plumber lucked out, then later on, when people touched the downstream plumbing and any ground—even a poor ground—while the power was trying to return, they got shocked. They also could have gotten shocked if there were a grounded-type appliance downstream on that circuit and they simultaneously touched its metal frame or chassis and a ground. A similar situation is described at the end of Chapter 2, "Troubleshooting," in the context of extending ungrounded circuits.

Four more incitements to premature balding

There are many, many other wiring peculiarities found in old buildings that are not necessarily a problem; some at least are not even technically violations. Before leaving the "omigawds," though, a few more of the nastier surprises deserve mention.

Buried splices

You will not find splices buried in the walls or ceilings of a modern, professionally wired and maintained electrical system. Old systems may be a different story.

Legally buried splices

Concealed knob-and-tube wiring was and is quite legal. Article 324 does not require that you install boxes, or self-contained devices such as are used in barn wiring, at knob-and-tube wiring's splice points. There are only two requirements for splicing it. A proper connector or a soldered-and-taped joint must be used, and each conductor must be supported near every splice.

What is the point in bringing up knob and tube wiring? Just this: if a concealed splice should go bad, you may not be able to locate it. It may be that the bad splice is in knob-and-tube wiring. Should you be unable to find an outlet box or junction box containing a bad splice, it does not mean that you have lost your touch. Nor is the experience absolute evidence that some carpenter buried a box, or that some amateur played loose and free with the NEC.

Legal or no, you may have a bear of a time finding the break. When you do find it, what then? Since extension of existing knob-and-tube circuits is explicitly authorized by Section 324-3 (1), repair certainly is legal. Before choosing that option, you have to answer two questions. First, do you want to be repairing a circuit that unequivocally is ungrounded and also whose insulation very probably is falling apart after so many years? Second, can you get hold of the necessary parts–knobs, leather washers, tubes, and loom?

When the customer decides that a circuit run in a knob-and-tube or open wiring is to be repaired, and you think the insulation is healthy enough to permit that repair, you would be well-advised to create a transition to more modern wiring. Bring the knob-and-tube wiring into a box, and splice a cable onto it in that box. To the judicious electrician the fact that the NEC grants permission to continue with knob-and-tube is not sufficient justification to do so. Change to cable, so as to protect your wires a little better, and also so as to start running a grounding conductor (provided that you can find a reliable ground).

There is one exception to this rejection of wiring with individual conductors as being antiquated and not up to snuff. In areas subject to severe flooding, open wiring retains an important advantage: it dries out faster than cable or raceway wiring. Note that the wiring in this case is neither decrepit nor ungrounded, unlike ancient concealed knob-and-tube wiring.

One last word about knob-and-tube. It was intended and permitted for use in free air, not buried in rock wool. The U.S. Consumer Product Safety Commission consulted a panel of experts about knob-and-tube wiring and thermal insulation. Based on the panel's advice, the CPSC recommends that where this wiring system is in contact with insulation, or even in a setting where insulation may be subsequently added, any #12-AWG conductors should be derated to 15 amperes.

Illegally buried splices

Other buried splices are just what you might fear: evidence that someone did what they should not have done. To locate such splices, look for new walls, added by the ignorant. Look for dimples in walls or ceilings, where the bodies (junction boxes) might be buried. Use a voltage tracer, if circumstances permit. In the worst case, disconnect the dead wires at the last box before the mysterious interruption, cap them, and run fresh cable. If you possibly can figure out which circuit is involved by consulting the panelboard directory, also disconnect and cap the wires at the last accessible box to the other side of the hidden interruption.

The reason for disconnecting and capping wires is twofold. First, the customer might discover another outlet that is mysteriously dead. Then, upon a little investigation, you may be able to say, "Whoa, that was no buried outlet that had me stymied, that was just an outlet located totally elsewhere than I was looking." Chapter 8, "Tearing it out and rewiring," discussed some reasons for that to be the case. The other reason for disconnecting the mysterious is that you do not want to feed or backfeed into a bad splice. It could arc.

Unexpected power

Ludicrous though it may seem, on any number of occasions you will find power when there should be absolutely none. Yes, that is the opposite of the problem that most commonly brings an electrician onto the premises—dead outlets. But it is very dangerous, very much a problem.

Before proceeding to the most likely ways this comes about, a warning: the cause could be even more outrageous. You may run across a self-sufficient customer with his own generator. With creative amateur rigging, there undoubtedly are photovoltaic systems, probably rotary converters, and even uninterruptible computer power supplies ready to backfeed into electrical systems without warning. See the section of Chapter 7, "Setting Limits and Avoiding Snares," called "The long and short of liability" for a discussion of "last person in" doctrine. It may apply when you encounter this sort of hazard.

With those warnings out of the way, there are two common situations in which power shows up where there should be none. One is where a meter is bypassed. Sure, that constitutes theft of power; certainly, the serving utility will thank you if you help them catch instances of power theft. If the condition predates your customers' tenancies even they probably will thank you, especially if you explain how dangerous it can be to steal power. But, in the meantime, that jumper or tap puts you at risk. This problem is more likely to be found in older buildings than in newer ones, because older systems tend to be less tidy. Power theft is quite infrequent, but it is still something for which you should be on the lookout. This is especially true when you are in the habit of pulling a meter in the expectation that all power will be killed—as the equivalent of lockout.

The other odd situation in which power appears where no power should be is where tenancies commingle. In a large industrial establishment, there often are multiple sources of power. Sure, each source has to have a permanent marking, identifying and locating the other power sources. This is not the issue that should worry you.

This should: you may find multiple sources of power in nonindustrial occupancies. In many apartment buildings whose tenants have separate meters, electricians have discovered power in apartments that was not controlled by their loadcenters. Sometimes this was because common building circuits entered their premises, circuits intended, say, for alarm systems or even for hall lights. On other occasions, electricians have found power from Apartment 24 serving loads in Apartment 23. Sam Levinrad, a vastly experienced consultant and retired inspector, says that he has found this situation to be quite common. The cause usually is hard to know. The most common reason probably is that when each apartment got its own panel, the person separating the circuits goofed.

Responding to the problem of finding power present when it should be absent

Whatever the reason that unexpected live conductors are found, the fact that they do turn up underscores one rule that is even more important to heed in work on old buildings than in new: you know that power is off only when and where a working tester tells you power is off. POW! Hacksawing through BX that was sure to be dead has burnt through many a hacksaw blade. Devil-may-care types flicking conductors against metal, in attempts to short them to ground in order to trip the overcurrent devices so the conductors can be cut safely, have often found POW! POW! POW! fountains of sparks because the circuits would not die. But some of those electricians have died.

Now that you have ample warning about the problem of encountering live wires where none should be found, the obvious question must be answered: what can you do when confronted with that danger? To be realistic, two levels of solution are offered. Most of the following presumes that you can work safely on live equipment.

At a minimum, notify and mark. In the case of apparent theft of power, this means notifying the serving utility—even if you hazard removing the jumper. Certainly the least you can do, when the customer does not authorize you to fix problems that constitute potential hazards to future electricians or tenants, is to permanently mark the service equipment with a warning.

If you do get the go-ahead for a real fix, here is how you can put things to rights. Start by removing jumpers or bypasses, and close any openings left by doing so. Disconnect power that originates in different occupancies, except insofar as it serves building loads such as common smoke detectors or exit lights. Blank off any outlets served by a foreign meter, rather than simply refeeding them from the proper loadcenter and capping the foreign wires. Even if there is sufficient volume at an outlet once fed from another tenancy for you to run power from the proper source, if you do so you risk that some future worker will ignorantly reconnect the capped wires. If you do decide to do so anyway, at least mark them to warn others of the danger they contain. Putting a note inside those boxes might be even more advisable, insuring against the possibility of paint or wallpaper covering a marking on the covers. Putting notes inside the blanked boxes is another wise move. Ideally, of course, you will kill them at their sources.

Example: mystery voltage To close this section, here is an odd little story about mingled electrical systems. An electrician was called in to troubleshoot at a townhouse whose owner had gotten a tingle in the shower. He detected no voltage using a solenoidal voltmeter (in this case, a Wiggy™). He switched to a wider-range multimeter, and found 10–13 volts AC between various points—the shower head, the faucet valve—and the metal drain grille. At first it seemed that the plumbing, perhaps the cold water line, was energized. Not so; a little further testing, using a properly connected three-prong extension cord for reference, confirmed that a potential difference was present between the drain grille and ground.

The mystery thickened. After pulling the meter—and confirming in the presence of a utility representative that no power was coming through that house's service—he continued to measure voltage between the drain and the extension cord's ground. Moreover, the drainpipe itself was plastic! The voltage was coming up through the sludge and slime coating it on the inside. So what was the story? The power apparently was coming from outside the house. Either the underground lateral had been pierced, or, more likely, the neighboring house had a leak or imbalance that was sending some voltage to ground through the plumbing system rather than, or as well as, through the main bonding jumper and the incoming service neutral.

What does this story have to do with old wiring? It just demonstrates that sometimes, in order to track a problem to its source, you have to go beyond the bounds of the tenancy or even the building on which you are working. In the case of this story, if the trace voltage had not gone away shortly afterwards, they would simply have told the utility to do something about it. Bringing the power company's representative in was an example of notifying those concerned. It would have formed the basis for a subsequent demand that the utility take the responsibility to remove the stray voltage.

UnListed equipment

Another problem that is not unique to old buildings, but that is found more commonly in older structures, is the use of unListed equipment. Chapter 10, "Historic Buildings," talks about some of the issues concerning unListed fixtures, including fixtures that antedate widespread Listing. Some people choose old buildings because they like the old and unusual. Still others live in old buildings simply because they have been in the family, others because, frankly, old houses are what they can afford. Especially in the latter cases, you may find that over the years lots of corners have been cut.

Inside meters

The last odd old design to be considered here is inside metering. In some cases, inside meters look just as safe and professional and guarded as anything modern. Their only disadvantages are that they may make it more difficult to provide the meter reader with ready access, and that it is harder for firefighters to disconnect power by pulling a meter before entering the building. In other cases of inside metering, open wiring, minimally protected by loom, dives through wooden boards on which the meters are mounted. From behind the boards, the wires go into a trough or wireway. The problem with these setups and some other types of old inside metering is that they offer no real mechanical protection for the wires where they emerge from the bottoms of the meters.

What is especially problematic about the latter design is that you find it in very old multiple-occupancy buildings. Each apartment might only have 30-amp service. None or few of the tenants may need more power than that provides. Therefore, it is hard to convince anyone to authorize the major effort that would be required to upgrade the building's service. On the other hand, in a building

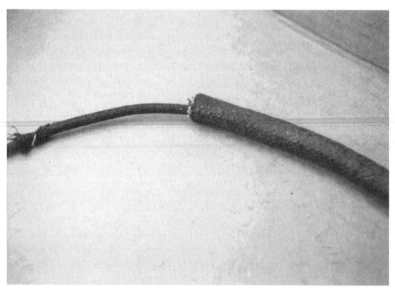

■ **9-4** *Open wiring protected by loom*

that old, a superannuated basement meter room probably does not pose the greatest hazard to the occupants. The more important task, upgrading the branch circuit wiring, is a far huger job. If that is ever authorized, upgrading the meters will be almost trivial by comparison.

Aluminum

The next big issue to cover is aluminum wiring. A few jurisdictions prohibit any use of aluminum conductors. Far more commonly, the use of aluminum is restricted in residences, and in smaller gauges. The larger conductors, it is generally agreed, are much less likely to cause problems. For that reason, this discussion starts by considering aluminum conductors that are likely to be found in commercial, institutional, and industrial settings.

Large aluminum cables

Many consider aluminum far more dangerous in branch circuit wiring than copper; such an attitude is much less present when electricians talk about heavier wiring. Here is one difference between copper and aluminum that is worth keeping in mind, where those heavier cables are likely to be found. When copper arcs to copper, the terminations frequently fuse and melt together, forming essentially a bolted short. When aluminum arcs to aluminum, on the other hand, the metal is far more likely to vaporize. This apparently is not so likely to produce an open circuit as it is to produce a somewhat high-resistance arc to ground or to any other conductors in the vicinity of the vaporizing metal. Be warned.

Before proceeding to a discussion of aluminum branch circuit wiring, there are other concerns that are particular to installations using larger aluminum conductors. Until the 1970s, electricians were not aware that lugs designed for copper conductors are not very compatible with aluminum and aluminum alloy conductors. Despite this incompatibility, you may encounter installations that have survived from the earlier period.

In such cases you once again face the issues of grandfathering and liability. Any aluminum conductor found in a lug that is marked "Copper only," or "Cu," or not marked, represents a violation. What should you do about it? Sometimes you can replace the lug. Sometimes you can install a copper pigtail, perhaps using a compression fitting, although at other times you will be stymied by insufficient room. Sometimes neither option is acceptable because the equipment, as opposed to the lug, is marked for copper only.

When you find an explicit marking to that effect, and aluminum in place, the original installer was either ignorant or a scofflaw. There may or may not be an onus on you to correct that installation, but it certainly is not grandfathered.

If you are repairing or replacing the cables or terminations, you must correct the violation. If your job does not actually require you to touch the terminations, the least you should do is notify the customer of the potential hazard. A superior option is to thermograph the equipment under load and record the readings. If nothing is running hot, that provides some reassurance—and you have established a baseline. What you should NOT do, all expert opinion agrees, is retorque suspect connections. Retorquing does little good and can do damage. If a connection is bad—has been running hot—it needs to be remade with a new lug, holding a newly stripped conductor end, rather than being retorqued.

Aluminum branch circuit wiring.

So much for feeders and large commercial, institutional or industrial branch circuits. Young journeymen may never have seen small-gauge aluminum branch circuit wiring; that qualifies it for this book because thousands of miles of the stuff remain in use. While aluminum conductors were first introduced in the 1940s, most aluminum branch circuit wiring was installed in the 1960s and 1970s. The U.S. Consumer Product Safety Commission's (CPSC) strongest warnings concern the original, unalloyed formulation, which was installed between 1965 and 1973. Even the later, alloyed version is something to be concerned about.

Small-gauge aluminum has not been in use for a few decades, and it is unlikely to come back. According to Ravindra H. Ganatra, a senior engineer with Alcan Cable, cable with #12 aluminum conductors probably would cost as much as or more than cable with #14 copper, which has the equivalent capacity. The reason is the increased quantity of insulation, paper filler (if used), and sheath required around fatter conductors.

Aluminum branch circuit wiring carries problems, but these problems do have solutions. First, though, you need to determine whether the stuff is aluminum.

It was not Al, Al, all the time

Lest they assume that there is more aluminum around than there really is, you need to get customers straight on one thing. Not everything that is a shiny silvery color is aluminum. No, you will never encounter cable containing silver conductors, not even where, as in the Depression-era Wobbly song, "the mills were made of marble..." (Silver conceivably may have been used, in the days when it was far less expensive than it is now, as a coating on copper fixture wires.) Plenty of armored cable was manufactured, though, with tinned (solder-coated) copper wires. Without that tinning, when wires were stripped the rubber had to be scraped off the copper because of gluing that resulted from catalytic action. How can you tell if you have tinned copper in your hand? Scrape the tinned wires and the copper shows. For a second test, tinned copper is stiffer than aluminum of the same diameter—stiffer than even hard-drawn aluminum.

The real thing

Having gotten that out of the way, it is time to discuss genuine aluminum wire. Not only are aluminum branch circuit conductors a source of concern, aluminum service and feeder conductors exhibit more creep than do copper wires. While you may never encounter a connector marked for use with aluminum only, there are plenty designed for copper only. There is no guarantee that such connectors, when used with aluminum, will make reliable long-term, low-resistance connections. In 1977, a lawsuit by the Consumer Product Safety Commission declared old-technology aluminum, the original version, "imminently hazardous."

Variations Aluminum wire comes in at least three flavors: regular, alloy, and copper-clad. Nickel-clad and tin-clad versions were Listed, but you may never come across them. The best advice is that those should be treated the same as the more familiar varieties of aluminum conductors. While they probably will work as well as copper-clad at terminals marked "copper or copper-clad aluminum only," using them there is technically a violation of NEC Section 110-3(b).

Copper-clad aluminum wires were manufactured, by Texas Instruments for General Cable, as a compromise between the better price of aluminum and the better electrical characteristics of copper. (In the early, early days of power wiring, copper-clad iron wire was used because it was cheaper than copper and perhaps also because it is stronger.) Copper-clad aluminum wire (henceforth just called, "copper-clad") has the ampacity of aluminum, but its ability to make connections at device terminals is based on the outer, copper cladding. No problems have been reported with its connections. There may never have been solderless connectors explicitly rated for use with copper-clad. While it may be technically a violation of NEC Section 110-3(b), generally accepted practice

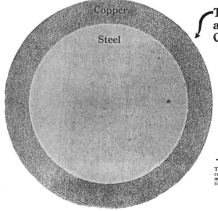

■ 9-5 *The other "copper-clad." You are highly unlikely to come across any of this.*

allows you to use any old twist-on connector with copper-clad, since what touches is copper—but see Section 110-14.

Relatively little copper-clad was installed; you may never encounter it. You do need to be able to recognize it. It feels more flexible, and is also more fragile, than copper of the same diameter. When you cut it, the center is silvery—the opposite configuration from tinned copper wire. An electrician's biggest danger in working with it is that he or she might mistake it for copper, and thus assume that a wire of a particular gauge has the ampacity of copper, rather than the lower ampacity of aluminum.

Taking on the bigger concerns

Having disposed of copper-clad, it is time to get back to aluminum and aluminum-alloy conductors. Whatever its metallurgical formulation, aluminum branch circuit wiring other than copper-clad needs to be treated with especial respect and caution. Two questions naturally arise. First, what, if any, is the problem? Second, if problem there be, what are the options for dealing with it?

To begin with, does aluminum branch circuit wiring constitute a problem you need to solve? There are arguments on both sides. The source of concern is twofold. First, aluminum has a higher coefficient of thermal expansion than copper, which means it flexes more as it heats and cools. Second, alum, oxidized aluminum, forms immediately on exposure of aluminum to air, and is a much poorer conductor than copper, aluminum, or even oxidized copper. These two factors, it is agreed, create the potential for overheating at every point where aluminum makes a connection. When connections show evidence of overheating, all will agree a problem exists. When they do not show such evidence, some say to leave well enough alone: "If it ain't broke,..." Others say, "Don't push your luck."

The discussion up to this point looked at aluminum wiring from an owner's or home inspector/safety surveyor's perspective. Now it is time for answers to more nitty-gritty, fix-it questions. If you need to replace a device or redo a splice where aluminum wiring is present, how complicated is your task?

Repairs To begin with, you have to use materials suitable for contact with aluminum. Another factor common to all work with aluminum wires is that they are more fragile than copper: as is true of copper-clad wires, they nick and break far more easily than do harder, tougher wires. Also, low-resistance terminations and splices are much harder to achieve. According to Dr. Jesse Aronstein, P.E., an authority who has extensively studied connections of aluminum branch circuit conductors, the conductors should be abraded (scraped) and immediately coated with a nonflammable antioxidant whenever they are stripped in order to be spliced or terminated.

Swapping in Leaving that basic advice, how complicated a task you face in repairing aluminum depends on what you find. Where you

189

are merely replacing a device such as a failed switch, you might be in luck. If the conductors still seem in good shape, and you need not touch any splices, you can simply replace the device with another that bears the Co/Alr or (if the device is rated over 20 amps) the Al/Cu mark.

Handling more complicated cases In many cases, however, you will need to make some change beyond just swapping a new device in place of the old. If the old device or wiring is not properly bonded to the enclosure by today's standard, you probably need to do some splicing. Likewise, if it is on a multiwire circuit, and the neutral daisy-chains through the device, you now have to change that arrangement to a pigtail. If one proposal for the 1999 NEC is accepted, you will have to pigtail all grounded conductors rather than daisy-chaining them, whatever the type of circuit. If a conductor looks nicked or fatigued, you will need to strip it back. If you see a splice that looks iffy, or if you need to make a new splice for any other reason, further repair work is required.

Systemic repair There is one unquestioned fix for aluminum branch circuit wiring, other than tearing it all out. This unique, patented system, Amp Corporation's Copalum™ uses special, calibrated compression connectors and tools to bond copper pigtails permanently to the aluminum conductors. You have to be trained and certified by the manufacturer to use the system, and the costs—both of the certification and of the tools and materials—are not negligible. They also are not overwhelmingly expensive. Surprisingly, the system is not Listed for making grounding connections, nor does the manufacturer anticipate seeking that certification. Nonetheless, that system is widely acknowledged to be the best solution you can get, short of rewiring. Highly-respected authorities call the lack of investigation for grounding a technicality. As of late 1997, Copalum™ was still the only approach recommended by the CPSC.

Solderless connectors There are several other repair options, all of which are held in lower esteem than the system just mentioned. One twist-on solderless connector, Ideal Corporation's Model #65 Twister™, is Listed for use with combinations of aluminum conductors, with the additional restriction that those splices must incorporate at least one copper conductor as well. The connector's Listing does include use for grounding connections.

As of late 1997, there still were major unresolved concerns about the testing program under which these connectors were evaluated. Nevertheless, given that UL and CSA both have accepted the results of additional testing, most inspectors will go along with your use of that Listed product—at least when connecting a single copper conductor with a single aluminum conductor. Whether the connectors will stand up over the years, in more varied types of use, is a question no one can answer. Because of the way these connectors were tested, you might want to hedge your bets if you decide to employ them. Use each of these connectors only to pig-tail a single aluminum conductor to a single copper conductor, and, if necessary, then splice copper to copper.

The #65 Twister has some across-the-border competition. Two corporations, Ideal and Marr, distribute solderless connectors Listed under the Canadian Electrical Code for use with aluminum. Some U.S. inspectors will accept the use of connectors that satisfy Canadian standards. Concerns have been raised in Canada, too, concerning the adequacy of the Listing standards employed. There, similarly, retesting has satisfied CSA, if not all challengers.

There are three more options. Some electricians still possess twist-on solderless connectors, rated for copper and aluminum, that were manufactured in the 1970s, before the UL standard changed. Listing is never "removed" from a product, even when the standard changes. While that technicality can be used to jus-tify employing products that no longer are up to current stan-dards, doing so is playing close to the edge.

191

Dr. Aronstein found excellent results from an alternative proce-dure. Unfortunately, it involved a solderless connector, 3M's Scotchlok™, whose design subsequently has been changed. If you are interested in a detailed description of the findings, including his proposed solutions and his warnings, try to find his report. As of the time this is being written, the report is available on the World Wide Web, at a posting listed in Chapter 14, "Resources." If it becomes unavailable at that site, the CPSC might be able to di-rect you to a copy. Again, do note that neither the manufacturer nor the CPSC recommends this approach, as it is not covered by UL's Listing for the connector.

Terminal strips The final fix you might consider is the use of splicing devices that consist of miniature terminal strips using setscrews. These have separate setscrew entries for each conduc-tor, so the connections do not depend upon contact between cop-per and aluminum. (There also are push-in versions, but many have less faith in them; besides, they are irrelevant here—they are Listed only for use with copper conductors.) Many terminal strips are Listed for use with grounding conductors. Most are Listed for use with aluminum conductors. This combination of characteristics suggests that they could fit the bill.

An issue that looms very large to some is that terminal strips have not been explicitly tested nor advertised as a solution to the problem of making reliable connections to aluminum branch circuit wiring. But their Listing does qualify them. The main practical installation problem is that they tend to be bulky. While the NEC does not account for splicing devices in box fill calculations, the introduction to Section 370-16 does require you to leave adequate space for the enclosed conductors. For this reason, when using this remedy you might have to add extensions to the boxes. It is ugly, but it appears safe, and your customer might be content because of its affordability.

Despite the overview of these various options, you have not been given an extensive, detailed, walk-through of the procedures; nor is there any great list of suppliers for aluminum repairs in Chapter 14, "Resources." That is because the best advice, if you run across an aluminum-wired home, is to either rewire or use Copalum™.

There are contrary opinions. The Aluminum Association does not see a need to do anything if the wiring has not been causing problems all this time. This is a judgment call that ultimately only the homeowner can make. It behooves you, though, to think out your position, as you are quite likely to be asked for advice.

An intermediate stance that some take, between the extremes of either dismissing the issue or else rewiring or remediating the aluminum branch circuits, is to set up a warning system. One manufacturer offers 6-32 cover plate screws whose heads darken if overheated. Another offers crayons whose marks do the same. The shortcoming shared by these intermediate-level answers is that color change has to be noticed or it is of no value. Unfortunately, it seems far more likely that an alternate notification system, the smoke detector, would offer the first warning customers would notice. It is conceivable that arc-detecting circuit breakers (AFCIs), described at the end of Chapter 6, "Safety Surveys," will provide an even better answer.

Relay switching

That is all the space available for discussion of aluminum conductors. The next rare beast is relay-operated switching. While many electricians never have encountered them, there are easily half a dozen such systems on the market. Relay switching is, and more commonly in the past than nowadays, was, installed for two primary reasons.

First, it permits more sophisticated control of lights, appliances, audiovisual equipment, and even alarms, than can be hard-wired for a reasonable effort. Second, in some forms it is considerably less expensive and messy, and in most versions it is more flexible than adding hard-wired line-voltage switches.

Class 2 wired relays

With adoption of the Consumer Electronics Bus standard in the 1990s, interchangeability is likely to be the way of the future. Not so in the past, and not so with relics of the past. The multitude of systems precludes comparing schematics or instructions in this chapter. Knowing that something exists, and having a basic idea of its operating principles, is a first step towards mastery. These basic principles will be enunciated shortly. When you need more information, you can order schematics from the same people who supply parts.

The Smart House™ system uses its own special cable, but most relay switching systems rely on regular Class 2 wiring. There certainly is plenty of installed capacity of General Electric low-voltage equipment, and every part of it is subject to wear and thus occasionally needs replacement. Cable sometimes suffers damage, and momentary-contact switches, relays, and transformers wear out.

Troubleshooting and replacement of this equipment call on the self-same skills you use to repair doorbell systems and fire alarms. Bear in mind, though, that not just the transformers but also the relays present the hazard of 120 volts. In this respect, working on relay-operated switching equipment is more like furnace repair than like working on other communications wiring.

In passing, take note of a dangerous item that is not at all hard-wired, and thus is less directly a concern for electricians. Still, be on the lookout for it—and, if you come across it, warn your customers. In a number of cases, Model 1C of the "Whistle Switch" by Mark Engineering (1-617-342-6034), also sold as Invento Products' "Sonic Switch," has overheated and ignited. The two-part unit consists of a plug-in adapter and relay, which is activated by a separate ultrasonic whistle operated by squeezing. The relay unit is beige and the whistle case is black. See Chapter 17, "Recalled Items," for more information on product recalls. Most of what you will find in that chapter is equipment you might be called on to install or repair. The "Whistle Switch" was mentioned now simply because this is the section on remote switching.

After low-voltage switching come systems whose installation takes no new cabling although some parts, at least, are more than simply plug-ins. Powerline carrier systems do not require Class 2 wire. X-10 (and compatible) transmitters simply superimpose signals over the 60 Hz waveform present in existing branch circuits. The manufacturers of X-10 produced equivalent equipment under dozens of other labels, including General Electric's and Radio Shack's Home Minder® system, Sear's Home Lighting Control®, Radio Shack's "Plug-N-Power®" and Stanley's system, including their "Light Maker®" infrared-controlled spotlight.

The X-10 factory also produced components for Leviton, which extended the system, creating a fluorescent lighting control that used the principle of X-10's appliance module, and also a locally operated dimmer. X-10 went for the homeowner market, whereas Leviton assumed that the installer of at least some of their equipment was an electrician—and charged considerably more money for equivalent items.

Some of the X-10 control units are battery operated and send out radio signals. Their receivers either plug into receptacles, screw into lampholders, or are hard-wired. The hard-wired units, with which you might be dealing, need to match the specifications of the loads they are to control. They are hooked up the same way you would wire the equivalent standard receptacles.

Each receiver needs to be set to the house code used by the transmitters, and each receiver must be given a unit number so that transmitters intended to control the receiver can turn it on and off, along with any other receivers given the identical unit number.

Certain problems with the system have been reported. X-10 has a master control (transmitter) that will operate any slave unit (receiver module) in the house by addressing its code. The only restriction is that the master and slave units must be fed off the same busbar, or a bridge must be installed to carry signals between the busbars. Unfortunately, the master might also operate any other X-10 unit in neighbors' houses that are fed off the same secondary distribution pole or even fed off the next pole within 300 to 400 feet if it comes off the same utility transformer, should the X-10 units happen to share the same code.

Another potential problem is that plug-in intercoms have been reported to interfere with the X-10. If, for instance, a customer left an intercom turned on in an infant's room for monitoring, the X-10 might not work at all.

Other control systems

Smart House Limited Partnership, Inc. was a bold venture into home control, with licensed installers, special panelboards, and whole-house control. In the several years since its introduction, though, the product has not taken off in the marketplace. This may be in part because Smart House Limited Partnership, Inc. expected appliance manufacturers to modify their products for use with the system. They have not scrambled to get on board. One interesting spinoff, from a Code standpoint, is Smart-Redi®. It is illegal for an installer to leave type NM or AC cable stapled or coiled in a wall, without an enclosure, against future need. The Smart House people, though, authorize just such a practice with their proprietary cable. Presumably, when a manufacturer comes along with an appliance that can benefit from the Smart House control system, it can be hooked up to that cable, without additional wiring. Some suggest that, while far from an old system, the Smart House™ is one step from being an orphan.

Square D was one of the panelboard manufacturers originally involved in Smart House Limited Partnership, Inc. Later Square D pulled back, perhaps because they have their own proprietary home-control system, called Elan. Elan is actually an audio-video-intercom system, but it does interface with the X-10 system, through either an RS-232 port or infrared.

Some of the proprietary systems lay claim to particular market niches. This does not mean their principles of operation are anything other than powerline carrier, Class 2, radio-control, or, within line of sight, infrared. Certain products are said to be particularly suited to serving the needs of persons with disabilities. They include the following, according to a disabilities activist: a security system by Honeywell, Inc. called "The TotalHome Home Automation System®," that controls the temperature, security, lighting, and appliances; and the "Mastervoice®" Home Automation System, the 1991 Easter Seals Society invention of the year.

Because relay-switching systems often were chosen, however long ago, by innovative customers, those same customers might very well present you with new cutting-edge challenges. Home automation is a natural candidate for computerization. Sony, for example, has a home automation system, their CAV-1, based on their PC Vision Touch® personal computer system. The IBM-manufactured

Aptiva™ computer supports the X-10 command set, and interfaces with the Home Minder™ system. The issue of installing high-tech equipment in old buildings is touched on below, as well as in Chapter 11, "Commercial Settings."

Thermostats

As with relay-operated home controls, thermostats come in too many versions for it to be feasible to take them all apart for you or present schematics. Programmable thermostats are a relatively recent development. There still are simple 2-wire thermostats that just contain a mercury bead in which a bimetallic spring makes and breaks contact as the temperature changes. Then there are more complex series, with separate heating and cooling mechanisms, and with adjustable lag.

As an example, here are a few types produced by Honeywell. The 24-volt Series 10 (with three internal wires) had a main contact and a holding contact, with in later versions a heater to cut it off sooner. The second or holding contact helped it keep from cutting off and on due to vibration. The holding contact established connection before the relay pulled in. Although it closed first, this holding contact would not turn on the furnace, but the furnace would stay on unless it released, even if the main contact had let go.

The 24-volt series 20 worked slightly differently. It had a common contact with one wire to turn it on and another to turn off. At first this required a motor that would rotate to turn it on and then start up again to turn it off. Later versions were manufactured with relays. The contacts to turn the furnace off served to short the holding circuit of the relay coil, (which had a resistor in series with it to take the shock) thus indirectly opening the contacts which kept the furnace on. The heaters were adjusted with a rheostat, or by screwing in replacement color-coded heaters. Aside from these, there was no adjustment to the differential (the difference between the cut-in and the cut-out temperatures).

The Series 80 used its own trick. It had a magnet for rapid make and break. Moving the magnet would adjust the amount of hang-in: closer to the armature and the relay would make sooner. The original version, like the series 20, contained a motor.

The bottom line with regards to replacement

That was just a taste of some old designs—to satisfy curiosity, not to help you make repairs. Whatever the brand, whatever the thermostat, you do not need to replace a thermostat with the same model, or even look up a replacement in an equivalency chart. What you need to know is the number of wires connected to it in the wall, the operations that the thermostat is to control—heating or heating and cooling—and what special features your customer desires, such as night setback or general programmability.

Caution: line-voltage thermostats

Line-voltage thermostats are another story. Far more common in Europe, they are two-wire devices, or four-wire if two pole, that are wired in series with heaters; no relays are involved. Beware: someone could get killed as the result of mistaking a line-voltage thermostat for a standard, low-voltage unit. Line-voltage wiring certainly could burn up a low-voltage programmable thermostat obliviously installed as its replacement. Even a single-pole line-voltage thermostat usually has line-to-line voltage available: one hot leg connected directly to it, the other available as a backfeed through the heater. If the existing thermostat seems to be fed by type AC or NM cable rather than by Class 2 wiring, you have an all-but-certain clue that it's connected directly to line voltage: 240 or 208 volts, commonly. Still, your voltmeter is the only absolutely reliable indicator.

Note that a two-pole line-voltage thermostat with a marked OFF position should by rights disconnect all power from the heater it controls, but it will do so only if wired for two-pole operation. Hasty electricians have installed such thermostats so as to interrupt only one of the two hot legs, and they have functioned fine as thermostats. Wired that way, even when in the OFF position a two-pole thermostat still feeds power from one hot leg to the heater or heaters it controls. This is misleading and potentially deadly.

The thermostat pictured in figure 9-6(a) offers no indication on the surface that it operates at line voltage, but you might suspect the likelihood, based on the fact that the thermostat and its fellows control a radiant heating system of resistance coils embedded in floors and ceilings.

(a)

■ **9-6 a, b, c** *Line Voltage thermostat*

Once you remove the cover, the manufacturer's sticker seen in Figure 9-6(b) clearly identifies the thermostat's mode of operation. Withdraw the thermostat from the wall enclosure, and you observe capped wires in Figure 9-6(c). This indicates that the electrician who last replaced these thermostats did not bother to wire them for two-

(b)

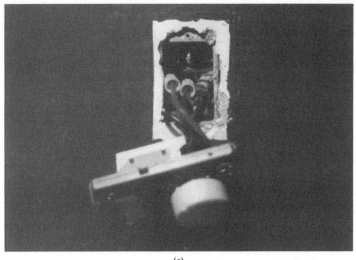

(c)

pole operation. They still work, but not quite exactly as intended. You should look out for evidence of other shortcuts. Indeed, some of the splices in this installation were rather sloppy.

Telephones

Returning to low-voltage systems, communications cabling has progressed so quickly, and can be such a bear to untangle and trace out, that the old wiring is worth little effort or discussion. Old four-hole phone jacks can easily be replaced with modular units, and hard-wired telephones likewise. The crimping tool and the terminals are readily available from your local supply house. Old one-line, four-wire cabling can easily be changed to handle two lines. When you encounter any commercial, complex phone or data system that needs significant repair or upgrading, though, the best advice seems to be to run new cables, and to document what you put in.

Having said that, if you want to get a rough idea of what is presently in place in a commercial setup, here is some guidance.

☐ Square connectors probably mean the system is IBM Type 1, Token Ring.

☐ Coaxial cable could mean an IBM mainframe installation, old Ethernet, or video lines.
  • Coaxial cable running radially means IBM.
  • Coaxial cable that is tapped means Ethernet.

Relatively Rare Situations

Warning: bonding cable

Here is a warning that rightfully belongs in Chapter 6, "Safety Surveys." It has been relegated to this later chapter because electricians, the main readers of this volume, are infrequently involved in cable television, and because telecommunications installations are more and more frequently being required to undergo municipal inspection. The warning is simple: whenever you have the opportunity, make sure that the low-voltage systems are properly grounded and bonded.

In one nasty case, omitting that step resulted in a fire that burned a number of contiguous townhouses. The TV antenna cable went to an ungrounded monument, and passed quite near power wiring—without the systems being bonded together. There was a power surge from the serving utility. It arced from the overhead conductors to the antenna cable, and then, inside the attic's thermal insulation, arced from the cable to a grounded branch circuit. The latter arc started the blaze.

Miscellaneous rarities

You will have to deal with a number of other odd items when working on old buildings. Unfamiliar overcurrent devices are an example of equipment whose repair requires no special skills but does demand knowledge.

Old breakers

One odd bird is the antiquated circuit breaker. Even so-called "interchangeable" circuit breakers are not interchangeable unless specifically Listed for use in the panels in which you want to use them. Loadcenters are evaluated by NRTLs in conjunction with certain circuit breakers, and vice versa. Even when they fit physically, breakers that are not mentioned in the Listing might not operate quite as they should.

The tripping curves of substitute breakers might not afford proper coordinated protection. In other words, the main might trip in place of the branch circuit breaker. Alternately, the branch circuit breaker might not be protected by the main tripping when it should in fact be superseded by the main because of the latter's higher interrupting rating.

Another danger is that an inappropriately substituted breaker will allow more circuits in the panelboard than the manufacturer intended, or will create overcrowding in some other way. An inappropriate breaker also could have a different response to ambient temperature, or even have different moisture resistance.

An exception to the noninterchangeability rule is that UL and some other NRTLs have Classified circuit breakers of certain brands for use with certain loadcenters as aftermarket equipment. Cutler-Hammer makes circuit breakers for Square D's QO panels; Trip-and-Lite and Thomas and Betts make circuit breakers Classified for quite a few brands. Moreover, many circuit breaker manufacturers have changed ownership over the years, so that the same breaker appeared under different names. The section of Chapter 13, "Dating and History," called "Begats" will help you trace this trail and thus will provide clues as to what circuit breaker to use in each antique panelboard.

From one perspective, liability is the main risk in using a nonapproved breaker, meaning one that has been neither Listed nor Classified for the purpose, nor given the nod by a fearless and helpful inspector as a reasonable substitute for a defunct model. If something should go wrong, it matters little whether the tripping curve was a mismatch or whether the breaker was in some other way not the equal of the OEM product. What matters in court is whether a lawyer, the manufacturer's or another's, can shove responsibility on to you. Contrariwise, in the event that something goes wrong, having used a Listed product in the application for which it was intended by the manufacturer and investigated by the laboratory offers you considerable protection.

For these reasons, as well as simply to do the best you can to protect your customers when a circuit breaker goes bad—a rare occurrence, but one you will run into—you want to substitute the correct replacement. Similarly, when you add a circuit to an old panelboard, you want to use a breaker Listed or Classified for the purpose. Finally, when you come on a breaker that was improperly installed, it is nice to be able to substitute the correct one. Nice? In view of "last man in" doctrine, doing so might be more than just "nice."

Note that you cannot determine the proper breaker by looking at what is already installed—nor by asking the customer over the phone what brand he or she has. There are many panels in which the majority of the installed breakers violate the panel instructions. You need to look at the legend, located with the schematic on the door or inside the cabinet. (Of course, if that is missing, you have to seek other evidence and go with your best guess.)

There are many panels around, especially older ones, that do not present this difficulty of determining which breaker is right. They simply require very, very different designs than the standard plug-in circuit breakers you might be used to. One is a breaker much longer than other brands, and with a different means of attachment: Zinsco. For a brief period, Zinsco private-labeled those breakers to Arrow-Hart as well. Another is the Bulldog or Pushmatic breaker, which has a pushbutton rather than a toggle handle. Federal Pacific manufactured the first plug-in circuit breaker. Their loadcenters are no longer manufactured, but the breakers are. They too are different, in their own way, from the most common ones based on Westinghouse's design. Wadsworth is another company that made slightly different breakers. Cutler Hammer is very much around. Even though they are united with Challenger under one ownership, they too are uniquely configured. Finally, the 1980s saw Square D's short-lived Trilliant line, which will no more fit into their QO panels than either will fit any other brand of panel.

Some of the oddest breakers are noninterchangeable with a vengeance. In the early 1950s, apparently in response to problems that came to the attention of Chicago's Chief Inspector, one and only one edition of the NEC required that breakers incorporate rejecting features at least equivalent to those of Type S fuses. Thus, not only did a circuit breaker of one brand not fit into a panelboard bearing another brand, a 15-ampere circuit breaker could not be replaced by a 20 of the same brand. In some cases, special tools were required to install and remove breakers.

Odd fuses and fuseboxes

You might also run into odd fuseboxes, although you are unlikely to still encounter any really odd fuses. Of course, what is familiar to some might be odd to you.

Old fuse types

Old, rare fuses are less of a problem than are old circuit breakers. It is unlikely that any reader will come across homemade fuses of wire solder still in use after a century and more. (In Europe, as of 1997, they remain quite common.) Substitutes are readily available for antiquated fuses that you realistically may encounter. They often represent a substantial improvement in function. A rebuildable cartridge fuse can be replaced with a one-time fuse of the same shape and ampere rating and the same or higher voltage rating and interrupting capacity. Where appropriate, a fast-blow fuse can be replaced with a more common time-delay fuse. An Edison-base fuse can be replaced with a Type S fuse, using the appropriate insert. A fuse of odd ampacity might not be necessary. You can safely, judiciously, replace many a 4-amp plug fuse with a standard 15-amp no-tamper fustat, based on evaluation of the wiring and the device being protected. That evaluation is not complicated. Simply check the #14 or greater AWG circuit to make sure that no equipment on it specifies protection at a lower rating than 15 amps.

Live front fuseboxes

Some outdated fusebox designs can be dangerous. One hazardous kind of antique fusebox is the live-front design. Open the door and not only the fuses but also the busbars are at your fingertips. On rare occasions, you might be able to cobble together a safety barrier. In doing so, you probably void the Listing. Whether to do so remains a judgment call, but one which a responsible inspector would probably support as enhancing safety. Here are some considerations:

☐ The barrier should be fairly rigid, and must maintain its structural integrity;

☐ The barrier should be nonflammable, which rules out wood and some plastics;

☐ The barrier should not significantly affect heat dissipation, nor impinge on the wiring space or the fuses and their terminals;

☐ The barrier should not itself pose a risk of injury, which rules out amateurishly cut and formed sheet metal;

☐ The barrier should not pose a risk of shorting.

A reasonable compromise in dealing with a live-front panel is to affix a permanent marking to the outside, warning users of the danger. The best answer, of course, is replacement—but that is not a call you always have the right to make for the customer, unless you perceive imminent hazard, say due to corrosion or other damage that creates faults or high-resistance connections.

Asbestos

Another odd style in fuseboxes is the enclosure with asbestos insulation. There is no need for panic if you encounter one of these. Asbestos is a wonderful insulating material, and it only constitutes a threat if breathed. Asbestos card is not going to find its way into your clothes or up your nose unless it's crumbling. There might, however, be a problem in safely disposing of such a box. Keep this in mind when you price an upgrade.

Switched fuses

Finally, some old fuseboxes have switches as well as fuses. Some have a master switch that disconnects the fuses before you open the box, as a safety factor, quite like a modern single-circuit fused safety switch. This was a great idea. Others like the one seen in Figure 6-2 had a separate switch for each circuit. These switches go bad; when they do, you are probably out of luck when it comes to replacing them. This might mean that it is up-grade time, at least to the extent of installing a subpanel. On the other hand, if switches for individual circuits fail in (low-imped-ance) conducting mode, this leaves you with a standard fusebox, in which circuits can be disconnected the usual way—by un-screwing fuses.

When the new is hazarded by the old

Before leaving the discussion of odd circumstances with which you might need to deal, it is worth touching on the problems faced by very modern additions to old buildings. Mismatches between new fluorescent lamp types and older ballasts are addressed else-where. Here is one more potential example: delicate electronics. This category includes home computers, and it includes electronic point-of-sale systems.

There is nothing intrinsic to old buildings to cause difficulties for such equipment, but there are troublesome tendencies. You are not likely to run into problems of triplen harmonics, except in commercial buildings that have had major equipment upgrades. You are more likely to be dealing with overloaded circuits, outages, arcing faults, and uncertain grounding. There is a greater likeli-hood that circuits will be confounded, so that you have not only multiwire circuits but also circuits that should not be connected nevertheless sharing identified conductors.

You certainly can find ways to add safety devices and high-tech equipment in old buildings. As an example, some of the problems mentioned above are also reasons you should be less likely to ex-pect GFCIs to remain untripped in an older building. The solution? Sometimes GFCIs need to be installed at the ends of runs, as pro-tection for single locations. If you do want to minimize unpleas-antness when installing delicate equipment in older buildings, dedicated circuits can be very helpful. More extreme solutions are presented in Chapter 11, "Commercial Settings."

One last caution regarding existing high-technology equipment is that it might have been installed early in the learning curve. Just as recently produced GFCIs and electronic ballasts are more sturdy and more reliable than were earlier versions, other equipment too changed for the better over time as manufacturers fielded complaints. Similarly, Code rules regarding new types of equipment evolve in response to field experience.

One example is the fact that until the 1990 NEC was adopted, conductors leading down from a photovoltaic array had less protection than they do now. If the conductors were sized so as to safely carry the maximum current the sun could generate from the panels to which they were attached, the Code did not require overcurrent protection. Unfortunately, when a fault occurred and the blocking diodes preventing backfeeds failed, those conductors could experience considerable overloads, resulting in fires. Just because an installation is cutting edge, like PV, does not mean that you can assume its wiring is in accordance with current safety standards. It might be just old enough to be dangerous: solar panels were being installed on houses as far back as the 1970s.

Historic Buildings

WORK ON HISTORIC BUILDINGS RAISES SPECIAL ISSUES OF safety, legality, value, and aesthetics. Those issues are not independent. On the one hand, the NEC and other building codes are intended to ensure safety, and fire does not care whether a building is steeped in history. On the other hand, if a structure has unique qualities, whether historic or architectural, some think it preferable to greatly restrict the building's usability rather than let it be "savaged" by trade contractors in order to make it safe. Of course, there are always people who want to have it both ways. You cannot always fully accommodate them.

This chapter speaks directly to these issues. It also touches on some of the codes other than the NEC that come into play when you work on old wiring, whether in historic buildings or elsewhere. Finally, it considers old and exotic fixtures, such as often are installed in and around classic structures.

Defining the subject

The term "historic building" can refer to two somewhat different kinds of beast. The first is simply a building whose owner values its charming traditional features. The second is a building that has made it onto the national registry of historic sites. The NEC does not distinguish between the one type and the other, or indeed between a building that is a decade old and another a century old. Localities, on the other hand, often impose legal restrictions on what you can do to an officially designated historic site, regardless of what you and the customer would like. Less often, they make special allowances for such structures.

What makes a building or a site of historic interest? How does this affect its value? The national registry of historic sites, of official landmarks, is maintained by the National Trust for Historic Preservation, 1785 Massachusetts Avenue, Northwest, Washington, D.C. Their telephone number is 202-673-4000. They are the people who best can answer those questions, which are beyond the scope of this book.

If a structure is not yet on their registry, your work on it could destroy the special quality that qualifies it, and, if so, might affect its value. To answer such concerns, talk to your local historical society or contact the National Trust. Fortunately, if your customers are seriously interested in such matters, they themselves very likely will have been in touch with preservationists.

Although you need to be aware that they come up, decisions involving property value are not yours to deal with, but your customers'. Some owners are troubled by the restrictions created by their properties' location in historic areas. The value of a property actually can be lowered because these restrictions make it less usable, less capable of generating income. Some resent this. Most owners, however, are quite enthusiastic over the possibility that their buildings may make it onto the national register. The reasons vary. Besides any finer feelings, there can be significant tax incentives for maintaining and restoring historic buildings.

Your responsibilities as a licensed professional put you in the position of having to please both your customers and the authorities. Even if a building is not on a historic registry, your customer might want you to treat it as though it were. How you respond depends on the local jurisdiction. The same rules that restrict what you can do to historic buildings sometimes also give you more leeway than does the current version of the NEC.

Other codes besides the NEC

The NEC is not the only law that rules your work, even if it has been adopted in your jurisdiction without a single amendment. While systems that were legal at the time they were installed generally are grandfathered under the NEC, you may be required to bring them into compliance with these other codes.

Every building code, while applying primarily to new construction, also addresses existing systems. The general grandfathering rule, regardless of which model code has been adopted in your area, is that within reason whatever exists can remain in use. The 1990 BOCA code has such wording in Section 103.0. The 1991 ICBO code has it in Section 104(f).

There also are more formalized requirements for existing buildings. The 1991 SBCCI code permits not only the use but also the repair and upgrading of existing installations, so long as that new work meets code. The SBCCI's Section 101.6 specifically references Special Historic Buildings. SBCCI's Standard Existing Building Code also offers guidance for electricians, albeit very general guidance. The BOCA Property Management Code has a section on electrical facilities. Its Section H602.0 has minimum outlet requirements that are different from the NEC's lighting and receptacle spacing rules. In habitable rooms and guest rooms, the BOCA code specifies two separated ("remote" in its language) outlets—not necessarily receptacles. In kitchens, it requires three outlets. In each case, one of them can be a lighting outlet, located on either the wall or the ceiling.

Many areas also have specific local codes. The 1990 California State Historical Building code includes electrical requirements. Other states with limited historic building codes, as of 1997, include Georgia, Wisconsin, Connecticut, Hawaii, Indiana, New York, North Carolina, and Pennsylvania.

Special codes

Besides the general building code in force in your area—CABO, BOCA, or whatever—at least one special code has been developed nationally to address historic sites. The ICBO publishes not only the Uniform Building Code but also the Uniform Code for Building Conservation. Section UCBC-4, "Electrical Guidelines," appears on pages 276–291 of the 1992 edition. If you are doing residential work, which is its subject, you will benefit from reading it whether or not it is in force in your area.

You may not endorse all of its recommendations. Nonetheless, besides offering an additional perspective to help you think things through, it does constitute the law in some areas. It also gives quite reasonable advice regarding inspection.

Until the mid-1970s, it was quite common for grandfathering to be determined by the "25/50%" rule. If work being done over a six-month period cost less than 25% of a building's worth (worth was calculated in various ways), the system was grandfathered with respect to building codes. Only that portion actually worked on had to be brought up to current code. If work performed over the course of six months cost over 50% of the building's worth, the entire building needed to be brought up to date. Jobs priced in the midrange, between 25% and 50% of total worth, were treated in assorted ways.

The fact that an installation is grandfathered does not mean that anything goes. Any inspector will insist that immediate, blatant hazards—for example, bare power wires—be corrected. Of course, whatever a building's stature, if it never has had an inspection, you may find various horrors. For a long time, some 120-volt circuits in the United States Capitol were carried on fixture wire and even bell wire run through chases and ducts, until a congressional appropriation in the late 1960s enabled the building to be completely rewired.

Where it has been adopted into law, the ICBO Conservation Code is enforced in two circumstances. The first is where a 25/50% or similar rule remains in effect. The second is where residential occupancy changes—say from single-family to two-family.

Just to give you a taste of the ICBO guidelines, they list a number of hazardous conditions that must be corrected. These include deteriorated equipment and loose receptacles (neither is defined); reversed polarity; receptacles without GFCI protection in baths, garages, and outdoor locations; and pull chain switches or brass lampholders within arm's reach in damp or wet locations.

In some ways, the ICBO guidelines are more liberal than the NEC's rules. The practices permitted by the guidelines include grounding an ungrounded system to the cold water pipe and installing a second, separate service in a residence simply because the existing one is inconveniently located. (The ICBO guidelines do require a permanent sign at each service giving notice of the other.)

In other ways, the ICBO guidelines are more restrictive than the NEC. They do not permit extending ungrounded circuits to kitchens, bathrooms, garages, basements, or other areas where grounded surfaces are readily accessible. Doing so, as permitted by the NEC, admittedly could result in some hazard. Under ICBO rules, adding a GFCI does not create an exception allowing one to use grounding-type receptacles in an ungrounded system. Of course, an inspector could, at his or her discretion, allow the NEC's permission to preempt that ICBO restriction.

Local codes are far more likely than national ones to specify what you can and cannot do. As a not-so-unusual example, in Prince George's County, Maryland, interior work on buildings in historic areas faces no special restrictions. Work on exterior features requires a special permit. There is an exemption for "ordinary maintenance or repair with no material effect on it as a historical, architectural, or cultural resource."

Urgent needs

Moving on from the subject of legalities, what aspects of historic buildings are likely to be unsafe? Government-sponsored research on "Common Code and Safety Deficiencies of Historic Buildings" included the following in its list: electrical equipment, fire alarm and detection, and lightning protection. The first fire sprinkler was patented in England in 1723, and alarms have been around as long as there have been bells. (Boston adopted electric fire alarms in 1851.) Westinghouse built a successful, nonarcing, lightning arrester in 1892. Nevertheless, many historic buildings have been lost due to inadequacies in these areas. You may just be there to look at the power wiring. Still, it is worth at least mentioning these other concerns to your customers, whether or not you are required by the jurisdictional authorities to install fire or smoke detectors or lightning arresters.

Old, odd, and unsafe equipment

Besides the more common safety defects, historic buildings present the challenges of unusual, unfamiliar equipment—some of which may be dangerous but cherished. Unfortunately, however long it has been giving satisfactory service, the old and unusual is not necessarily safe.

Great-great-granny's floor lamp is no concern of yours, unless you are asked to repair it. If you do consider taking on that task, you are strongly advised to check for hazards such as rotted internal wiring—even if you are just replacing a switch. Also look for problems such as inadequate insulation between a metal body and a metal lamp shell. Consider issues such as lack of a grommet and strain relief where the cord exits. Once you fix something, you may be considered responsible if it subsequently causes injury.

Portable fixtures, however, are not your major concern. When asked to install an antique luminaire, a work of art, or an import—whether historic reproduction or avant-garde—be very, very leery. Artists do not know UL standards. European manufacturers, unless working to U.S. standards, often assume higher voltage and correspondingly lower ampacity, and thus use lighter-gauge wires. True, NEC Section 110-2 says that any fixture can be installed so long as it is acceptable to the Authority Having Jurisdiction. But why should an inspector bestow his or her blessing on unListed equipment?

What can you do when your customer wants iffy equipment installed or reinstalled? There are two options that often will serve to cover both installer and inspector. The less-certain one is to hire a licensed professional engineer to sign off on any unusual installation. The far more reliable one is to request the services of an NRTL. Even the oldest of them, UL, has a Field Services division. As of 1996, the evaluation and labeling of an individual piece of equipment could, in some circumstances, with some NRTLs, cost less than $50.00. See Chapter 14, "Resources," for a partial list of NRTLs.

Other unusual antiques

Odd electrical fixtures are only one problem. Where the original lighting system involved gas fixtures, you may very well encounter leftover connections to the gas system.

That statement encompasses a number of terrors. One is electrical outlets that are hung from, meaning actually supported by, old gas pipes. The characteristic appearance is old black iron conduit capped not with an electrical capped bushing but with a plumbing cap, complete with sealing compound. Do not assume that such a pipe is empty of gas. (There is no need to be terrified, either; residential and even much commercial gas service is supplied at low pressure; you can restore the cap if you inadvertently remove it.)

What should you do about connections with old gas pipes? Nothing. If a sconce supported by such pipes is working, and the wires are good, there is no reason to mess with it. If you are upgrading, or if the customer is too concerned to leave it be, then by all means move the wiring out of that box. There is no reason not to bury a pancake box, for example, once you relocate the cables that entered it to an adjacent, more capacious box. A box without wire, cable, or conduit is just a chunk of formed steel. Do leave any gas pipe termination accessible (unless a plumber is called in to remove it entirely).

Another hangover from gas days is the reconfigured gas light fixture. There are Listed electric post lights that are designed specifically for retrofitting gas lawn fixtures. These too are perfectly okay to leave in place or to install. (How historically accurate such fixtures are is not your worry—that decision lies with the customers or with their preservationists, unless the local jurisdiction has something to say about it.)

If you fail to find a Listing mark on an existing, retrofitted fixture, evaluating or repairing it becomes a little more complicated. Examine the fixture and decide whether its design and installation look professional. If these satisfy your judgment, and the fixture does not show signs of deterioration or other hazard, you might well choose to consider it grandfathered and not worry about the lack of a label. You also can bring such fixtures to the attention of an inspector, if he or she comes by the job before you have to make your decision, and request an official decision.

Satisfying customers' aesthetic requirements

Moving on from nonstandard fixtures, it is time to consider ways to install more conventional wiring. Historic buildings present you with many decisions based on the intersection of law, safety, and aesthetics. What kinds of building features do people cherish? What kind of changes may be restricted, and how can you deal with such limitations when the existing wiring is shot?

To answer these challenges, it is worth looking at what some experts have suggested. One conservator concerned with visual quality has recommended that alarm bells be located just above baseboard to reduce their visibility. Another, museum lighting specialist Gersil Kay, suggests mounting emergency lights remote from their sources of power so that only the heads are visible, rather than displaying the complete units including battery boxes.

213

She also fervently believes, and has documented, that a vast proportion of historic buildings are designed in such a way as will allow the concealment of building services. This extends beyond the mere capability to fish vertical runs of conveniently flexible electrical cables. She also has managed to hide plumbing pipes and HVAC ducts in hollow spaces of wall and ceilings. "More thought, less damage" might be her motto. She thinks little of electricians who cut into decorative moldings and sculpted surfaces for their runs, rather than moving over a few inches and cutting into plain plaster, which is easier to restore.

Minimizing visual pollution through relying on unusual locations is a common theme. One authority, historical preservation architect Alfred M. Staehli, has asserted that when a historic building's walls would be "marred" by switches and receptacles, floor outlets can be the solution of choice. John Fidler, a renowned British architectural conservator, agrees with that unusual suggestion in an article referenced in Chapter 14, "Resources." A second odd option apparently favored by some pundits is installing floor pedestals near the wall. A third is recessing receptacles and switches behind baseboard, with flaps or doors providing access.

You have to evaluate these possibilities in terms of your understanding of the NEC. The suggestion that you recess receptacles within the baseboard could be considered in violation of NEC Section 370-20 by those who prefer a narrow interpretation of the NEC. From a broader perspective, if the box and cover plate form a complete enclosure, and the contents of the box are fully accessible, Sections 370-20 and 410-56(e) are satisfied. You are merely redefining "wall surface." A good carpenter can do that. (Alternately, a casuist [a.k.a. wisenheimer] could argue that baseboards are not walls or ceilings, so those sections simply do not apply.)

Incidentally, for many decades, Edison-base outlets were used as receptacles, and appliances were fitted with free-swiveling screw-in attachment plugs. At an intermediate stage before that system was totally outlawed, when installed in the baseboard such outlets were restricted to ones with protective covers—a bit like that third option just mentioned. As of mid-1997, a house wired in the 1940s still had one outlet harking back to that practice. It consisted of a duplex flush receptacle on an enclosed porch, with a weatherproof cover, the type having two round doors that snap shut when nothing is plugged in. The most unusual feature was that the receptacle, and the "doors," were designed to accommodate not modern three- or even two-prong cord connectors but Edison-base outlets.

Larger upgrades

In a historic building, carefully thought-through creativity is important. If you fail to evaluate the implications of your planned upgrade, you may have to backtrack—expensively. While HVAC work is slightly beyond the scope of this book, here is an example from that realm that could easily be applied to equipment falling within an electrician's sphere—equipment such as a step-down transformer.

Putting a central air conditioner on the roof can sometimes provide a happy solution to the problem of how to serve comfort while maintaining historic appearance. There are two potential problems with that fix. The first is visibility from the street. If the unit is an eyesore, will it mar the historic lines of a building. The second is support. While installers surely will assess the structure's ability to handle the unit's weight, a related but less obvious issue is vibration. A historic building may contain relatively delicate fittings that could come to grief.

Service changes

What about service upgrades? What do you do when you find deteriorated service equipment, or calculate the need for a higher class of service to feed new loads? Here you are getting into changes that could affect the external appearance, and thus the historical value, of the property. Fortunately, utilities may be willing to accommodate your customer's special needs. PEPCO, the utility that at the time of this writing serves much of Maryland, has a rule that all new meters must be outdoors. Nevertheless, they will not insist that the meter be moved outside when you upgrade, so long as their workers will have no problem regularly gaining access to the meter. Nor will they insist on changing to underground service in their underground area, or even on running triplex. They will replace a rack if that is needed for historicity, and bring open conductors to it from the pole.

Utilities are not alone in this considerateness. When dealing with historic structures, inspectors are likely to honor the principle of reasonable accommodation, by extending special permission when necessary and safe.

215

Commercial Settings

WHILE MOST OF THE CHALLENGES DISCUSSED IN EARLIER chapters can be found in commercial, institutional, and industrial installations as well as in residential, certain conditions are far more likely to be encountered in stores and factories. Two situations are addressed in this chapter. In one, you are confronted with types of old equipment that you can expect to find only in commercial and industrial premises. In the other, you are asked to accommodate new demands on old wiring in a commercial, institutional, or industrial setting.

The presenting problem might take the form of a customer who says, "What will it cost me to repair or replace this motor?" or "What will it take to replace this with a more efficient setup?"

Note that industrial settings permit some exceptions to Code rules, provided that operations are under the supervision of qualified persons. As the glossary points out, "qualified" means something specific here; having an electrician on-site, even a master electrician, is not enough.

Phased-out systems

An electrician whose experience has been purely residential could easily be fazed by polyphase systems. Almost all residential wiring is single-phase. Even in multifamily dwellings served by three-phase, four-wire, 208Y120-volt power, the wiring in each occupancy normally is single-phase. Edison's original AC system was single-phase. Soon, though, because of the need to handle the heavier demands of commercial installations, polyphase power and higher voltages became far more common.

Relatively few readers will encounter the very unusual old systems discussed next, so the first part of this chapter will be mainly of historical interest—except to those readers who experience an "aha," an epiphany: "That's what that puzzling setup was about!" Feel free to skip through the next few pages, if the designs that follow are not present at your customers' premises and you are looking for practical advice immediately .

Synthesized three-phase

If you are used to efficient three-phase industrial equipment, served by efficient three-phase utility power, certain old systems may surprise you. Yes, the motors are three-phase. No, the incoming power is single-phase.

There are two ways to convert single-phase to polyphase power: rotary and static. The first employs either a rotary phase converter or a motor-generator. A rotary converter is in essence a three-phase motor designed to run on single-phase power; its effect on the power system enables you to run a regular three-phase motor or motors in parallel with it to operate utilization equipment. A motor generator is, essentially, a single-phase motor turning the shaft of a true three-phase generator. The latter tends to be rather more expensive.

The second approach is to simulate a third phase using capacitors— either with capacitor-relay type synthesis, or with a solid-state device that produces three-phase power by rectifying single-phase and synthesizing the desired wave forms. There is no need to go into greater detail about either approach. Few of you will encounter these, and there are experts out there ready to serve you, if needed. Incidentally, phase conversion does not make it wonderfully efficient to continue using the old, odd equipment; doing so by using phase converters is equivalent to keeping an old gas-guzzler on the road by pouring lead substitute into its tank at each fill-up.

Phase converters do have a place. Normally, in locations with only single-phase power available, you find single-phase motors. Some utilization equipment, however, is only readily available with three-phase motors. Also, a good three-phase motor may have a significantly longer life span in some applications. According to phase-sythesis expert Gary Werner, when powered by single-phase motors 7.5 HP or higher, heavily used air compressors in body shops last a year or two. They last seven or eight years powered by three-phase motors of the same size—utilizing single-phase incoming utility power, adapted by a static phase converter. Also, municipalities often find it worthwhile to use converters to enable them to equip sewage lift stations with three-phase pumps, when three-phase power is not conveniently available.

This example should not be generalized to other existing single-phase motors. Werner warns that converters (his company's sole product, so he is certainly not biased against their use) are there for when you must serve three-phase equipment. In other situations, installing three-phase motors plus a converter or converters is not economically competitive with using single-phase motors. He also mentioned that new high-efficiency motors are less friendly to phase converters than were older ones, whose steel laminations had lower silicon content. With the newer motors, the phase angle tends to collapse, unless you specify larger rotary converters than you would have formerly, and rely less on capacitors.

One grave drawback to the use of static converters is that they will trip phase-loss detectors. This is even true of some rotary converters. Therefore, as of 1997, there is no way to ensure that three-phase motors fed through static converters are protected from burning out due to single-phasing. If a proposed Code change to require phase-loss protection is passed, there could be a severe increase in the cost of utilizing three-phase motors where only single-phase power is supplied. Motor-generators could be required. UL Standard 508C already requires that polyphase drives incorporate phase-loss detectors.

Other manipulations to put off changing the class of service

There are a number of other ways to finesse the supplied power to achieve the equivalent of a different class of service. Buck-boost transformers are an example. You will encounter systems using autotransformers to step voltage up and down between, say, 208

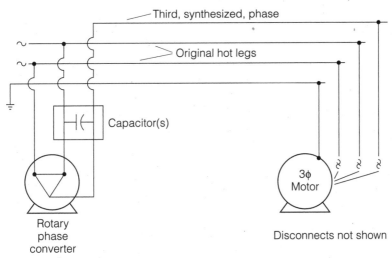

Third, synthesized, phase

Original hot legs

Capacitor(s)

Rotary
phase
converter

3ϕ Motor

Disconnects not shown

■ **11-1** *Rotary phase converter.*

and 240 volts; that is a standard practice. The 1999 NEC should explicitly authorize the use of an autotransformer to buck or boost between 208 and 240 volts to supply either an individual load or a feeder, in either old or new work. Some systems utilize 208 volts, single phase, obtained between the high ("red") leg of three-phase 240-volt delta wiring and the grounded midpoint of the other two phases. For the application of Code restrictions, such as the grounding requirement in Section 250-42(f), do note that this 208-volt power is over 150 volts to ground, unlike that obtained from 208Y120-volt wiring.

Why would people do these things? Money. It might seem far more efficient to order the right equipment and the right utility service, so that they match each other from the start. That option, alas, is not always available. It certainly may be far more expensive, at least in the short term, than one of these creative kludges. If the options that were chosen violate neither manufacturers' (and Listing agencies') specifications nor the NEC, it is your job to maintain them. It even can be argued that you have a responsibility to recommend such approaches when they could save the customer money. The comparisons involved in that determination can be complex. Life-cycle costs have to take into account at least the following considerations:

☐ up-front costs, including the difference in value between money spent now and money saved over time;

☐ the costs of training associated with making any change;

☐ the costs of power, both in terms of kilowatt-hours and of peak demand and power factor penalties;

☐ differences in equipment lifespan:

☐ costs associated with differing maintenance requirements;

☐ and costs of having to increase, or savings associated with being able to put off increasing, electrical system capacity.

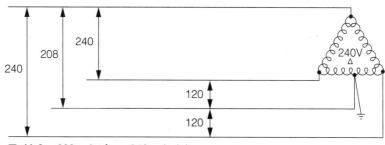

■ **11-2**   *208 volts from 240 volt delta.*

Other considerations to keep in mind with regard to creative solutions is whether they will deceive users into misapplication. Burning out a few motors, or starters, or even lights, on the wrong voltage can very quickly wipe out initial savings. And then there are the "we can't get this to work" call-backs.

In terms of the overall electrical system, one important thing to keep in mind with a three-phase converter is that it draws the amount of power required by the equipment (volt-amperes times 1.73 because of the three-phase demand), plus some additional, though not at all large, losses associated with the converter itself. You can severely overload a two-wire, single-phase circuit feeding a three-phase motor by unthinkingly performing current calculations based on three- or four-wire three-phase service. The NEC will steer you safe. Section 455-6 (exception) requires you to multiply the full-load three-phase amperes of all your converter-operated equipment by a minimum of 250% to arrive at the single-phase ampacity.

Updating class of service

Because of this inherent complication, and inherent inefficiency, it is worth backpedaling just a bit. If you are called in because phase conversion equipment is reaching end-of-life or rebuild time, you might research the possibilities for upgrading the service—especially if the present electrical system is deteriorated or borders on the inadequate. There may be advantages to considering a different class of service. The serving utility may have changed policies, making that more attractive than in the past. Utility deregulation may result in wider choice for at least larger users. Alternately, if the machinery is out of date, new equipment that matches presently available power may prove to be a cost-effective investment.

This is the way the representative of a large northeastern utility described the economics of single-phase versus three-phase service, in late 1996. Basically, he said, it comes down to what is available. Utilities charge by the kilowatt-hour, plus demand factors for larger customers. Phase is irrelevant to the utility in terms of pricing. It may not be irrelevant in terms of what your customer uses and therefore pays; three-phase, being more efficient, sometimes leads to slightly lowered total and peak loads.

If three-phase power is available near your customer's site, offering three-phase service just means running two more wires pole to pole. As of late 1996, utility installation of 13-kV wire, carried in 1/0 aluminum (which, the representative observed, suffices to serve any reasonably-sized service) cost a customer less than $5.50 a foot. There was no charge for new transformers. If three-phase service was not available in the area, though, the numbers were very, very different. In that case, the customer had to pay a considerably higher fee, equal to the cost of setting additional poles. At the time, this came to about $10,000 per pole.

To that utility, incidentally, three-phase power means three-phase, four-wire, 208Y120-volt service. While their transmission lines carry three-phase delta, they do not want to stock transformers to supply that configuration to customers.

Odd old services

From phase issues, the discussion will move on to even more exotic matters. Some readers are bound to encounter certain systems that generally were phased out long, long ago.

DC

The first system that the above-mentioned utility offered customers was DC. If your customer is one of the few still using utility-supplied DC, he or she certainly can be expected to tell you. (Here, as everywhere else, expectations are no substitute for using your tester.) DC was supplied to Washington, D.C. customers who lived near substations supplying the trolley system, when trolleys were around. That system is long gone. Because of the investment tied up in rolling stock, the rails are an excellent example of a system that changes slowly. Electric rail systems in the Northeast Corridor distribute DC power because they own so many DC locomotives; they invert it in the more modern locomotives that run on AC. Nowadays you will find utility-supplied DC serving a few, rare industrial customers. Some older buildings in large cities may also have elevators run by DC motors. Warning: in faulting, DC can draw particularly fat and hot arcs.

Freaky frequencies

60 Hz, and 50 Hz in Europe, has not been the standard forever. In London, England, in 1950, there were substations within one block supplying 600-volt traction power at 16.66, 25, 33.3, and 50 Hz—the odd frequencies were derived from their standard 50 Hz via motor-alternator sets. The first high-tension line went from Niagara Falls to Buffalo, New York, carrying 25-Hz power. It has not disappeared; 25-Hz power is still supplied to some large old industrial customers

all around this country. Apparently, 25-Hz power is terrific for incremental control of motors such as are used for elevators.

Forget the ancient history. Union Carbide's Electric Welding Products department was still building 25-Hz equipment for customers as late as 1965. As recently as 1992, the Firing Circuits Company of Norwalk, Connecticut, would get an occasional inquiry for 25-Hz motor starters. And as of mid-1997, Amtrak and NJ Transit operated sections of electric traction rail equipment under 12-kV 25-Hz single-phase power. Out in the boondocks? No—in New York City's Pennsylvania Station.

The 25-Hz utilization frequency has for a long time been derived from utility-supplied 60-Hz power. Conversion originally relied on vacuum tubes—later, diodes—and now uses SCRs. Motor-generators might seem a reasonable option for accomplishing this purpose, but greater maintenance costs cause that option to be a noncontender.

Two-phase

Two-phase systems, some three-wire, some four-wire, and some five-wire, are still to be found, because people still have motors designed to work on two-phase power. The phases are 90 degrees apart, as opposed to open delta, which is still three-phase, and thus has phases 120 degrees apart. It is possible that there no longer are utilities that still supply two-phase power, but it can be derived from three-phase. A Dr. Scott developed and patented the Scott-T transformer in England in the 1950s to convert three-phase power to two-phase. Plans are still available.

While rare, two-phase is something a few readers will need to troubleshoot. The NEC recognizes two-phase, in places such as Table 430-149. In the late 1960s the Niagara Mohawk utility in upstate New York sold the transformers it had used to derive two-phase to some large customers, weaving mills. As of 1997, the transformers were still in use, to operate the many, many two-phase motors the mills own. Those motors apparently are very durable.

Because of their rarity, when these two-phase motors burn out, most customers should be happy to replace them with three-phase. Given that three-phase is more efficient, there is no obvious reason not to make that change rather than try to find someone to rebuild two-phase motors. A potential problem when there are many two-phase motors, and only some are replaced, is that the customer could end up having two-phase and three-phase motors working side-by-side, requiring separate feeders.

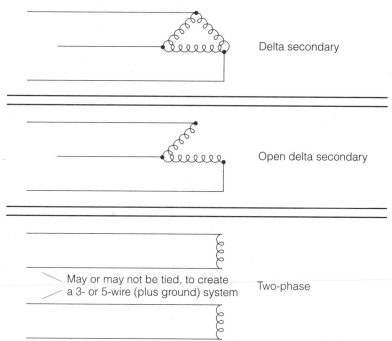

Delta secondary

Open delta secondary

May or may not be tied, to create
a 3- or 5-wire (plus ground) system

Two-phase

■ **11-3** *Two-phase power versus open-delta three-phase.*

In some cases, therefore, you may be called upon to maintain a two-phase supply. Motors are not the only use. Apparently, two-phase power has some advantage in arc welding. Using two AC electrodes in parallel from separate phases of a three-phase supply will produce a rotating movement in the liquid metal in the weld pool. When using two-phase power in contrast, the arcs will "cross" one another, one tending to oscillate along the weld seam while the other oscillates crosswise.

Plain old motor replacement

This is not a place to learn how to rewind or replace burned-out motors, though it is worth mentioning that a motor rated at less than two horsepower probably is cheaper to replace than to rebuild. Regardless of a building's age, when you are called in to deal with a sick or dead motor you probably will send it to a motor shop. Does it make sense to rebuild it? They will handle that evaluation. Is it a lost cause? They will be able to read the voltage, power, frame size, speed—all that needs to be known for a match.

In tight quarters, motor replacement may demand equipment redesign. For safety's sake, the 1996 NEC increased the size of motor terminal enclosures. This meant that replacements for some tightly wedged motors no longer fit in the same spaces. There was an attempt to reverse this decision for the 1999 NEC, but that effort appears likely to fail.

Presuming that it does not make sense to suggest that your customer consider buying a replacement motor of different voltage or phase, you still might ask what a higher service factor would add to the price of the replacement. One reason to do so is that older facilities are prone to heavily loaded feeders, resulting in voltage sag. To a motor, that can be equivalent to overload, which wears it out faster, and can melt the thermal protectors and interrupt its operation. This is exactly the kind of stress that a higher service factor helps a motor withstand.

Motor standards continue to change. Before the 1965 NEC was adopted, the standard industrial motor design was the U-frame, most commonly rated at 440/220 volts, three-phase. It was replaced, in equivalent installations, by the T-frame, which has a smaller enclosure, and is rated at 460/230 volts. Design E motors are on their way in as replacements for existing three-phase motors, although as of 1997 few of them were in use. Installing one of them means making some adjustment. With one exception, all high-efficiency motors, Design E or no, have much higher inrush current than do older designs. Peak current can be from 10 to 18 times FLA. This means that you need to check Code requirements carefully with regards to overload and ground fault protection. Furthermore, as mentioned above, there is a movement afoot to require that new motor installations, especially of Design E motors, be installed with phase loss or reversal protection. This requirement probably will be applied to replacements as well, in many jurisdictions that adopt it.

The exception to the rule that high operating efficiency means high inrush current is the written-pole motor. Instead of relying on fixed-position magnets in its rotor, this type of motor "writes" poles in the magnetizable (and erasable) material of its rotor, at optimal positions, as the motor comes up to speed—or as it responds to a power disturbance. The written-pole motor has two advantages: very low inrush current and the ability to protect sensitive operations from the speed fluctuations that otherwise would result from irregularities in supply voltage or frequency. A premium, external rotor version can have very high rotational inertia to enhance its ability to ride through brief outages. Finally, for other applications where an old business with chancy utility power is installing high-tech equipment, a motor-generator can use written-pole technology. The 1999 NEC does not specifically recognize it because written-pole production had not reached mass-market levels at the time revisions were being proposed. It has, however, been in use, in rugged applications, for several years.

Finally, here are two facts familiar to electricians who have worked with motors in newer or older buildings. First, while it can be wise to install heavier services and feeders to allow room for growth, that generous principle does not apply to the motors themselves. True, it is not a good thing to load a motor to within an erg of its capacity. On the other hand, oversizing a motor by a third or more results in much lower efficiency than does matching it more closely to its load. In older buildings where the system is already near capacity, you certainly do not want to specify an energy-wasting replacement, even if the motor itself would have a longer life.

Second, confirm that your replacement motor turns in the right direction. Overlooking that one test can result in a silly—and potentially painful—call-back. Whether you check rotation electronically or by jogging (presuming that the driven equipment can be safely jogged), do check before you leave.

Leave it functioning and safe

You dare not assume that you can repair motors and go merrily on your way, content in the knowledge that the motors are working. You may very well be landed with the responsibility to see that the installation provides a modern level of protection, at least of personnel protection, when you restore it to operation. This depends on the local inspection department and on how much change you have introduced into the electrical system. There almost always will have been changes in NEC Article 430 and other relevant code sections since the initial installation. You need to protect yourself from claims that you should have known enough to bring the installation up to modern standards of safety.

Personnel protection

Keep in mind such requirements as suitable switch enclosures, conductor protection, and, most especially, compliance with rules concerning disconnects and controllers. Some very hefty OSHA penalties have been assessed when workers were inadequately protected around high-tension equipment. In some cases, customers—and inspectors—might be open to alternate means of guarding. One example is light curtains, which automatically shut down electric-braked machinery when anyone interrupts infrared beams (perhaps visible ones, as well) aimed at alarm-system type detectors.

Warning signs A change associated with increased requirements for guarding is the need to mount warning signs. A notice "Danger: High Voltage. Keep Out!" may not have been present when you arrived to work on the equipment, but it needs to be there before you leave high-tension equipment that is accessible to people who should keep out. (If it is locked or otherwise guarded, "Keep Out" may be unnecessary and even confusing.) Incidentally, there is very solid research saying that the more authoritarian a sign, the more likely it is to inspire defiance—to boomerang. Therefore, when you find polite, informational wording that is clear and definite enough to pass legal muster, it is more effective.

For additional discussion of the liability you take on simply by walking onto a job, see the first and last parts of Chapter 7, "Setting Limits and Avoiding Snares."

Motor protection

Even if a requirement for phase-loss and phase-reversal protection fails to make its way into the 1999 NEC, such protection is very much worth suggesting to customers with polyphase motors. If one old fuse blows and the result is loss of that phase, your customer could lose a motor. If phase loss is the result of a utility disruption, your customer could similarly lose the motor—and the utility will not accept liability. Overload protection—even internal thermal overload protection—does not normally protect motors from these dangers.

Panelboard arrangements

Since most three-phase panels are in commercial or industrial establishments, this is the best place to bring up the issue of changes in phase arrangement. Delta services providing line-to-ground power by means of a center tap have one phase "high," or further from ground. It most commonly is called the "red leg," though other names are also used. Current Code says that the high leg must be connected to the "B" or middle busbar. This requirement only arrived in the 1975 NEC, with a new Section 384-3(f). Therefore, different arrangements will be found in many older panelboards. The exception to Section 384-3(f), exempting a switchboard or panelboard section containing a meter that the serving utility has connected differently, is yet more recent.

Should you encounter an older phase arrangement, and it appears to have been professionally installed, it certainly will be grand-fathered, and not something to concern you. The only caution is that you cannot make assumptions about which phase is the red leg. This is of particular concern should you need to add a sub-panel, or a line-to-neutral circuit.

Factory lighting

It is time to look at lighting systems. Old factory lighting is less likely to be esoteric than is old machinery. Still while it may amply reward upgrades, the upgrades can present complications.

History of modern industrial lighting

Early American industrial plants used gas lighting. From that system, they were converted to incandescent fixtures, not arc lamps. Mercury vapor lamps did not become efficient enough to supplant incandescents until almost World War II. As the country moved into the 1950s, it became apparent that high-bay factories were relatively ill-served by incandescents. As efficient high-intensity discharge lamps became available, they took over for that application, while fluorescents began to light workplaces whose ceilings were at more moderate heights.

Voltages have changed considerably. Lighting in factories used to be operated at 115 volts, stepped down from motor and process voltages of 440 or 460 volts. Sometimes, the lights operated at 220 or 230 volts. Then 480-volt three-phase power supplanted the slightly lower previous supply voltage, and this new supply voltage was stepped down to new utilization voltages for lighting: 240/120 volt, single phase or 208/120, three-phase four-wire. As of the early 1990s, many industrial plants retained lighting at the latter voltages. Newer plants, then and now, use 480Y277 systems.

Energy efficiency

Is it worth your while to change a factory's lighting voltage? Unlikely. Sure, if plant management wants to massively rethink the lighting system—or you can convince them with a payback analysis that the energy savings will make it worthwhile—you might throw more convenient voltages into the stewpot along with new controls, circuiting, placement, and light sources. On the other hand, ballasts for electric discharge lighting come in many voltages, and sometimes in multiple voltages. This fact means that you generally can upgrade to more efficient lighting systems without changing the supply voltage.

Old industrial plants certainly present great opportunities for win-win lighting changeovers, perhaps even more so than old office buildings. To be explicit, "win-win" means you can generate work that pays you well, gives your supplier a nice bit of business, pays for itself quickly, and then continues to save your customer money. More efficient lighting can save power usage (an out-of-pocket *and* ecological benefit), reduce peak charges, reduce power factor penalties, and improve the work environment. The latter benefit can both increase productivity and make OSHA happy. Major luminaire manufacturers often will assist contractors in putting together proposals for relighting large plants.

The discussion of motors was rather specific to industrial settings. Opportunities for lighting upgrades are found everywhere but in the smallest occupancies. Such improvements can be designed and installed well or poorly. Follow Illuminating Engineering Society guidelines and respect the buildings' architecture, the areas' users and their needs, and the equipment's specifications.

229

Strength in numbers: large installations mean fast payback

Numbers add up very differently in commercial, industrial, and institutional than in residential settings. Mother may have told you to turn the light off when you left your room. Frankly, the penny an hour that your parents saved, counting both energy use and lamp wear, was trivial. It may have been good as character development, but it was irrelevant as an economy. Not so in a factory, where the security force does not need the entire shop floor brightly illuminated after hours. Not so in an office complex, where the cleaning crew can turn lights on and off as they enter and leave a room. That having been said, if the entire plant is on the go 24 hours, with the exception of one or two offices that keep an 8-hour day, installing sophisticated occupancy controls for the shop floor, or even the offices, is very possibly the penny-pinching equivalent of listening to Mother's fussing.

Specific modifications

This discussion will not go into excessive detail about particular efficiency enhancements, because they are not specific to old buildings. For those who use this book as a guide to "old work," though, as opposed to genuinely old buildings, it is worth at least mentioning some additional opportunities.

Exit signs Replacing exit signs can be a simple source of savings. Since the signs must shine continuously, a considerable amount of energy can be saved by replacing even low-wattage incandescent lamps with fluorescent; doing so also extends relamping intervals considerably. Retrofit kits utilizing LEDs offer even greater energy and labor savings.

Fluorescent fixture inserts Even when there is no change in wiring, a retrofit kit, like any other Classified equipment, must be applied in accordance with its Listing. Such is the case with specular (extra-shiny) reflectors, designed to reduce the amount of light lost before it even leaves a fluorescent fixture. An NRTL evaluates and classifies a reflector kit for fluorescent fixtures as one of two types. Type I is a simple reflector, which either sticks to the fixture's existing reflector or else replaces the original reflector. Type II is more sophisticated. Besides increasing reflectance, it often reduces the number of lamps; doing so may require reballasting or changing lamp locations in the fixtures.

The way to proceed will vary. Some of these fixtures are Classified for use with specific fixtures within particular manufacturers' product lines. Others have been investigated for use with, for example, all 2'-by-4' troffers. Refer to the directory of the relevant NRTL—UL's green book, for example. Some jurisdictions require permits for such retrofits; others do not.

The SWD rule

Before leaving the subject of lighting, take warning. If the old lighting system is incandescent, controlled directly by circuit breakers, beware of changing over to fluorescent lighting without upgrading the controls. Even relamping with compact fluorescents can bring in the requirement that the circuit breakers used to turn them on and off daily be SWD-rated, in the opinion of many, including Code expert Charles Eldridge.

Age rarely bars high-technology improvements

Most of the changes proposed in the name of energy-efficient lighting are quite blind to the age of the building or of its electrical system. These include the following:

☐ occupancy sensors;

☐ time controls, including daylight sensors and time clocks;

- [ ] delamping;
- [ ] task lighting;
- [ ] installation of more efficient reflectors;
- [ ] fixture cleaning (simple, but not to be discounted!) or lowering;
- [ ] purchasing higher-efficiency or reduced-wattage lamps of the same type.

So long as you encounter no deteriorated equipment or materials, there is no difference between making such changes in an old building and in a fairly new one. Of course, the old building's conductors are likely to be rated for 60° C, but that is likely to be an issue when you change fixtures, not just lamps or controls.

When system age is a factor

Most but not all improvements in efficiency not only will save energy, but will also extend the life of the wiring system. Certain changes, though, can create problems for old wiring and service equipment. The reason for both facts has to do with the far greater likelihood that in a commercial or industrial setting, old wiring is loaded near its ampacity. On the one hand, this means that reducing the load reduces the operating temperature of the wiring system. On the other hand, because commercial and industrial installations tend to use multiwire circuits, you may run into the problem of overloading the neutrals—and sometimes the transformers—with triplen harmonics (waveforms that add in the neutrals of multiwire circuits, rather than canceling each other out). Ill-planned use of electronic ballasts is notorious for causing this problem. A knowledgeable, responsible lighting representative is not likely to put you in dutch with your customers by carelessly cookbook-specifying a retrofit plan. Still, if you know that a customer has many multiwire circuits, you would be wise to keep the issue in mind.

As a side note, you may find knob-and-tube wiring utilizing open conductors in noncombustible locations in some industrial settings. Without dielectric insulation to deteriorate, these could last essentially forever, even at the elevated ampacities that were permitted. It was okay, for example, to protect #12-AWG conductors at 30 amperes. This could be woefully dangerous in the event that conductive or combustible materials are moved into those spaces, and serious overheating could result from the addition of thermal insulation. The worst scenario would be where blown-in insulation became combustible. The fireproofing on cellulose insulation, for example, can deteriorate in well under a decade.

The electronic age

A ballast is less likely to cause problems for its fellow ballasts than is digital equipment such as a computer or a programmable logic controller, the latter two being types of equipment that can give each other and their fellows the heebie-jeebies. This is not the place to go into solutions for such problems as waveform distortions. It is, however, worth pointing out that while high-efficiency innovations reduce the overall load, this does not necessarily mean you can blithely rely on existing circuits without the nasty (if it is unforeseen) need to add accessories such as waveform-smoothing inductors or tuned filters.

Other risks associated with old buildings

So much for circuit protection. There also are personal safety problems associated with old buildings, as discussed in Chapter 17, "Hazards and Benefits." Old industrial and commercial buildings probably pose greater risks of exposure to hazardous materials than do old residences. From asbestos and rock wool to solvents and process chemicals, you can find many toxic challenges to worker health on the job — and no Material Safety Data Sheets. If you are not sure what you are dealing with, assume the worst until you find someone who does know for sure.

Increased risks also are associated with the fact that some commercial, and many industrial, occupancies use high voltages and currents. One serious danger at industrial sites is that, even where the initial installation did not include any misapplications, over the years the serving utility may have upgraded its transformers and lines. This can increase the available fault current to a level that exceeds the rated interrupting capacity of disconnects and overcurrent devices, in violation of NEC Section 110-9. Another danger is that lugs may have been overtorqued or subjected to repeated torquing. The problems that they can create will be discussed in a few pages. For both reasons, some very knowledgeable persons recommend standing off to the side, "out of the line of fire," when operating disconnects, as partial—but only partial—protection in the event that they explode and rupture their enclosures.

Examples: ill-considered upgrades Another kind of risk associated with modernizing older industrial buildings is that if you try to update without considering all aspects of your customer's needs, your improvements can backfire—and this can mean, at the least, that you have to eat the expense of doing things over. Three examples follow. The first two are cases where the upgrades involved not industrial motors, not lighting, but HVAC controls. Both were upgrades to food warehouses.

The first warehouse, in Hershey, Pennsylvania, stored confectionery products and previously had been kept much cooler than the products required. A modern, more-sophisticated temperature control system was installed. While it allowed the temperature to climb higher than before, it was guaranteed to keep the stored candies from rotting even at the high end of its range. Unfortunately, one factor had not been considered: bloom. When chocolates repeatedly warm up and then cool down, their fat blooms (migrates to the surface). This results in unappealing appearance and texture; even without hazardous spoilage, the product becomes unusable.

The second warehouse had a different problem. Again, considerable energy wastage seemed evident, because the goods were kept much cooler than was warranted by product characteristics and length of storage. With modernized controls, the temperature was still kept well within the acceptable limits for the foods themselves. As the temperature rose, though, vermin attacked the foodstuffs. A more modern warehouse would have handled the higher temperature just fine, but the old building was just too porous to bar the bugs.

The third case was a little different. A cheese factory was inspected in September 1994 by the Food and Drug Administration and found to be very unsanitary. According to the report, one of the problems involved a cutting room so dank that condensation dripped from the ceiling onto the cheese (which was not covered). The inspectors returned in May 1995 to find that a fan had been installed to eliminate that condensation by drawing in fresh air. Unfortunately, the fan drew air from the direction of the company's sewage treatment plant. While ultimate responsibility for the plant's further history does not rest on the electrician, nor was it a direct result of this blunder, the company did go out of business.

Opportunities for upgrading low-voltage systems

Industrial buildings constructed up into the 1950s were built when energy was much less expensive. They have high ceilings, a characteristic that allows renovation designers to hide wiring in raised floors and dropped ceilings. This is useful when replacement seems a better alternative than repair.

Many commercial buildings require the latest in power and tele-communications wiring, without fudging. Old telecommunication wiring, especially, often is worth abandoning. While most electrical contractors do not pursue this work, increasing numbers do. The EIA/TIA have introduced designs that do not tie users to a particular manufacturer, designs whose organization is relatively consistent. There are Listed cables capable of significantly faster transmission than what is likely to be in place. Old cabling fails to even approach these standards, so it is rarely worth ringing out the existing runs when an upgrade is sought.

Suppose you are interested in becoming a preferred source of ser-vice on old buildings, but you are not one of the contractors who specialize in telecom. Manufacturers may be willing to assist you with specifying and estimating jobs in this and other low-voltage areas. For instance, while some alarm system manufacturers are vertically integrated and only sell through their captive installers, others will meet you more than halfway if you come to them with a chance to win a sizable job.

Assorted errors to watch for

This chapter started out discussing some legitimate but highly unusual types of industrial equipment. It closes with certain types of installation you are likely to encounter in commercial and indus-trial settings that are not legitimate; in fact, they are simply wrong.

Overly isolated grounding

One is the isolated ground circuit whose ground is isolated all-too-sincerely. NEC Section 250-21 authorizes disruption of some grounding paths in the case of objectionable current. As recently as the mid-1980s, some manufacturers—or at least distributors—of electronic cash registers advocated bonding their equipment, and the circuits feeding it, to grounding electrodes that were com-pletely isolated from those of the services or derived systems. Consequently, Section 250-21(d) now clarifies the intent of its authorization to disrupt grounds; for its purposes, currents that cause noise or data errors are specifically excluded from being classified as "objectionable." This means that you have no right to remove any part of the grounding electrode system because of stray currents. You can install isolating transformers, but they isolate the power, not the ground. You can use dedicated circuits, and isolated-ground receptacles, but their grounds must be bonded to the system ground at the source of power. Any installation you come across that was done differently is a candidate for correction.

The danger associated with a completely separate grounding path is this. In the case of a ground fault, to complete the circuit so as to blow the fuse or trip the breaker current has to travel from the service equipment to the outlet; then out, through the faulty equipment, and back to the outlet's isolated ground; from that to the earth; then through that high-impedance path back to the service grounding electrode and the main bonding jumper. The impedance of the earth between the two electrode systems often is high enough to forestall operation of the overcurrent device.

While on the subject of isolated ground installations, be on the lookout for the following violation in this area, newly defined in 1997. In

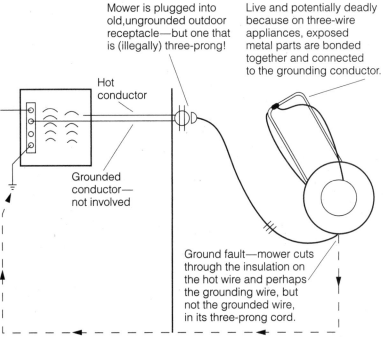

Mower is plugged into old,ungrounded outdoor receptacle—but one that is (illegally) three-prong!

Live and potentially deadly because on three-wire appliances, exposed metal parts are bonded together and connected to the grounding conductor.

Hot conductor

Grounded conductor— not involved

Ground fault—mower cuts through the insulation on the hot wire and perhaps the grounding wire, but not the grounded wire, in its three-prong cord.

Long, high-impedance return path; unlikely to trip circuit protective device

■ **11-4** *Fault current trying to flow through the earth. Why show a lawn mower as opposed to, say, a cash register? Electricians recognize the danger of using ungrounded tools outdoors. Yet, an indoor cash register with an overly-isolated ground may actually have a higher-impedance fault-clearing path than the lawn mower, because the earth through which it runs is drier.*

order to harmonize U.S. and foreign standards, an isolated ground receptacle now must have an orange triangle on its face. Ere now, an orange face served the same purpose. Few inspectors will require changing the receptacles that are in place in existing installations, but this new rule could create problems when new certificates of occupancy are requested. This is a matter of fussy detail, rather than a serious danger. Your inspector may not sweat this one.

Too much tightening

Finally, you need to think about the issue, touched on earlier, of overly torqued connections. On commercial jobs, with relatively high ampacities, there is some temptation to retorque connections. The best advice on this issue is consistent: "Don't." When you are concerned about the condition of lugs, run a heavy load, and thermograph all connections. If you find hot spots, troubleshoot.

Retorquing has the effect of extruding (elongating and narrowing— think of a taffy-pull) lugs and bolts, and in most cases does nothing to improve the connection. Sure, if local law demands it, you have little choice. For example, in adopting the 1993 NEC in late 1996, the city of Annapolis, Maryland, introduced such an amendment. Switchboards having a capacity of 1,000 amperes or larger in Annapolis are to be cleaned and their connections torqued to specifications every three years. As of mid-1997, Arlington County, Virginia, had a similar requirement.

Over time, a perfectly well-made connection will relax, so that a torque wrench will show it to be below specifications. This relaxation, however, does not indicate that the connection is worse; its resistance is lower, if anything, than when the termination was initially torqued! According to the best counsel, retorquing can be of benefit when—and only when—the original installer totally neglected to tighten a connection. (In that instance, it is not really "retorquing.")

Thoughtless risk-taking by customers

Some other dangers you ought to point out, even if you do not automatically correct them, are found in offices as frequently as in homes. One reason for pointing these out is that you are a responsible person on the spot. Another is that remediating the following dangers can mean added business.

Spaghetti

The first bad'un is the cord run under the rug, or behind the furniture. Many businesspersons and homeowners have expressed surprise when an electrician explained that extension cords are for temporary use, and that cords hidden out of sight can cause fires. Major manufacturers, such as Wiremold, sell cord protectors that neatly route equipment cords or extension cords out of the way. At least some such products have no NRTL marking. It would be reasonable for a layperson purchasing a cord protector that does boast a NRTL mark to presume that using the protector made it safe to run a cord under, say, a rug. How should they know any better? Putting aside whether they happen to be overly impressed by a brand-name product, many office managers are insufficiently aware of the distinction between "it works" and "it's safe."

Protection rejection

The bypassed overcurrent device is a similarly widespread danger. The NEC makes special allowances for certain processes that can be extremely hazardous to interrupt. For instance, if an electromagnet used for lifting should fail because its circuit breaker tripped, or if a fire pump should cease pumping because its fuse blew, that consequence could result in greater danger than the risk of overheating conductors that the overcurrent device was protecting. This type of situation justifies a variance; others do not.

Hodge-podged bypasses are employed to save some commercial, institutional, and industrial occupancies not from danger but from the inconvenience of having to replace a fuse. You are likely to find this sort of criminal creativity most frequently, albeit not exclusively, in some old shop that has plenty of scrap materials on hand to use in creating a substitute for an out-of-stock or frequently blown fuse. A short length of copper plumbing tubing is probably the most common substitute for a cartridge fuse with cylindrical ends. One benighted soul fit a rifle cartridge in a fuseholder!

Electrician errors

That sort of bunged-up job is strictly amateur. Other problems are created in commercial occupancies by careless electricians.

Multiwire misapplication

Confused by crowded panels and subpanels, unthinking installers working on commercial jobs are at least as prone as on residential jobs to misapply "multiwire" circuits. Article 605 forbids using multiwire circuits to power cord-and-plug connected office partitions. A far greater concern, though, than the illegal use of real multiwire circuits is the following. Two or three spare breakers are combined in one cable with a "neutral," willy-nilly, for a use that does not forbid multiwire circuits—but those two or three circuits are combined without confirming that they originate from different busbars. Commercial installations (to some safety-oriented professionals' surprise) do not require handle ties on breakers serving multiwire (as opposed to polyphase) circuits. For that reason, it is far easier in commercial panels than in residential to accidentally share a grounded conductor between two or three hot conductors on the same phase. This can mean a badly overloaded return conductor, in cases where it is not indeed a neutral.

Rule breakers—breakers do not always rule

The next problem you are apt to encounter more frequently in commercial establishments than in residential—inappropriate protection—is likely to arise when you upgrade service equipment or loadcenters. And it really can sneak up on you. It is easier to detect and handle, without damage to your pocketbook, when installing new utilization equipment, and is more likely to sneak up on you when you are just changing a panel.

The problem is improperly feeding equipment whose instructions include wording such as, "Protect with a time-delay fuse rated at no more than 15 amperes." When manufacturers employ such wording on their equipment, they do not mean, "Protect with a 15-ampere or smaller fuse or circuit breaker." Such a misreading is a common mistake.

If you run across equipment with some such legend, and a circuit breaker loadcenter will be supplying it, you do not have a situation that is insoluble, or even difficult to solve. You simply need to add fuses as supplementary overcurrent protection between the source and the equipment. If you unthinkingly bid the work, or pulled a permit for the work, without allowing for an additional fuseholder, adding it may sting. It need not be terribly expensive: depending on the characteristics of the utilization equipment, the supplementary protection that you must add may be as simple as a fuseholder mounted in a handi-box. (When adding new equipment, it is easy to solve this sort of problem by installing a fused, rather than plain, local disconnect.)

Anonymous installers: look out for trouble created by them

You should be fully prepared to find the interior wiring of commercial, industrial, and institutional facilities in violation because it has been repaired and upgraded over the years by maintenance personnel, people who never studied the NEC; who may never have had to earn or maintain electrical licenses. Master electricians intimately familiar with two major institutions in Washington, D.C. mentioned in mid-1997 that, over many years, they had never seen an electrical permit pulled for either minor or major changes to the electrical systems of these institutions. They were referring to work that included hospital renovations and the creation of new equipment rooms!

You might expect such buildings' incoming services and exterior wiring, on the other hand, to be in compliance, being so much more in the public eye than the interior wiring—perhaps even more so than the equivalent wiring at dwellings. Such assumptions can be dead wrong.

Example: a bad sign

For an example, consider the repairs required at a fifty-year-old church in Washington, D.C. This discussion focuses on a few aspects of how the church's internally illuminated street sign was wired.

Parenthetically, after fifty years, the rubber-insulated armored cable was badly deteriorated at many outlets, even where there was no reason to assume that circuit overloading, or the nearness of heat-producing equipment, had contributed to degrading the insulation. Fifty years is pushing the limit, or, truthfully, beyond the limit, for rubber.

You can reasonably expect the body of such a sign to constitute a moisture-proof enclosure, rated NEMA 3R. This one fell short. Perhaps some part of it had failed over the years; more likely, the way the supporting angle iron was brought into the top of the sign when it was first installed was what had provided an ample path for rain, whenever there was a driving wind. This is what the electrician concluded. The interior showed clear evidence of moisture, and the terminals for its two six-foot fluorescent lamps had corroded.

The original installer had wired the sign with lead-sheathed armored cable, run out through the wall of the building, along the supporting angle iron, and into a conduit body at the bottom of the sign through a standard BX connector. The conduit body was secured to the sign not with a chase nipple but with another BX connector passing through from inside the sign. So there were long-standing violations:

☐ While BXL, or ACL, was permitted in damp locations, and for use where exposed to weather, there was no evidence that the opening through which it left the building was ever sealed.

☐ Indoor connectors are just for indoor use.

☐ BX connectors are not listed as substitute chase nipples.

☐ Conduit bodies are not an approved means for making a transition from armored cable to other wiring methods.

To complete the story of the sign's woes, there were a few more problems:

☐ Its ballast appeared to have cooked;

☐ One of the wires in its fixture channels was badly charred;

☐ The sign ran off a time clock that was not working.

☐ Also, in violation of NEC Section 110-17(a)(2), the clock was missing the protective interior shield that guards the live terminals from contact when you swing open the front cover to reset the clock, or to turn the load on or off manually.

How did the sign come to be installed in so questionable a manner, right on the front of a stately church? A reasonable guess is that this came to be because the sign is a good 20 feet in the air, and the inspector was not eager to clamber up there, even assuming that a ladder was at hand when he came by. Inspectors cannot catch everything. Besides, the sign may have been installed after the church was complete. Sign installers are not the most by-the-book trade with regards to pulling electrical permits and calling for inspection.

Here is what was done to remedy the situation. First, the electrician had to decide how to get the sign working at all. While he could have tested the ballast and replaced the lamp terminals, it seemed wiser to replace the entire weather-worn fixture, consisting of one ballast and two 72-watt lamps and lampholders. There was ample room for two high-output, 4-foot, 2-lamp fixtures with 0° ballasts.

These were not outdoor fixtures. Recall that he had determined how moisture had entered the sign at the top, where a hole had been cut to permit the angle iron; he caulked that hole. He then removed the BXL running outside the building, and (although not in this sequence) terminated it in the back of a bell box he installed right on the outside wall. He wired the sign using UF cable, run from that box. On the bell box, he mounted a photocontrol, rated for use with ballasts, well over the wattage of the ballast installed in the sign.

Inside the church, he bypassed the time clock, properly mounted and enclosed a switch to serve as the disconnect, and pulled back the BXL, cutting back to where, to his surprise, he found perfectly good conductors. These he fed into the bell box, into which he had screwed a BX connector. After tightening the setscrew against the

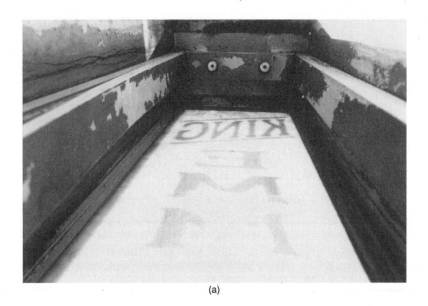

(a)

(b)

■ **11-5a, b** *11-5(a) Enough corrosion to call for gutting rather than repair. 11-5(b) BXL cable leaves the church, then enters the sign through a BX connector and conduit body.*

armor, he pulled the cable back so as to recess the connector into the wall, as best he could, before securing the box—and then caulked the dickens out of the space behind and around the bell box, so moisture could not weasel its way either into the building or into the connector, and thence into the box.

Other options would have been to use conduit, EMT, or flex. Sealtite would have been a fine option, but given that the knock-out was on the underside of the sign enclosure, even regular flex would have met Code.

Changing rules—neon

A different sort of problem to watch for derives from changing Code and Listing requirements.

Neon-type lights are bent, and installed, by specialists. These entrepreneurs may be willing, even more than willing, to sell self-contained luminous signs for use in residences, especially when configured as works of art, despite the normal presence of over 1,000 volts at the electrodes—which makes their use in residences illegal. Still, these folks tend to know what equipment is suitable for use with their installations. They are no more likely to ignore Listing requirements than is an electrician or HVAC installer.

Even so, problems can arise if you assume that old outline lighting fixtures comply with modern standards. It is likely that, around the time that the 1999 NEC is adopted, a new UL standard for neon transformers and neon power supplies will be enforced. UL Standard 2161 will require that equipment to incorporate secondary ground fault protection.

This change will have two consequences. First, if an electrician, who need not be a sign expert, is called in to replace a burned-out transformer, installing the identical unit no longer will be appropriate. (This will be true even if supply house personnel still have the old unit in stock and ignorantly insist that it represents the current standard.) Second, a few years further ahead, electricians could be at risk when they assume that the equipment they are working on has secondary ground fault protection, which would to some degree protect them while working on it.

Sometimes, on the other hand, Code proposals alert you to problems that may require remediation in existing wiring. For instance, Section 600-32(a) of the 1999 NEC may forbid the use of non-metallic raceways to enclose high-voltage secondary conductors. Data indicate that such raceways are subject to physical damage due to the corona associated with cold cathode secondaries, and may not contain high-voltage arcs.

The risks associated with making assumptions about existing wiring extend way beyond the area of signs, and way beyond commercial settings. A dangerous assumption that can prove false in any type of occupancy is that the three-prong receptacle you want to use for your extension cord is grounded (or GFCI-protected, even though not so labeled).

Example: It is worth taking the trouble to find out what the customer wants—and needs

One final example will demonstrate a general issue that applies to commercial jobs as much as to any other. A customer may call on you to perform a simple change, a straightforward, make-this-place-a-little-safer matter of convenience, and be quite unprepared for its implications. When those implications mean too much expense, this can put the kibosh on the whole project. When the implications are more a matter of requiring significant planning and figuring, you retain a chance of selling the customer on that effort.

In this case, the owner of a high-tech service business asked what it would take to equip his newly acquired warehouse with power to the workbenches. He and his partner probably would be the only employees, but they would be servicing a fair number of items at any one time: mostly consumer electronic equipment. The electrician suggested running cable under the workbenches, popping it up to feed receptacles. The customer suggested surface metal raceway, by which he actually meant a multioutlet assembly. This made sense.

The electrician told him that he would be glad to do the job, and that pricing should be straightforward once the customer knew how long he needed his runs and how many circuits were required—presuming his existing panelboard had the capacity. The number of circuits, he told the customer, legally depended on how many items might be simultaneously worked on, or in the process of burning in.

You may wonder why the electrician was making life so difficult, rather than pulling an estimate out of his hip pocket. For one thing, the customer had ballparked the length of multioutlet assembly he wanted at 75 to 100 feet. Twenty-five feet is a considerable difference. More important, though, is the fact that without some careful figuring, NEC Section 220-3(c) Exception 1 could present a challenge to the amount of space in, if not, realistically, to the capacity of, his panelboard.

Take the worst-case scenario. 100 feet at 180 VA per foot gives 18 kVA. Divide that by 2400 (120 volts×20 amps) and you get 7.5, meaning 8, circuits. This would call for a 20-ampere circuit serving every 12½ feet of work space—considerable overkill for two men's work, unless they would be simultaneously, continuously, powering equipment that has a significant number of heating elements or fair-sized motors.

Take the other extreme. Run one circuit in each direction from the panelboard and assume that each man will work on only one piece of equipment at a time. (It is not credible that only one partner would work at any one time.) Now the first part of Exception 1 permits you to consider each 5 feet rather than 1 foot of multioutlet assembly as a 180-VA load. (Every foot is a 36-VA load, if you ignore fractions shorter than 5 feet.) Now you can feed 66 feet of multioutlet assembly in each direction.

Unfortunately, it is hard to believe this scenario too. Sometimes, each man will be working on a piece of equipment, and each will have a few other items plugged in and drawing power at once: for instance, a 120-volt tester and a power tool; or a couple of modules being burned in. Even if the receptacles in the assembly are duplex, which, in other cases, allows the two receptacles sharing a yoke to be counted as one outlet, there is little reason to assume that the Code panel meant for that definition (multiple receptacle) to apply to Section 220-3(c) Exception, concerning multioutlet assemblies. Does this mean you should return to the first, more stringent calculation? Not necessarily.

There are two ways to avoid that; either would require consulting further with the customer. Multioutlet assembly can be purchased with outlets spaced over a foot apart. Then, instead of applying the exception, the general rule in NEC Section 220-3(c)(7) of calculating 180 VA per outlet can be applied, bringing the spacing up to 20 feet of multioutlet assembly per 20-ampere circuit. This might well work, although it would not be a useful option where the customer wants an outlet every 6 inches.

A second possibility would be to ask, "Is there any section of workspace where, in the ordinary course of events, you expect to just store equipment rather than work on it?" The answer may be, "Can't say." It might be, "No, I meant 75 to 100 feet of raceway with outlets spaced throughout it." However, if the answer is "Yes," that less-used section can either be run in surface metal raceway with an outlet or two, or run in a multioutlet assembly with its load calculated at 180 VA per 5 feet.

There are a couple of other possibilities. One is the option of cutting the number of home runs, in the event that eight circuits are required, by using multiwire circuits. Multioutlet assemblies do come in a two-circuit configuration. On this particular job, a judicious electrician would be reluctant to employ multiwire circuits because of the risk of the power supplies' overloading the neutral, and also the risk of sending noise through the neutral from one piece of equipment to another.

A final consideration is this: if you do your calculations and your installation based on a two-person shop, what happens if the owners are successful and hire more help? This could mean more intensive use of the workspace, and much heavier electrical loads. Sometimes, saving the customer money up-front is no favor. Other times, it makes sense because when they are ready to expand they will have the capital to pay for additional work. Where everything is being installed on the surface, as in the job just discussed, coming back to do more work does not create the disruption that it would if walls had to be ripped open.

Inspection Issues

THIS CHAPTER IS FOR INSPECTORS AND FOR ELECTRICIANS who are concerned about permits for and inspection of work on old systems. To start out, if it is a toss-up whether you, a contractor, need a permit and inspection, it can be very much worth your while to arrange for inspection. This is certainly the case where jurisdictions in which you work employ qualified inspectors who perform their work responsibly. The inspection means that one more person looks things over, perhaps catching problems you missed. This is well worth a few dollars and some inconvenience. An irresponsible inspector is of less value. The owner of one inspection company reportedly confided that he will look for major violations because "I don't want anything to burn down," but he doesn't spend a lot of time looking for small problems so that (when something does blow up and) he goes to court he can say, "While I was there I can honestly say that I did not see any violations." He also told our source that he keeps a minimum amount of records so that he can say, "I have no record of that and I don't remember that." No, he is probably not doing the world a lot of good.

Even the best inspection cannot be exhaustive, but has to be a spot-check, a "best-guess" evaluation. Still, however sporadic the examination, having your work approved by the AHJ is, if nothing else, an indication to the customer that you have nothing to hide.

Judgment calls

Of course, there are questions and controversies. Here is a sore area, one that generates many appeals from contractors to chief inspectors and to Code authorities.

When does neatness count?

The first sentence, or "general-duty clause," of Section 110-12 is one of the NEC sections most easily abused by poorly grounded (meaning in this case inadequately trained) inspectors. "Neat and workmanlike" is such a very subjective call. This Code proviso should not serve as a surrogate for "Do it the way I would." Nor have inspectors the right to act on the pop-psychology assumption, "If he cares this little about making the job look good, who knows what corners he cut. I'd better find some technicality to nail him on, so he knows I take my job seriously." Still, authorities do need to apply their judgment, to wing it if you will, to fill gaps—places where the NEC fails to specify details. Here is where a double standard is in fact appropriate. As a conscientious jurisdictional inspector, it is your duty to enforce the minimum requirements explicitly set forth in the NEC: no more. As a conscientious contractor, it is your duty to yourself as well as to your customer to go a little further; not because anybody is forcing you to, but because it is the right thing to do.

Filling in the holes in NEC coverage

When the NEC does not provide clear guidance as to the appropriate way a piece of equipment is to be secured or mounted, an electrician must extrapolate. The specifications that are most suitable to guide the installation of exterior UF cable, for instance, might be the $4\frac{1}{2}$-foot support rule in Section 336-18 for inside NM cable, supplemented by the "closely follow the building" rule of Section 336-6(a).

That was guidance for installers, not inspectors. As an inspector, you should apply a rule that is at the same time looser—more generic—and tougher. The purpose of the NEC is safety, not attractiveness. Therefore, where no detailed guidance is offered, you must require installations to be sufficiently "workmanlike" to avoid hazard. You must red-tag installations that are so sloppy as to create risk—and pass those that merely offend the eye. Is the appearance of the installation messy? Oh, well. That is between the customer and the contractor. Is the wiring likely to get torn loose? That issue falls in your bailiwick. If the cable or conduit is so poorly secured, insist that they add some straps! How about the completeness and legibility of the panel directory? That unquestionably is a far tougher judgment call. You are being extremely generous if you accept "plugs and lights" as sufficient legend to identify several circuits. Ask yourself: does this meet the intent of the NEC requirements? Does it sufficiently serve the safety goal?

When are permits needed?

In rewiring, one fuzzy area, alluded to at the top of this chapter, is whether permits and inspection are required. Most jurisdictions will not demand permits and inspection when a contractor simply performs repairs, even repairs that extend to replacing rotten bits of cable. Conversely, when a building is gutted for rewiring, the job generally will be treated very much like new work. Contractors should expect to need both a concealment inspection and a final sign-off. In a sense, the customer is getting a new certificate of occupancy, at least insofar as the electrical system is concerned.

When the job is extensive but not a gut, you have a more complex situation. Suppose the customer chooses to have one room redone at a time. Should the inspector show up to approve each one, once before the holes are patched and once again before the new wiring is energized? That would be absurd. Generally, when the walls and ceilings have not been completely opened in rewiring, there is only one inspection, at the end.

How does this jibe with the idea that when fresh wiring is run, it should be inspected before final energizing, before the customer gets to utilize it? It really does not. Contractor and inspector need to put their heads together to find a reasonable solution. When a building is rewired piecemeal, because it is in use, there is bound to be some accommodation to avoid unnecessary hold-ups. At the same time, paying for a permit does buy the right to get work checked. Competent, intelligent, sensible inspection is a benefit, not a burden.

Can the installation be grandfathered?

It would be nice if there were a rule book that explained how to apply the Code to old buildings. Although there is no general inclination to adopt it into law, there is an NFPA document on electrical inspection of old work, a "Maintenance Code." Its scope is 1–2-family dwellings and mobile homes, and its language is mandatory: "Thou shalts" and "Thou shalt nots." There is general agreement in the electrical industry with the document's principle that items that never were legal, such as zipcord wiring, must be corrected. But very little is specified. Will an old system have to be brought up to current Code? No. To the standards in force when the place was built? No. What, then, will be required? It leaves plenty (inspectors discussing this say, "too much") up to the judgment of the inspector at the site. For instance, the fact that it uses mandatory language, saying that receptacles must grip cord connectors firmly, at the same time not defining "firmly," makes it far more a statement of principles than the practical guide that many may seek.

To even consider grandfathering an installation, you need some assurance that it started out legal. Chapter 13, "Dating and History," can provide some evidence as to whether an old, odd installation simply represents the way things were done—and approved—way back when. That chapter does offer a number of cautions. Underline these two: first, even if something did not comply with the wording of the NEC, "Special Permission" may have been granted; second, Chapter 13 gives a greatly shortened version of Code developments—you need to look up articles for yourself in an old Code book to be sure what the rules said.

When an electrician works on equipment that was legal when installed but is no longer up to code, it is a harder call than finding clear violations. There are many gray areas at the interfaces of the old and the new. Here are some once-legal installations that may or may not be grandfathered. Those making these decisions cannot always put a finger on why they choose the way they do. A contractor certainly should check with the inspector, where available, before going ahead with the choices described as "reluctant." Inspectors might check with their chiefs.

Very rarely will you find written guidance to help you make decisions such as those that follow. All of these judgment calls are based on the assumption that related wiring is not dangerously deteriorated—rotten insulation is a "gimme," forcing rewiring.

☐ Replacing a switch or receptacle in a now-undersized box: sure, within reason.

☐ Replacing a light fixture at an undersized box: usually yes, depending on crowding; be very concerned about pinching conductors.

☐ Extending a circuit (and therefore adding conductors) from an undersized switch or receptacle box: not without installing a box extension or replacing the box.

☐ Adding circuits to split up loads from a now-undersized loadcenter, or a split-bus service: sure, unless it shows damage; even adding circuits to a service with only 60 amps incoming and more than 6 branch circuits can be acceptable.

☐ Adding branch circuits from the main section of a split-bus main panel: no way.

☐ Adding major loads to an undersized service: nope. Of course, "major" is itself a judgment call.

☐ Adding a subpanel, run from the main section of a now-undersized fusebox: usually, unless the subpanel is needed because the customers are adding major loads beyond what Article 220 says the service can support.

- ☐ Replacing a Carter System-wired multipoint switch: no, ma'am. (Others would call this one differently.)
- ☐ Adding a circuit to a live-front, ceiling-mounted (unswitched, cartridge) fusebox with an empty fuseholder: yes, reluctantly.
- ☐ Adding circuits to loadcenters in other illegal locations, such as over the kitchen sink: usually yes, reluctantly.

It is not hard to explain and justify these decisions, but reasonable electricians also see arguments for the opposite choices. There is no way to prove one right and the other wrong. These are the stands a careful but not totally conservative person will take as a contractor. In some cases as an inspector you might be more liberal, unless particular guidelines had been established as consistent policies in your jurisdiction, and contractors working in your area had been informed.

Can antique equipment be reinstalled?

Installation instructions are another troublesome issue. In light of NEC Section 110-3(b), in new installations an inspector quite reasonably can demand to see manufacturers' instructions. With old equipment, whether located in a remodeled area or whether being reinstalled in a new location, such a demand is not reasonable. This leaves inspectors up in the air. If there is no evident Listing mark, or if there is reason to suspect that the installation is not in accordance with the makers intended use, what is an inspector to do when confronted with no-longer-manufactured equipment? Knowledge of the NEC and of accepted wiring practice is an inspector's basic area of competence. ANSI standards are way beyond that.

An inspector need not take it on faith that equipment is suitable for a particular installation. When there is a serious question, and an absence of appropriate documentation, he or she can demand field labeling or certification by a competent authority. That authority could be a licensed professional engineer with the appropriate expertise. Better yet, a hands-down favorite is the field inspection service of NRTL. A number of companies are recognized by the Department of Labor as qualified to investigate electrical equipment. Many of them not only work with manufacturers but also perform single-unit investigations. See Chapter 14, "Resources," for a partial list.

Courtesy inspections—Easing citizens' concerns

When an electrician, or a homeowner, or even a tenant, calls the county or the city and says, "I think what's here was done illegally," how should a jurisdictional inspector respond? Conscientious inspectors tend to steal the time to come out and look at the situation. Sometimes, though, there are impediments.

When they are constrained by local ordinance not to come on a job unless there is a paid-for permit, inspectors usually can find a way to come out anyway, provided that they are there to nab someone for working without a permit or to gather evidence against a scofflaw installer. Where the perpetrator is unknown, this loophole is absent. When local inspectors are constrained by such a rule, one common way to get an inspector onto the job is to have some licensed electrician apply for a permit, charging the minimum fee. See Figure 7-2.

Jurisdictions that contract out their inspection services to private agencies also tend to limit the services available. One can always hire such a third-party inspection agency to come out anyway, though the visit probably does not have official standing.

What should a municipal inspector do in the course of such a courtesy consultation? A full-scale safety survey, such as is described in Chapter 6, is definitely beyond the scope of the job. Looking for deterioration in old wires is certainly not a jurisdictional concern, not unless the building department is cooperating with the fire department in a home safety campaign.

An inspector asked for a big-picture assessment actually has a very tough job, by virtue of being the voice of authority. An electrician can say to the customer, "You ought to change this; what say I go ahead and do it?" and leave it to the homeowner to decide how much work to authorize. The words of an inspector carry much more weight because he or she has no financial stake in the outcome. Also, in extreme cases, an inspector has the responsibility and power to notify the serving utility that it cannot continue to safely supply power. For that reason, and because the jurisdiction's attorneys may have something to say about any pronouncements offered by public employees, inspectors need to weigh carefully the conclusions they draw and the comments they make. At the same time, there are no points to be earned for a poker face. Taxpayers are inspectors' employers and have the right to benefit from their expertise.

The ultimate job any inspector is there to perform is ensuring safety. This means that, most especially on courtesy inspections, determining whether something is a violation may not be essential. Here is an example. Before the days of the obelisk in Article 310's ampacity tables, rubber-insulated #10-AWG conductors in knob-and-tube wiring, used to be rated at 40 amperes, and #12 AWG at 25 amperes. If an inspector encounters such antique installations, there is no reason to assume that they are in violation, being grandfathered. But they are probably unsafe, and that is what should be communicated: "Get yourself an electrician to replace them."

The best inspectors work with, rather than against, reasonable, conscientious contractors, as described in Chapter 7, "Setting Limits and Avoiding Snares." Incidentally, cooperating with contractors an inspector trusts does not mean neglecting to inspect them properly. Consider this: an electrical permit pays for having a second competent person check the work. Failing to inspect Joe's work, because Joe is known to be conscientious, means giving Joe, and Joe's customers, less value in return for the permit fee.

There is no point in discussing unethical behavior such as bribery, or third-party agencies that might compete for business by being easier on installers—more lax in their enforcement of the NEC. Corruption is corruption. It is hard to imagine that individuals who engage in such behavior would have enough professional commitment to spend their time reading books such as this.

Making old buildings safe: the ultimate, problematic, goal

The numbers tell us that something needs to be done about bad old wiring. More than 40,000 fires are caused each year by problems with home electrical systems. They kill close to 400 people and cost $2.2 billion annually. Perhaps 300 people more die annually of other injuries caused by electricity, such as electrocution. The CPSC has data showing that homes over 40 years old are over three times more likely to suffer electrical fire than are newer homes.

Bad wiring in homes may even become an affair for OSHA. Increasing numbers of employees working for major companies, such as DuPont and ARCO, telecommute, with the employers providing office equipment for use in the employees' homes. With homes serving as extensions of workplaces, employers have increasing reason to be concerned about the adequacy of home electrical systems.

Homes are not the only worrisome area. There are numerous other examples. Although no one may sleep in them, places of assembly are particularly problematic in part because so many people can be at risk, and in part because coming down heavily on community institutions can be politically hazardous.

(a)

■ **12-1a–c** *The basements of some churches (and presumably temples and mosques and synagogues) suggest that no one has ever broached the idea of ready access to, or clear working space around, electrical equipment. Requirements such as the need to cover panels and to limit the height of switch handles seem not to have been included in the education of the member who installed much of this church's wiring. Many inspectors will bend over backwards to accommodate community institutions. Still, it would be hard to grandfather some of this work, clearly performed without a permit; or to smile on the practice of storage in the electrical room.*

(b)

(c)

Beside homes, there are other examples of places where people may be at risk from old wiring while they sleep. One that may never have crossed your mind, unless you remember being there, is a college. Old dormitory buildings, housing youngsters who have not previously lived on their own, can be quite dangerous. Pre-1998 halogen torchiere floor lamps pose significant fire risk when bedding, laundry, even Nerf™ balls land on them. Harvard University outlawed the use of such lamps in their dormitories and provided safer torchieres, utilizing compact fluorescents, in their stead. Students have been badly burned while cooking in their dorm rooms, perhaps using a hot plate plugged into the one overloaded outlet that also served the clothes iron or hair dryer. In most jurisdictions the dormitories, being preexisting buildings, are more or less off-limits to inspectors. A carelessly-written proposal for the 1999 NEC would have added an article on child-care facilities. These are other worrisome settings.

Should reinspection become the rule? In the Cleveland area community of Shaker Heights, Ohio, reinspection is required, and despite old housing stock, they have one of the lowest fire rates in the country. New York State has held hearings considering similar legislation, and many other jurisdictions have talked about it. Think about it, and make your voice heard.

Also think about whether you want to support The Inspection Initiative (1-800-647-3156, extension 521), the national movement to oppose downsizing and otherwise strangling inspection departments, often while maintaining or even raising permit fees. Some inspection departments have offered rudimentary training to inspectors with no background in electrical work, and subsequently authorized them, as combination ("multihat") inspectors, to pass or fail electrical jobs. They can be quite lost looking at old work, having some idea of current Code requirements, but lacking the perspective of people with a real grounding in the electrical industry. Some jurisdictions have forced inspectors to "tighten their belts" and make up their minds about each installation in five or ten minutes, no more. Some jurisdictions allow contractors to certify their own installations as code-compliant. That is convenient, but it also removes one level not only of safety but of legal protection.

To some degree, the fight to have work inspected by someone at least as expert as the installer is a losing battle—though it is far from lost. As of mid-1997, the hotline associated with The Inspection Initiative was not even seeking information on shifts to combination inspection for residential jobs, only commercial and industrial.

A report in the *NFPA Journal* (1-2/97), said "In Texas, for example, insurance companies have made several recent attempts to sue local fire departments, citing failure to properly inspect and warn building owners of fire hazards." Insurance agency involvement is evidence that The Inspection Initiative is more than a self-serving effort to protect electrical inspectors' jobs.

IV

Supplementary Material

Dating and History

THIS CHAPTER IS SUPPLEMENTARY MATERIAL, RATHER than an essential part of the book. Knowing what used to be done can be helpful, but it is not of primary importance. The latest version of the NEC embodies the best information about wiring safely that is readily available. To put that more bluntly: anything that falls short of the current NEC skimps, to some degree, on safety. And the NEC is a minimum standard.

Age versus condition

Age is not really a safety issue. At various points in this book you have been presented compromises that you can offer customers who want to salvage older systems but who still seek a degree of safety. You also have gotten pointers on how to judge whether wiring is deteriorated. Whether an item was installed a week ago or a century ago does affect its likely degree of deterioration, but you judge safety by evaluating the deterioration, not the age. Antiquity of design does not necessarily indicate how old or how worn a particular item is. The looks of porcelain lampholders, for example, have not changed significantly over many, many decades.

Dated can indeed mean tired

Notwithstanding all that, some equipment generally can be assumed to be hazardous or borderline-hazardous simply because it is so old. Rebuildable receptacles are an example, mentioned in Chapter 9. Receptacles lose spring tension in the course of use. It has been so long since rebuildable receptacles, or their replacement parts, have been produced that unless such a receptacle has been idle for 50 years it is a prime candidate for replacement.

Good condition may not make up for design shortcomings

Other old designs have to be judged individually. A rebuildable light fixture may consist of a nineteenth century skeleton whose lampholders and internal wiring—the entire guts—were replaced in 1993. Maybe, if the fixture is a portable lamp, despite the overhaul it still lacks a strain relief bushing to protect the cord where it exits the base. Quite commonly, with both portable and hardwired fixtures, there is no label giving the maximum safe wattage. On the other hand, you may find no appreciable deterioration. The two design lacks may be acceptable to the customer. If you are in any way involved, you should nonetheless point them out. And since anything you actually install carries some liability, when you are asked to reinstall or move an old, hard-wired fixture that may not be quite up to par you have a judgment call.

The value of antiquarian expertise

So your primary concern is the safety of equipment design and condition. There still is value in being able to date systems, if only the value of enabling you to demonstrate expertise. Dating installations is a better way of establishing your authority than the use of scare tactics. "Look at this! Do you realize how close you came to having a fire?" can appear pretty self-serving, as it directly implies that your customer should have called you in sooner. (It still can be worth saying, when you do uncover an "omigawd.")

Contrast this with a far gentler way to warn customers. "This beauty has been going strong for over half a century" still leaves room for you to say, "We do need to look at the wires feeding it. They are probably way past their design life." To make such statements with authority, though, you need to know some chronology.

Besides, playing with the "Wayback Machine" can be fun. It is an innocent enthusiasm you can share with your customers. Moreover, here and there you will find tidbits that help you not just to date and evaluate antiquated systems, but also to work on them.

Grandfathered? Grandfather never wired like this.

Grandfathering is another reason to know some history. It is especially relevant if you are an inspector, but is of benefit to other readers as well. If a device or fitting or piece of equipment has an NRTL mark, you know that it is okay (unless you have reason to believe it was installed in a manner inconsistent with its Listing). Old equipment frequently lacks such a badge, if only because the UL label has dried up and fallen off.

If you can determine when a now-illegal type of system stopped being legal, you may gain a clue as to whether the installation you are looking at was originally done to Code. As Chapter 12, "Inspection Issues" points out, the fact that an installation is old and not imminently hazardous is not reason enough in itself for it to be considered grandfathered.

The situation with respect to grandfathering gets even hairier when you are looking at an installation that may have been worked on over the years since it originally was installed. In the 1970s, for instance, conduit fill rules depended on whether the conduit was preexisting or new. If there are 20 #12s jammed into $\frac{1}{2}$" trade-size conduit, you know that it was and is in violation (not to mention in-credible). At more plausible levels of fill, though, it is hard to know just which year's criteria to apply.

The dating game

The chronologies that constitute the latter part of this chapter describe what was commonly accepted in the field during a particular period. This will be more useful than just specifying the legal standards defined by successive editions of the NEC—although a certain amount of that is included as well.

Although not exhaustive, this is solidly researched. Besides calling on one middle-aged electrician's knowledge of history, well over half a century of electrical magazines have been leafed through to see what products were advertised at different times and what issues of work and workmanship were being wrangled over. The Library of Congress's collection was employed, as well as some of the holdings of the Smithsonian Institution's libraries. Best yet, a number of old-timers were consulted during the preparation of this chapter, as they have been tapped repeatedly over several decades. You should take advantage of similar resources yourself, when you can create the opportunity to do so.

263

This offers guidance only, not precision

For a number of reasons, most of the dates should be considered approximate. First off, this book is not being prepared by or for historians. It is possible that some of the sources, other than the old Code books, were not consistently correct. Furthermore, the date of an item's invention is not the same as the date of its introduction. Here is an example. Alexander-Edmond Becquerel discovered the photoelectric effect in 1839; Einstein explained it in 1905; yet you find not even one photocell installed before the last few decades of the twentieth century.

Even more to the point, the dates that a product was designed, patented, first manufactured in quantity, and eventually acknowledged in the Code are all different. Edison patented a fluorescent lamp in 1896. Even so, the first commercial neon was installed only in 1923, and the first commercial production of fluorescents was not until 1938. Thus, the earliest relevant date you could conceivably apply to an old fluorescent is more than 40 years later than the date you might unearth in the Patent Office. Sometimes it is the other way around. The laser patent finally granted in 1977 was based on 1957 laboratory notes. Meanwhile, lasers were being busily developed in the 1960s.

Even when you can identify when a practice first was authorized or forbidden by the NEC, this does not tell you much about a particular installation. Local dates of Code adoption differ, as do local amendments and even codes. As early as 1881, 16 years before publication of the NEC, the New York Board of Fire Underwriters adopted its own "Standard for Electric Light Wires, Lamps, etc." In 1925, for example, New York City prohibited open surface wiring, supported by cleats and knobs, in loft buildings. This makes dating New York City installations different from dating surface wiring elsewhere.

A finer-grained example concerns grounding. In California in the early 1940s, ungrounded, two-wire NM cable was supplemented during installation by an external ground wire. Yet in Illinois in the early 1950s, ungrounded, two-wire NM cable was installed in new houses without such supplementation.

Another consideration can be even more important than local amendments or dates of NEC adoption. Even if a design was supplanted in, say, 1970, and stopped being manufactured, this tells you only so much. When did jobbers run out of their supplies of the product? When did the last distributor sell the last box or coil off the shelves? When did the last old-timer, clearing out his truck, install the last forgotten sample? In most cases, inspectors permit electricians to use up existing stock.

The changing NEC affects accuracy and ease of dating

Especially when there are legal ramifications (and grandfathering is not the greatest of them), this chapter can only point you in the right direction; you have to track down the date a particular language was incorporated into the Code. The NEC is always changing—and for good reason. It includes new types of material and new systems in its coverage, eliminates old ones very occasionally, and continues to get much, much more specific.

Searching through old Code books can be a chore, both because the NEC is not the most readable book, and because its organization has changed and changed again. Until 1914, when the first edition of a proposed National Electric Safety Code appeared, the NEC covered both premises wiring and utility work. You are just as well off that utility issues were removed that far back—the NEC is quite fat enough! Then, in 1959, the NEC was again redesigned: the order and the numbering system were changed. In 1966, Walter Stone was hired to do a massive rewrite, to make the book easier to read. This was not a success. Technical Committees accepted many of his changes in 1971, but in 1974 rejected the rewrite, considering too many of Stone's changes substantive rather than merely editorial. In the mid-1990s, there was again a move for editorial revision to make it easier to locate rules in the NEC, and to make the book more understandable. Few of the changes proposed in response to this initiative have been welcomed. Electricians have enough trouble navigating through the Code, and do not eye changes in a friendly manner unless they seem very important.

Here is a simple example of how NEC changes keep everyone on their toes, be they antiquarians or not. In 1965, the NEC had about two pages on health care facilities. In 1996, Article 517, Health Care Facilities, ran to more than 30 pages. If you are looking at an older hospital, you must refer to the Code that was in force when it was wired to determine whether the automatic lighting control was properly installed, or whether the redundant grounding plausibly can be left alone. So dating is important. In any case, it is patently unfair to judge the professionalism (other than workmanship) of even a 20-year-old installation by current standards. Its safety, yes. An anesthetizing location brought up to the standard of the 1999 NEC will be far safer than one whose wiring remains in accordance with a much earlier version of the Code.

Here is another twist: occasionally, restrictions come back after years and years of absence. In 1996, the NFPA removed the exemptions in Section 250-60 for electric dryers and ranges from the rule in Section 250-57 and Section 250-59 that grounded conductors not be used for equipment grounding beyond the service. In 1935, that practice required special permission. The World War II-era official exemptions lingered for half a century. With the 1996 NEC, the practice was again outlawed for new installations. But there are many thousands of old NEMA 30Rs and 50Rs out there that are perfectly legal not only to use but also to replace, under the current version of Section 250-60. This complication makes it yet harder to identify just when an outlet was installed. The fluctuation in this Code section, which took so long to reinstate a restriction that enhances safety, also highlights the fact that installers can create safer installations by heeding some old requirements that have been dropped. As an example, it can do no harm, only good, to obey the 1928 NEC requirement that white conductors in switch loops be recolored.

The dates

What follows next is not exhaustive linear table of progress, but rather a series of snapshots. Before going into the details of when important individual rules entered the NEC, you will get a look at two time capsules from the middle of the twentieth century. The first simply describes good professional practice.

Those fabulous fifties

After World War II, the Industry Committee on Interior Wiring Design offered residential wiring recommendations that look not too alien 50 years later—nor half bad. As it did not have the force of law, you will not find this model of wiring showing up in every residence of that period or even most of them. This fact lets it serve as an indicator of quality: more conscientious electricians and more forward-looking customers would have been likely to push wiring that meets these specifications.

Lighting is a good place to start. Most habitable rooms were to have wall switches controlling lights. Except for a dining room or bedroom, if a room's longer dimension was greater than twice its shorter, that room was to have two such lights—regardless of overall room size. Recreation rooms were to have wall switches controlling one lighting outlet per 150 square feet of floor area. Basements and utility spaces also fell under the 150-square-foot rule, but, there, lights could be operated by pull chains.

Receptacles are next. Today's 6-foot spacing rule for convenience receptacles was augmented with the recommendation that at least two were to be controlled by wall switches. One outlet was to be installed flush in a mantel shelf, where such a shelf was present. Receptacles were not required in isolated lengths of wall 3 feet long, rather than the current 2 feet. The 6-foot rule was relaxed to 10 feet for dining rooms, and kitchens were exempted altogether from that particular recommendation.

Special spaces had their own guidelines. A kitchen, kitchenette, or pantry needed a switch-controlled ceiling outlet, as nowadays. It also was to have a receptacle for each 4 feet of working surface, though it was not necessary to space them at 4-foot intervals. A bathroom needed a lighting outlet, controlled by a wall switch, at each side of the mirror. A ceiling outlet controlled by a wall switch was called for in any bathroom larger than 60 square feet and in any completely enclosed shower compartment. One convenience outlet was to be located near the mirror. Halls were to have convenience outlets and switch-controlled lights. Staircases, except for stairs leading to storage attics, were to have multipoint-switched lights. Larger closets were to have lights. Additional outlets were recommended almost everywhere, to handle possible use of electric heaters.

There is plenty more. The point is that, while the NEC has continued to grow fatter and fatter, well-wired houses of 50 years ago were wired very similarly to modern ones. No GFCIs, of course. No smoke detectors. No central air conditioning. But they had reasonable outlet spacing, if not broken out into as many circuits as one might wish. Of course, the standard materials and methods used then were a far cry from those of today.

One other time capsule Ideal specifications are one thing. What every electrician needed to know is a more basic perspective. A 1948 (first edition) handbook, *The Electricians' Pocket Companion*, still taught the use of open wiring, knob-and-tube supports, and soldering as the method for connecting conductors; and it talked of rubber as the insulation one dealt with. On the other hand, it did acknowledge the existence of solderless connectors and of thermoplastic insulations. Without any other discussion of aluminum, it mentioned aluminum conductors as having 84% the ampacities of equivalent copper conductors.

Backing up further

Now you will have a chance to leaf through "snapshots" that will help you assess much older systems. One purpose of these snapshots is to show how long some old systems of wiring persisted. Rules of particular importance are presented in UPPERCASE. To provide a complete list of important changes would take a separate, very fat, chapter; an exhaustive list would take a separate, thick book. Hence these snapshots. Most are derived from Code responsa, articles in electricians' magazines of the time in which senior inspectors and other Code experts discussed the practices being introduced and the violations causing installations to be rejected.

The early 1920s In the early 1920s, according to one report, the following developments were current:

☐ Soldered splices started to be phased out, though it took over 30 years for them to be fully supplanted.

☐ Using bent-over nails as a means of support was frowned upon and sometimes rejected.

☐ Two-wire Armored Cable was deemed unacceptable for 3-way switching, on the grounds that the NEC required all circuit conductors to be run within the same armor.

☐ Armored Cable bushings were brass thimbles. While bushings were not explicitly required by the NEC, many inspectors demanded their use as necessary for safe and workmanlike installations.

☐ Two wires could be terminated under one screw, if soldered together or installed with washers.

1920 Now for a closer look at year-by-year changes for a decade or two. The NEC introduced the following changes in 1920:

☐ Low voltage was set at 600 volts—there had been agitation for 750 volts.

☐ A requirement was added specifying that the identified conductor at an Edison-base outlet go to the screw shell.

☐ A requirement was added that grounded conductors of #12 and #14 AWG be identified.

1921 As of one year later, the following requirements were enforced, and the following practices continued to be required or permitted:

☐ EACH OUTLET NOW REQUIRED A BOX, CABINET, ROSETTE, FIXTURE BLOCK, OR EQUIVALENT IN ADDITION TO THE DEVICE ITSELF, EXCEPT IN SOME CASES WHERE KNOB-AND-TUBE WIRING WAS USED.

☐ Stranded conductors, other than those in flexible cord, had to be soldered before being tucked under a binding screw.

☐ Arc lamps were still covered in the NEC.

☐ The following ampacities were assigned to #14 conductors: 15 amps, when rubber-insulated; 18 amps, cambric-insulated; and 20 amps, other-insulated.

☐ Equipment had to be grounded if the circuit feeding it was over 25' long; or if the circuit was less than 25' long but in contact with grounded surfaces or metal, and not insulated or isolated.

☐ Gas pipe was only legal as a ground for equipment, not for services, and only when water-pipe grounds were not available.

- [ ] Damaged building surfaces had to be repaired.
- [ ] Holes in boxes had to be closed.
- [ ] Outlets could not be recessed, except for the same $\frac{1}{4}$" maximum in noncombustible surfaces that is now permitted.
- [ ] Circuit breakers were permitted as switches only where used under competent supervision.
- [ ] 440/220-volt three-wire circuits in wooden raceways were prohibited.
- [ ] Service switches were knife switches in closed cabinets. A snap switch was not forbidden as a service switch, but using one for that purpose was a novel idea.

1922 Here is a snapshot of 1922:

- [ ] In early 1922, a two-wire circuit to a lampholder still required a fuse in the grounded conductor.
- [ ] Edison-base outlets were still being accepted as receptacles in baseboards, provided that they had hinged doors covering them. They were being phased out.
- [ ] A number of manufacturers introduced a "flat" (actually oval) armored cable. It was used for under-plaster extensions, and required connectors or box openings and clamps designed specifically for it.

269

1923 The NEC made a few significant changes in 1923, including the following:

- [ ] GROUNDED CONDUCTORS NOW WERE REQUIRED TO BE WHITE OR GRAY.
- [ ] Minimum service conductor size grew to #10.
- [ ] "Readily accessible" was first defined. By 1927, consequently, cutout boxes on ceilings were rejected by two-thirds of inspectors polled.

1925 In 1925, there were further important changes:

- [ ] LAMP SOCKETS' CENTER CONTACTS WERE TO BE SWITCHED, AND NOT THE SHELLS.
- [ ] Switches and circuit breakers had to be not just enclosed but externally operable.
- [ ] Edison-base "receptacles" less than 4' AFF became illegal unless intended for lampholders.
- [ ] The rule that receptacles be loaded to no more than 660 watts and that no more than 16 outlets be installed per circuit was

changed, with the 660-watt restriction dropped, and circuits further limited to no more than 12 outlets.

The late 1920s There are three more snapshots from the period preceding The Great Depression. Here is 1926:

☐ In the NEC Supplement, there was one curiosity of note: in the course of a conduit run, you were allowed today's four quarter bends, plus one more bend at each end.

☐ You could simply bring loom out through the wall to the back of a switch mounted on the surface; compare this to 1921.

☐ Polarized fixtures were required, but not polarized receptacles.

☐ Three-prong plugs were ready for market. It took half a century, though, for them to supplant two-prong receptacles in new construction.

In 1928, a big year for changes, the trade left legacies such as these in the buildings they wired:

☐ GROUNDING WAS REQUIRED FOR EQUIPMENT IN CONDUCTIVE LOCATIONS SUCH AS BASEMENTS OR IN WALLS CONTAINING METAL LATH, EVEN IF THE EQUIPMENT HAD BEEN FED BY KNOB AND TUBE WIRING, CONDUCTORS IN WOODEN MOLDING, OR NONMETALLIC CABLE LACKING AN INTERNAL GROUNDING CONDUCTOR.

☐ THE GROUNDING CONDUCTOR NO LONGER COULD BE SOLDERED TO THE GROUNDING ELECTRODE.

☐ ELECTRICIANS WERE ALREADY REQUIRED TO USE WHITE AND GRAY FOR THE GROUNDED CONDUCTOR. USE OF WHITE AND GRAY FOR OTHER CONDUCTORS WAS NOW MADE ILLEGAL. SWITCH LOOPS HAD TO HAVE THEIR WHITE CONDUCTORS REIDENTIFIED AT BOTH ENDS.

☐ A NEW REQUIREMENT WAS INTRODUCED: WHERE POSSIBLE, BORED HOLES WERE TO BE KEPT 2 INCHES AWAY FROM EDGES OF WOODEN STRUCTURAL MEMBERS.

☐ THE NEC REQUIRED A BARE COPPER GROUND WIRE IN NEW NONMETALLIC CABLE. IN MANY JURISDICTIONS, THOUGH, THIS SIMPLY DID NOT HAPPEN.

☐ THE 40% MAXIMUM FILL RULE FOR CONDUIT WAS CHANGED FROM A RECOMMENDATION TO A REQUIREMENT.

- [ ] Conductors with varnished cloth insulation were still legal and employed in dry locations.
- [ ] The NEC said, in rule 1403(f),"Where a gas pipe...is present, the fixture shall be attached thereto." (In New York City, hundreds of fixtures supported that way were still in use as of 1997.)
- [ ] Grounding to water pipes, which had previously merely been permitted, was redefined as essential, where such pipes were available.
- [ ] Where no water pipe was available, gas pipe could be used for grounding of either systems or equipment.
- [ ] Open knife switches were still legal.
- [ ] Snap switches fed by knob-and-tube wiring, without outlet boxes, no longer were legal.
- [ ] Demand factors were introduced into the NEC.
- [ ] The minimum service conductor size grew to #8 copper.
- [ ] A requirement to use running boards or guard strips for armored cable was introduced.
- [ ] The NEC required that NM cable be fastened every 3' where run exposed.
- [ ] NM cable could be fished.

271

- [ ] The support requirement for concealed NM cable stretched from 4' to 4'6".
- [ ] A requirement that NM cable have braided cloth protection over each individual conductor was removed.
- [ ] Rubber-insulated conductors still were required to be tinned, to prevent chemical interaction with the copper (but the requirement was moved out of the NEC to the UL rules).
- [ ] "Outlet plates" were made illegal (although their use still is mentioned, as a normal procedure, in 1932 publications).
- [ ] Outlet boxes had to have a minimum of $1\frac{1}{2}$" depth, except where using boxes this deep would cause damage to structure—not just to plaster—in which case boxes of $\frac{1}{2}$" (not $\frac{5}{8}$") minimum depth was acceptable.
- [ ] Wooden molding was still an approved wiring method, although frowned upon.
- [ ] Three-prong equipment began to be introduced into industrial establishments.
- [ ] Use of Edison-base outlets other than as lampholders was on its way out.

□ Multifamily dwellings could have four service disconnects with no single main switch, provided that they had four separate meters and cutouts.

The following snapshot dates from 1929:

□ Somewhat over two-thirds of U.S. homes had electric wiring, and almost as many had telephones.

□ There was plenty of rubber tape in use, and some combination (cloth-and-rubber) tape, although the latter was unListed. Friction tape was not available in a Listed version. This means that scrupulous installers would not use it to insulate splices, only to protect the rubber tape that did the actual insulation.

□ Some still used wooden plugs for mounting equipment on masonry walls, although this practice was frowned upon and often—but not always—rejected by inspectors.

In the late 1920s and early 1930s, there was some dispute over where conductors in conduit required lead covering. Were the conductors likely to be exposed to moisture, and therefore in need of the added protection, when conduit was installed underground, outside but above-ground, or both? The NEC soon clarified requirements.

The 1930s Progress continued throughout the 1930s. In 1930, this was the state of electrical work:

□ A COMMON NEUTRAL, APPROPRIATELY SIZED, WAS STILL ACCEPTABLE FOR ANY NUMBER OF CIRCUITS; THEY DID NOT HAVE TO ORIGINATE FROM DIFFERENT BUSBARS.

□ Transformers were replacing the batteries used to operate doorbells and telephones.

□ EMT still was not recognized as acceptable for grounding.

□ Dimmers were permitted on grounded conductors; they generally did not have a "full off" position.

□ Single-conductor armored cable was introduced—in #8 AWG, it was legal to use as a grounding electrode conductor.

□ French hooks were accepted as part of the grounding path.

□ The TEFC induction motor was introduced.

The 1931 NEC includes the following:

□ THERE STILL WERE NO RESTRICTIONS ON BOX FILL.

□ THE RULE REQUIRING SIX INCHES OF FREE CONDUCTOR WAS INTRODUCED.

- ☐ Single-family dwellings fed at 70 amperes or less no longer needed a main service switch, provided they had no more than six circuits.
- ☐ Two-wire services were still common—and, in buildings fed by such services, inspectors forbade circuits with common neutrals.
- ☐ Refrigerators still could be connected to lighting circuits.
- ☐ Underground conduit, except for service runs, required lead-covered conductors—except in the case of standpipes that dipped underground for a few feet as they entered a building.
- ☐ Services serving no more than one branch circuit (1,000 square feet or less, maximum) could be fed by a pair of #12 service conductors.
- ☐ Bare service neutrals were allowed.
- ☐ Service conduit no longer could contain wires other than service conductors.
- ☐ Service conductors no longer could be spliced before the service switch. They could, however, be tapped at the switch.
- ☐ Multiconductor armored service cable was introduced.
- ☐ Threadless couplings now were allowed on service conduit.
- ☐ Canopies now had to fit snug against the edges of boxes.

The 1933 NEC allowed multifamily dwellings to have up to four main disconnects, regardless of the number of meters.

A highly significant change began in the mid-1930s. Until then, many farms relied on windmills and batteries for power and were wired at 32 volts, DC. The Rural Electrification Administration began extending utility wiring to those areas in 1934–1935, and they gradually converted to standard utilization voltages. There was an effort to remove Article 720 from the 1999 NEC on the grounds that it has not been needed since then. You may, however still find remnants of the 1920s wiring, intended for lower voltages, in some rural systems.

The following 1935 practices seem noteworthy:

- ☐ LOCATING CUTOUT BOXES OVER TUBS AND SHOWERS WAS ADJUDGED OKAY, SO LONG AS THE ENCLOSURES WERE WEATHERPROOF AND READILY ACCESSIBLE (AND THEIR DOORS WERE SECURED WITH ADEQUATE LATCHES).
- ☐ Grounding a range frame to the grounded conductor was not accepted, except by special permission.

- [ ] Nonmetallic surface extensions had to be secured every 6 inches.
- [ ] Lightning arresters could be grounded to the grounded conductors of services.
- [ ] Load calculations employed a 500' rule—one 15-ampere circuit with any number of outlets could serve 500 square feet of space.
- [ ] BOX FILL RESTRICTIONS NOW WERE IN PLACE—FOR INSTANCE, A MAXIMUM OF EIGHT #14s WERE PERMITTED IN A $3\frac{1}{4}$" BOX.

The 1937 NEC allowed any service to have up to six disconnects.

The World War II era In many ways, permanent changes at home were put on hold as America geared up for, and then went to, war. Just as the gender barrier was breached because of the necessities of mobilization, other changes took place.

Funny things happened in the 1940 code: a number of wire sizes were dropped and ampacities were changed. Here was the problem: copper was scarce during World War II. And here is one consequence: while the 1940, 1942, and 1943 codes were basically adopted, due to local amendments the 1937 ampacities, which had been supplanted in those later editions, continued in force in many localities. It seemed to work. Another apparent consequence was the permission in NEC Section 250-60 to ground electric ranges and clothes dryers to the grounded conductor.

How about insulation types? Plastic came in during World War II, and the rubber got thinner and flimsier. In itself, thinner insulation should both reduce voltage rating (and, arguably, ampacity) and at the same time enable more conductors to fit into conduit. Not so: it was not until 1958 that the type of insulation began to affect ampacity. Incidentally, the 1940 code did not require that sockets be polarized. They could either be polarized or else be fused on both sides.

An alternate vantage: equipment by category

So far, you have a lot of dry dates. The preceding lists correspond to the earliest wiring you are likely to have to evaluate. In contrast to such early wiring, if you have to investigate some World War II era or postwar installation and decide whether the original work was legitimate, you have a fighting chance of obtaining the Code book that was in use. This is one reason the text does not inch forward over the subsequent half-century.

The following part of this chapter is organized differently. If, instead of finding out whether this 1925 house is all-original, you want to determine whether its wooden "wiremold" is likely to have been installed legally during grandfather's Depression-era renovation, you can go right to that category and find out.

Basic materials The first category that will be covered is the materials used in making electrical products. The logical place to begin is way back, with nonconductors and material used explicitly for insulation. Such materials were needed well before electricity was used for light, heat, or power.

Most but not all of the materials employed in the very earliest days, like wood, are still in use. In some areas, wooden fuseboxes installed in the early 1900s are still in use as junction boxes. Wood still is legal as part of fixtures, and is explicitly mentioned in at least the 1996 NEC. Porcelain was another early insulator. If unglazed, it required a test of absorbency because if it could take in moisture too readily, it would not insulate very well. Paper was an even earlier insulator. In the 1890s, paper-insulated mains were used.

Cloth was used from the beginning, and jute fibers still are used as filler in flexible cords. At one point, a major nuisance at telephone office central switchboards was that lint from textile fibers used in insulation would float free and foul relays. Hemp, while less flammable than cotton, and stronger (having much longer fibers) wicks water more wickedly than does cotton. For this reason, the hemp used to insulate the first transatlantic cable was sealed against moisture with tar. Hemp also was used in capacitors and eventually in making cellulose-based plastics. Cotton too was used as a filler, and also in cloth coverings and sheaths.

All of these materials and others needed some treatment to reduce moisture absorption. Receptacles incorporated sealing wax into the twentieth century. Some of the earliest insulation was paraffin-dipped cotton. When it got too hot, it burned. The paraffin was replaced with lead in "underwriters' wire"—which still burned. (It received the uncomplimentary nickname, "undertakers' wire.") Oily varnish not only was used in original cable manufacture but was field-applied, to the point where a long-ago inspector complained about the oily pool he found at the bottom of a long vertical run.

Rubber goes way back as an insulator. Vulcanized rubber cable insulation was invented in 1860. In the 1880s, copper conductors began to be tinned to prevent them from causing deterioration of their rubber insulation. In 1932, 75°C rubber insulation was introduced, and in 1935 a thinner-walled rubber insulation came on the market.

By the end of the 1960s, *The Graduate* told Dustin Hoffman and America that the future was in plastics. The electrical industry saw the light well over a decade earlier. Bakelite, the first modern plastic, was patented at the end of 1909, and in use in the electrical industry by the second decade of the twentieth century. Vinyl resins were introduced in 1930, and neoprene and fiberglass in 1931. Vinyl chloride insulation was introduced in 1937, and its uses have grown and grown:

☐ house wiring in 1939;

☐ yellow! electric tape in 1946;

☐ telephone switchboards in 1946;

☐ phone cords in 1954;

☐ commercial and industrial power cable in 1955;

☐ rigid conduit, underground, in 1956;

☐ rigid conduit, indoors, concrete-encased, in 1971;

☐ and plenum cable in 1982.

Plastic did not take over from rubber immediately upon being introduced into the marketplace. As of 1950, Type T was so unfamiliar that it was not accepted for use in service drops. UL first listed what was later called THWN in 1951. Nonetheless, THW was far more common than nylon-jacketed wire as late as the early 1970s. In houses wired during the 1950s, it is common to find nonmetallic-sheathed cable whose sheath is cloth rather than vinyl, but whose internal conductors are plastic-insulated. Warning: very commonly, this cable does not include a grounding conductor.

A few stranger materials have been employed by electrical manufacturers. Leather washers or bushings were used under the nails that secured knobs in knob-and-tube wiring. Asbestos was used for a long, long time both as conductor insulation and as enclosure lining. This does not even consider products such as rubber-asbestos tape.

Wiring systems The next category after raw materials is wiring systems—mainly raceways and cables.

Conductors The actual wires are the place to start. Both the sizes and the makeup of wire have changed over the years. Hard-drawn copper wire was developed in 1877, supplanting iron wires. During and shortly after World War II, iron or steel wires were temporarily brought back into use as conductors, due to copper shortages. While most electricians are unfamiliar with copper-clad steel conductors, they still are mentioned in the 1999 NEC, for use in CATV and fire alarm wiring. Figure 9-5 shows an earlier type. Tinned copper was used through a large part of the period when rubber-insulated copper was employed.

Aluminum goes back quite a way. Bare aluminum conductors were used in bus and in high-tension lines from the end of the nineteenth century. A 1948 article in one of the trade magazines discussed the fact that, to their advantage, aluminum conductors are not corroded by the sulphur in rubber insulation. It talked about soldering them, for splicing or pigtails. The 1960s and 1970s were the period of major aluminum use in branch circuit wiring. After the 1970s, aluminum wire of 8-12 gauge no longer was made. There is little point in describing changes in wire sizes, but, for example, while readers are fairly unlikely to run across #5-AWG conductors, that size was listed in the NEC until 1947.

Raceways Now it is time to shift to the wiring system proper: that which surrounds and contains conductors.

Wood Wooden cleats initially were used to support wiring but were outlawed by the beginning of the twentieth century. Wooden molding, containing chases for individual conductors, was used for quite some time after wooden cleats were banned. Theoretically, wooden molding was a two-part hardwood construction. Flimsier versions, lacking the base section, were used into the 1920s, but became illegal in most jurisdictions by the 1930s. There was a long period when "molding" could mean either surface metal raceway or wooden molding. Metal moldings, incidentally, first were Code-recognized in 1907.

Wooden molding was not so bad as one might think. Its good features are that it was nonconductive and that it stood up well, so long as it was properly sealed against moisture. It had two bad points. The obvious one is that it was quite combustible. A less obvious problem stemmed from a characteristic of plaster and gypsum block walls: they hold nails poorly. If a homeowner wanted to hang pictures, the common answer to that problem was to install horizontal, wooden, "picture rail." Affixed to the wall, it alas looked quite similar to wooden electrical molding. In consequence, wooden molding, along with the enclosed conductors, was regularly penetrated by nails driven by ignorant occupants decorating their walls. In this way it was not unlike cable or EMT passing through an unfinished basement, on which some homeowners dry laundry or hang storage .

Conduit On to more familiar types of raceway. Conduit has undergone a number of changes over the years. Some early rigid metal conduit was made of thin brass. Early rigid steel pipe, unlike conduit as you know it, required an inner lining of resin-impregnated paper. In 1927, both enameled and galvanized conduit were going strong. Enameled conduit was readily available at least into the 1950s. While black enameled conduit was still being installed in California up to about 1950, most conduit was galvanized for some time prior to that.

Some early rigid nonmetallic conduit was made of resinated vegetable fiber. Rigid nonmetallic PVC conduit was introduced in the market in 1962, nonmetallic flexible conduit in 1981, and nonmetallic tubing in 1983–4. Code acceptance was another matter. In 1981, there was an effort to bring rigid nonmetallic conduit, PVC, into the NEC. The effort was narrowly defeated through expensive efforts on the part of the steel conduit manufacturers. In 1988, the U.S. Supreme Court upheld a judgment for over $11 million dollars against Allied Tubing in a lawsuit alleging anticompetitive practices, namely their having purchased NFPA memberships for their employees to make the 1980 vote turn out right. PVC did not have to wait quite that long. RNMC found limited acceptance in the 1984 NEC, and full acceptance in the 1987 NEC.

While brass tubing was one of the earliest raceways, and even speaker tubes (yes, like on an old submarine) were converted for wiring use, thinwall conduit came a bit later. EMT was code-recognized in 1923—for exposed, dry use only, in $\frac{3}{8}$"–1" sizes. In 1928, was recognized in sizes up to 2"; in 1965, $2\frac{1}{2}$" to 4". EMT was restricted to use with 1/0 or smaller conductors until the 1946 NEC. Amusingly, the UL Standard for Safety for Electrical Metallic Tubing, UL 797, was not issued until 1942, 19 years after EMT appeared in the NEC. The use of EMT for circuits over 600 volts has varied a great deal, with general acceptance taking place in the period after 1970. By 1995, both the NFPA and UL acknowledged the legitimacy of such use.

Fill Before leaving raceways, it makes sense to take a look at their capacities. Until 1953, conduit fill (except in special circumstances such as runs between a motor and its controller) was restricted to nine conductors. The 1959 code allowed more, in large conduit, but still restricted $\frac{1}{2}$" conduit to three #12s, and $\frac{3}{4}$" to five! In rewiring, one was allowed additional conductors. The 1968 Code recognized that thinner insulations made room for more conductors.

Thus far, electricians pulling another circuit or two into existing conduit were permitted more crowding than electricians installing conduit for the conductors. It makes sense on a gut level to give people a break on fill calculations to prevent their having to run a second conduit rather than stuff a little more in the existing one—going, perhaps, a single conductor beyond what the calculations say is okay. On the other hand, it obviously is far easier and less likely to damage insulation to pull, say, nine conductors into an empty raceway than to work two additional conductors into a raceway already filled with seven. Whether this was the logic that determined the decision, or whether it was simply that heat dissipation depends on physics, not on convenience, the rule changed. The 1971 Code was the first to stop differentiating between conduit fill for old and new work.

Cables Cables were introduced a little later than raceways.

Armored Cable Conduit and armored cable are almost the same age. In 1899, the NEC removed the requirement that conduit be lined. About the same time, it recognized armored cable, invented by Gus Johnson and Harry Greenfield. (Flexible conduit, still known as greenfield, after Harry, was not recognized until two years later, although it had been in use before the turn of the century.) Use of armored cable did not become widespread until 1910, and it did not achieve major popularity until the late 1920s or the early 1930s. In 1959, aluminum-armored cable was introduced.

"BX" was never a Code or UL or ANSI designation. As of 1932, armored cable was officially called Type AC, although BX (the trademark of cable made by G.E.'s Sprague Electric division) was then and is still the colloquial term. In the early 1970s, apprentices were taught to think of "AC cable" as "BX with an internal bonding conductor." The bonding wire in armored cable was first required in the 1959 NEC. That skinny strip of aluminum made a big difference in reliable operation of overcurrent devices in the case of ground faults.

At one point, armored cable came in strange versions. Before the development of armored cable with the internal bonding strip, and before double armor, which is discussed next, there was cable armored without any interlocking strips—just a wrapping of four galvanized wires, two round and two square.

An intermediate product intended to overcome the inductive impedance of armored cable's long steel spiral was introduced by 1929; rumor says that it was in use before 1920. Double-armored

"BX"— a G.E. product —"BX"

There is only *one* "BX"

One, and only one, flexible armored conductor can rightly be called **"BX."** This is the product made by the General Electric Company—the pioneer product of its type and the established standard of quality for twenty-five years.

To be sure of getting genuine **"BX,"** look for and find the letters **BX**—a registered trade name of the General Electric Company—at regular intervals on the armor. The distinctive orange and blue label is another means of making sure that you are getting the real thing.

*"BX"—a trade name of a
G. E. Product.*

GENERAL ELECTRIC

GENERAL ELECTRIC COMPANY   MERCHANDISE DIVISION   BRIDGEPORT, CONNECTICUT

■ **13-1** *A rearguard effort to protect the term, "BX."*

cable (previously mentioned in Chapter 5, "Accept, Adapt, or Up-root?" in the section called "Grounding and Bonding Replacements") was produced by the National Metal Molding Division of National Electric Products Corporation. It had both an inner and an outer coil of steel. You still will find some of it in place, though it has been out of production for decades. It is remarkably tough—those two coils do not appear appreciably reduced in thickness compared to the single coil in modern cable. Installers could double-dare-defy any rodent to chew its way through them, and they give drywallers' nails considerable resistance as well.

It is not the easiest material to cut. The spirals do not line up exactly, and the inner spiral snuggles rather intimately against the conductors. To salvage installations using this type of armored cable, you probably will need to cut back the armor to reach sections of wire whose insulation has not dried out and cracked. In cutting back this armor, you need to be extra-careful not to damage conductor insulation.

Nonmetallic sheathed cable NM cable was first recognized in the 1926 NEC, although it was used considerably earlier. You will recall, from earlier in this chapter, that it required a copper grounding conductor as early as 1928. Nevertheless, ungrounded NM was used as late as the mid-1960s. The insulation on its individual conductors was changed from rubber to plastic after World War II, though there was a transition period when its outer sheath remained cloth. That version was installed on East Coast jobs as late as 1962. Plastic-covered Type NM Cable appeared in California about 1960.

Groundless worries In the early days of its use, electricians did not know what to do with that extra wire in nonmetallic sheathed cable, its grounding conductor. As a result of ignorance, installers often would clip off the ground wire. By the 1940s and 1950s more conscientious installers would turn the grounding conductor back, and spiral-wrap it around the sheath, so that the cable connector or clamp bore against it. (They used a tight spiral, like a motor winding, unlike the spiral created by laying armored cable's bonding conductor back in the grooves of the armor corrugations.) The 1962 NEC was the first to require installation of a screw in each box for termination of that equipment grounding conductor, a screw to be used for no other purpose.

While wrapping the grounding conductor back under the clamp or connector is not legal today, most of those installations do offer reasonably good ground continuity. The effect is roughly equivalent to that of the old practice of relying on the mounting screws to ground/bond a flush-mounted receptacle or switch.

Nevertheless, when you rework outlets originally wired that way, and you can finesse access to these grounding conductors, you will do a better job if you unwrap the grounds, splice them, and bond them to the boxes using ground clips or separate screws. Watch out for installations where, far later, someone unfamiliar with this early practice added a new cable and, not noticing any existing grounding wires, said, "What's the point?" and left his or her ground wire unattached—or snipped it.

Weirder wiring: beyond raceways and cable Next come some other interesting systems, which, however, are a century or more old—definite "tear 'em outs." They include individual conductors embedded in plaster, illegal in the United States by 1901, but still common in Europe and South Africa; individual conductors fished in walls (likewise); and individual conductors encased in circular loom (semirigid cloth) fished through walls (a system that remained legal in the United States into the early twentieth century). Regardless of the condition of their insulation, the fished conductors may still be conducting just fine, with no interest in shorting, because air is a fine insulator. Still, you may not want to put your guarantee on a job in which you had so much as touched equipment connected to anything that old, with insulation so deteriorated.

As mentioned in Chapter 9, "Relatively Rare Situations" in the section, "Buried splices," wiring with individual conductors need not be antique. Porcelain and glass cleats, as well as knobs and tubes, were used for open and concealed single-conductor wiring from the 1880s, but this does not mean the knob-and-tube wiring you find is necessarily so old. Knob and tube was the most commonly used residential wiring method in San Francisco in the 1940s. Frequently flooded areas in Louisiana were still being wired using open conductors in the 1980s, according to one report.

Enclosures Wood already has been mentioned as a material used for enclosures, and asbestos as a liner. By the early 1920s, asbestos-lined wooden electrical boxes no longer were being installed. Instead, electricians sweated up 12" × 12" or 12" × 18" boxes of lead sheet in the field.

Most familiar types of enclosure have been around forever. Earlier, this chapter mentioned Code changes that began to require enclosures, and then later to require enclosures of at least $1\frac{1}{2}$" depth. Nevertheless, one can purchase shallower utility boxes. Pancake boxes $\frac{5}{8}$" deep still are being produced, sold, and legally (as well as illegally) installed as of 1998.

Round—not merely octagonal—boxes in the 1½" deep size were frequently used in the early 1940s. The first galvanized boxes, according to one report, appeared around 1935.

The rules about box fill have changed over the years. Sometime after 1971, the capacity of 8B boxes was reduced to the present 15.5", permitting their use with no more than seven #14 conductors. Each device they contained counted as one conductor until the 1990 NEC, which boosted devices' volume allowance to two conductor equivalents. There have been numerous other minor changes, including the ways grounding conductors, clamps, and other equipment are counted.

Plastic boxes' volume is less standardized. You must rely on their markings. Union Insulating, now part of Challenger Electric, was making Bakelite enclosures in the 1950s. Carlon introduced PVC boxes in the late 1960s. In the mid-1970s they introduced larger enclosures of PVC for industrial use, wherever corrosive environments demanded them.

Connectors and clamps have not always been required where cable and conduit enter enclosures. Consider the box used for early lighting outlets, designed to slip over existing gas pipe. Not only was there no clamped connection to the pipe, a U-shaped piece was bitten out of opposite sides of the box to accommodate the gas pipe. Toward the latter part of the twentieth century, though, connectors or clamps have been required, except in specific circumstances. (This ignores the open bottoms of switchboards.) Single-gang nonmetallic boxes with nonmetallic cable entering were exempt from the clamping provision, provided that the cable or cables were secured within 8" of the box. Sometime in the 1980s, the precise year depending on the jurisdiction, that exemption was narrowed to just one circumstance: single-gang nonmetallic *switch* boxes. Until that time, nonmetallic round boxes also had been covered under the exception.

A piercing oddity So much for enclosures. No history of covers is broken out here, although some of it is covered in earlier chapters. The same is true for devices—except for GFCIs, whose chronology appears a little further on.

One rather peculiar outlet design is worth mentioning. Floor boxes, at one point, contained receptacles that had shutters closing the slots for the connectors' prongs. Cords hanging down from dining tables had connectors (plugs) with sharp points that would pierce carpeting and then push aside these shutters to complete the circuit. Although it is highly unlikely that any examples are currently in use, you may encounter some outlets of this design that were never removed. It will be a feather in your cap if you recognize them. More important, being aware of them and considering the possibility that you have encountered one will increase the chance of your solving the mystery of interruption to a circuit that passes through one of them, despite the fact that the box is concealed under carpeting.

Power delivery This section describes how power got from the serving utility into premises' branch circuits. It is strictly of historical interest. Circuit breakers are covered separately, later, due to unique complications. Part of the section on circuit breakers is of considerably more than historical interest.

What power companies chose to provide has changed over the years. The first centrally distributed power was two-wire, 120 volt. Edison quickly realized the economies of center-grounded three-wire circuits. Edison patented his three-wire distribution system in 1882, and first installed it in 1883. Three years later, in 1886, the first commercial AC systems were installed—by Westinghouse and Brush in Buffalo, New York, and by Stanley in Great Barrington, Massachusetts. In 1926 or 1927, 208Y120 3-phase, 4-wire power was introduced. At that time, incidentally, many farms still had to generate their own power. Very commonly, they produced and utilized lower voltages such as 32 volts.

The transitions between different systems have been interesting. In the late 1920s, New York's Consolidated Edison was converting Manhattan from DC to AC. As they worked their way north through the island, Con Ed replaced all appliances affected by the change, at no charge. (The Hotpoint line of appliances first was produced for this replacement program.) If a customer did not want to change over to AC with his or her neighbors, fine: Con Ed would continue to supply DC. When such customers later applied to Con Ed for conversion to AC, they were stuck with replacing their own appliances.

One utility employee's job was to take customers' DC appliances that Con Ed accepted in trade in South Manhattan, and clean them up. They then were resold cheaply in North Manhattan. As the conversion progressed north, Con Ed replaced the same appliance several times! The French provided a similar exchange program for their citizens in the 1960s, when their supply was converted from 120 to 220 volts. In the 1940s, Con Ed still offered both AC and DC service in some places, often going to the same building, and fed to indistinguishable receptacles!

Who bears responsibility for what part of a service installation has also varied. In the 1930s, in at least part of Maryland, power companies required electricians to install and ground only the service panels. The power companies did the rest of the service work, from installing poles and meters to running wiring all the way to the panel. Whatever the switch size, they ran #4 copper, and later #2 aluminum. Around 1970, inspection authorities told the utilities that wiring into panels had to correspond to switch size. The utilities then turned responsibility for installing conductors at the premises, from the weatherhead on, over to the contractors.

Here are two final notes about changed requirements related to bringing power into buildings. First, the 1940 NEC required that flammable oil-filled transformers mounted on roofs be in vaults; the 1946 NEC eliminated that rule. Second, in 1926, Washington, D.C. required separate electrodes for the neutral and the ground, and they were not bonded together. Surely other jurisdictions had similar quirks. When you can find a source of information on local history, such factors provide additional means for dating systems.

Overcurrent devices Moving on from power delivery to premises wiring, a natural first stop is the history of overcurrent devices.

Fuses The first distribution fuse was a length of 1/0 cable that Edison spliced into his 500-kcmil underground feeders.

The first premises wiring fuses were bare wire or solder stretched over wood. Weights were suspended on the solder or copper wires to break them when they softened under overload. Next, springs were introduced to supply the tension in place of the weights. Initially the springs were part of the current path. They lost tension when subjected to even low-level overload, because of the associated heating. As a result of this metallurgical change, they failed to do their job and quickly break the circuits as needed. In fairly short order, therefore, springs were moved out of the current path, reducing their heating.

Later fuses consisted of wire or solder enclosed in ceramic holders or between asbestos pads. The Square D company patented an enclosed fuse in 1905. Cartridge fuses and more-or-less modern Edison-base plug fuses took some time to develop. Type S fuses were mentioned first in the 1940 NEC.

Fuseboxes certainly were common at least into the 1960s, and not only in low-end houses. Split-bus panels were installed in residences until about 1974. By then, though, most residential loadcenters being installed were circuit breaker panels.

Circuit breakers Circuit breakers are the logical next subject. A company that is now part of Siemens may have invented the first circuit breaker, which looked like a telegraph switch. In 1888, partners Henry B. Cutter and A. Edward Newton established the Cutter Manufacturing Co. to produce circuit breakers in Philadelphia. In 1904, théy developed the Inverse Time Element or ITE breaker, whose speed of operation is proportional to the amount of overcurrent. In 1928, finally bowing to trade usage, they changed the company name to I-T-E. At that time Westinghouse, at least, was selling breakers that made and broke their internal arcs at carbon pads to spare the copper contacts. In 1932, Westinghouse put their modern molded case air circuit breaker on the market. Square D introduced a panelboard with plug-in circuit breakers in 1951 (though FPE may have been first).

There is a long tradition of less-alert electricians, and less-knowledgeable or -scrupulous sales personnel, who figure that if a breaker fits, it will do the job. This presumptuous presumption does not result in reliable and safe wiring. Yet if the manufacturer listed on the panel door is out of business, what should you do? Sometimes you can find a breaker Classified for use in that panel. Other times, you can contact the manufacturer's successor. At worst, you may have to initiate a hunt among the purveyors of outdated equipment mentioned in Chapter 14, "Resources."

Quality On the other hand, perhaps some breakers fail because they never were wonderful, from the beginning. The bad rap on some brands can be reason enough, for some electricians, and some customers, to replace panels. It is hard to evaluate such stories, which could after all represent isolated experiences rather than shoddy design or indifferent quality control. What makes a company untrustworthy? In the 1950s, for instance, when four of the major breaker manufacturers were fined, and some of their officials were imprisoned, for price-fixing, that did not necessarily mean their products were unreliable. The fact that the former manufacturer of Federal Pacific breakers was at one point caught defrauding UL regarding their current withstand testing does not necessarily mean that the FPE breakers protecting a particular house's wiring are not up to snuff.

Similarly, even the most reputable brands have manufactured low-end products. An old-timer who said that when he was an installer Eagle and Royal were known for making "cheap" switches and receptacles acknowledged that the brands he respected for quality (Leviton, Arrow Hart, Pass & Seymour, and Hubbel) produced low-end as well as premium products. Such variability is even more likely to be present when different brands are manufactured under a single ownership.

The begats The most important part of this section is the "begats." This almost-biblical record of circuit breaker manufacturers and designs is a good part of what you have to rely on in dealing with old and unfamiliar panels.

Crouse-Hinds magnetic-hydraulic breakers, with silicon-bronze guts, were very highly esteemed. They originally were manufactured by Heinemann. Moisture getting into the arc-chute vent would not cause them to rust, as it would equivalent breakers with steel innards. The magnetic-hydraulic design also made the Crouse-Hinds units relatively insensitive to variation in ambient temperature, compared to thermal-magnetic breakers.

When the original patent expired, Murray adopted the design. (Thomas Murray, an early electrical inventor, had over 2,000 patents.) According to some knowledgeable sources, the Crouse-Hinds design was a great improvement over the less-respected breakers Murray had been selling. Arrow-Hart bought Murray. Crouse-Hinds then swallowed Arrow-Hart. In 1983, Siemens bought the low voltage I-T-E electrical product line from Gould. In 1992, Cooper Industries bought Crouse-Hinds and sold a portion to Siemens, which called it Murray Electrical Products. Siemens eventually changed Murray breakers, alas, to the standard steel thermal-magnetic design.

The trail gets still more complicated. In 1991, Hubbell bought Bryant. Subsequently, Cutler-Hammer and Westinghouse and Challenger merged; the new organization included Bryant and Zinsco among the subsidiaries. The Federal Trade Commission required the company to spin off much of Bryant. Thomas and Betts (T&B) acquired Bryant loadcenters and Zinsco breakers. They redesigned the loadcenters to take Zinsco main circuit breakers, but Bryant branch circuit breakers. Bryant circuit breakers continued to be made by Westinghouse. T&B now also owns the Challenger line.

Is your head spinning yet? When you are faced with the problem of replacing a circuit breaker in an old panel, there are practical answers. For instance, if the panel requires Bryant "BR" breakers, you still can obtain breakers with that designation, even though the freestanding Bryant Electric Company that may have manufactured the panelboard is but a memory. Which company owns the right to employ that mark is nothing for you to worry about, unless you have reason to believe someone is trying to sell you a counterfeit—a thoroughly unlikely occurrence.

Some brands are no more. Panels manufactured under the names Frank Adams, General, and General Switch are orphans. They seem to have been manufactured, under license, to Westinghouse specifications. Therefore, using "interchangeable" breakers in them is possible but iffy. If such a breaker explicitly says that it is NRTL-classified for use in your particular panel, you are, of course, on solid ground. The circuit breaker in your hand may indeed have a different internal design than that which was originally intended for the panel, but there are a manufacturer and a testing laboratory behind you asserting that its tripping curve is equivalent to the original, and its interrupting rating is as good or better.

Square D's plastic-enclosure, cut-rate loadcenter, the Trilliant line, was available briefly in the early 1990s, but production was discontinued because of low profits. Parts, however, are still available through Square D's authorized distributors. FPE panels no longer are made, but another company acquired the right to make the circuit breakers, which continue to be available—although rather pricey.

GFCIs To complete the discussion of protective equipment, it seems reasonable to shift a little further downstream and look at ground fault circuit interrupters. GFCIs go back further than many realize. Of course, reliable GFCIs are a relatively new development, vast progress having been made between 1980 and 1990. GFCIs were used in Europe beginning in the 1950s. The GFCI was marketed here in 1962 as a self-contained unit, to be wired in series with a load. The first circuit breaker type was introduced around 1968, and the first receptacle type in 1972. This gives you the early end of the range to use in dating GFCI installations.

Next comes the history of Code requirements for GFCI protection, to help you with the question, "Was this [non-protected outlet] legal when originally installed?" The following is far from a complete list:

☐ GFCI protection for underwater pool lights first was required by the 1968 NEC.

☐ With adoption of the 1971 NEC, it was required for storable pools, for pool receptacles, and for some feeders.

- In 1971, it became an option for protection of receptacles at construction sites.
- In 1973, it was required for outdoor receptacles at dwellings.
- With adoption of the 1975 NEC, it was required for receptacles in dwelling bathrooms, and for fountains.
- With adoption of the 1978 NEC, it was required for receptacles in dwelling garages; for bathroom and outdoor receptacles in mobile homes and recreational vehicles; for receptacles in recreational vehicle parks, marinas, and boat yards; and for receptacles in health care facilities.
- With adoption of the 1981 NEC, tub motors and portable signs were included.
- The 1984 NEC explicitly covered replacement receptacles, and also added hotel bathrooms.
- The 1987 NEC added receptacles in commercial garages, dwelling basements, and boat houses; spray washers, and receptacles serving kitchen counters near sinks.
- The 1990 NEC added photovoltaic roof arrays, receptacles serving kitchen islands, receptacles in crawlspaces, and receptacles near sinks in mobile homes and recreational vehicles.
- The 1993 NEC added receptacles in nondwelling bathrooms, on rooftops, in elevator pits, and serving wet bar sinks in dwellings.
- The 1996 NEC added receptacles in trailers, receptacles serving escalators, electric vehicle chargers, receptacles in sheds, balcony receptacles, and receptacles serving previously exempt kitchen counters.
- For readers of future editions, the 1999 NEC may add neon transformers' secondaries, and also accessory buildings with floors at or below grade that are not intended as finished, habitable spaces.

Lights The rest of this chapter is primarily of historical interest, with little direct practical application: something like what a journalist might call, "deep background." Still, a customer might be interested in knowing that, as mentioned below, her wall sconce with integral receptacle might date back as far as 1927.

To start, here is a brief history of the first type of equipment to utilize power wiring—lighting.

Gas lighting There was a considerable overlap between electric lighting—especially arc lighting—and gas. Oil lamps were replaced by gas lights over a period extending from the 1830s to the 1840s. The gas supply initially depended on water dripping onto calcium carbide in acetylene generators located in basements. Gas mains came later.

Before commercially generated power became widely available to serve gas igniters, a pull cord would open and close the gas cocks, while also striking a spark to ignite the jet. Later, inventors developed a battery-operated pushbutton that performed the same functions, and thus "electrified" the gas lights. One leg of the battery was grounded to the gas pipe; the other connected to the buttons. The white button used an electromagnetic vibrator to ratchet the valve open and spark ignition; the black button pulled the valve closed.

In the 1890s, gas lights were overtaken by electric lights: not the lighting with which you are familiar, but arc lighting. If you have ever had the unnerving experience of seeing electrical equipment arc, or if you've arc-welded, you know how bright such an arc can be.

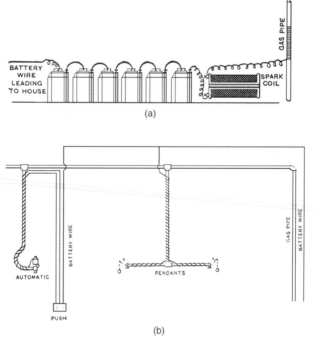

(a)

(b)

■ **13-2a,b** *Gas-lighting circuits.*

In 1876, the year after Samuel F. O'Reilly invented the electric tattoo machine, Paul Jablochkoff invented an alternating current arc lamp. Its two carbons burned evenly, unlike the electrodes in the DC version, whose cathode was consumed more rapidly than its anode. His arc lamp was installed in Paris in 1878, with the invention of an efficient AC generator. The year before that, 1877, marked the invention of hard-drawn copper wire, which began to replace iron wires. About this time Hertha Ayrton figured out how to shape the carbons in DC arc lamps to minimize flicker. This earned her election as the first female member of the Royal Society and the first female Fellow of the Institute of Electrical Engineers. Incandescent lighting did not take over from gas till about 1910. The arc lamps then in use were ten times as energy efficient as the initial incandescent lamps.

Between 1880 and 1910, gas and electric lighting coexisted. In 1905, the downward-pointing, direct-light gas mantle was invented, giving gas lighting its last wind. Dual-source fixtures provided the reliability of gas to back up the uncertainty of electricity generators.

Gas lighting still is in limited use. According to one source, street lights in some Texas, Missouri, and Louisiana towns were using gas as of the late 1980s. In the northeast, gas companies still maintain residential lawn lights using gas mantles. Normally, these lights are unmetered, and customers are charged a fixed monthly fee.

Lights as you know them Here is a laundry list of dates and facts related to modern lighting.

- ☐ In 1879, Edison lit his first tungsten-filament lamp, according to a history by Arthur Ferguson.
- ☐ It was not until 1907, surprisingly, that the first tungsten filament lamp was patented. General Electric bought the rights from an Austro-Hungarian company.
- ☐ In 1911, ductile tungsten filaments were introduced.
- ☐ In 1927, lighting fixtures were appearing with integral receptacles, wired along with the lampholder in #16 or #18 AWG.
- ☐ In 1931, mercury vapor lamps were patented.
- ☐ In 1933, sodium vapor lamps were introduced.
- ☐ In 1945, the first circline fixtures were marketed.
- ☐ In 1949, Soft White lamps were introduced.
- ☐ In 1959—yes, that long ago—halogen lamps were introduced.

- [ ] In 1960, the development of longer-lived mercury vapor lamps led to their adoption for street lighting.
- [ ] In 1961 metal halide lamps hit the market.
- [ ] The sulfur lamp, induction lighting, LEDs, and tritium fixtures are all too recent to be of value in dating older systems.

Here are a few perhaps-surprising facts about the history of light bulbs. Before 1892, light bulbs were created individually by glass blowers, without molds. In 1922, a patent was granted for a light bulb without a tip (tips were used for the vacuum suction). It was not until 1925 that the first practical inside-frosted bulb was produced. Before then, they had been outside-frosted.

Miscellaneous devices, tools, and equipment This section offers a laundry list of isolated facts and dates.

- [ ] Lightning rods in some form go back a long way—to ancient Rome and Palestine, by some measures. (A curious quote from mid-18th Century Moravia, where scientists were experimenting with lightning rods, goes: "Profesora gromom sachiblo"—the professor's been rubbed out by lightning.) UL developed America's first lightning rod standard in 1915, and began field inspection and master labeling of lightning rods in 1923.
- [ ] In 1890 the megohmmeter was invented.
- [ ] In 1896, Harvey Hubbell patented the pull-chain porcelain socket. He followed that with the separable plug. It did not have flat prongs like the cord connectors now used, but round pins.
- [ ] In 1917, snap switches were replacing two-button push switches. (Push the top button in for ON, and the bottom button pops out; push the bottom button in for OFF, and the top button pops out.) That does not mean that, if you find an installation with a two-button push switch, it necessarily dates back to the teens or earlier.
- [ ] In 1935, the household disposal was introduced.
- [ ] In 1936, the electric blanket was introduced.
- [ ] In 1940, home window air conditioners were introduced.
- [ ] Around 1948, Square D and GE introduced low-voltage switching.
- [ ] In 1949—yes, that long ago—heat pumps were field-tested.
- [ ] In 1957 machine tools were controlled by paper tape or punch cards, descendants of player piano rolls and punched-paper controls for machine looms, and antecedents of modern programmable controllers.

292

- [ ] In 1961, Nabisco used computers in a baking plant, and in 1962 they were used to run steel foundries.
- [ ] In 1958, room temperature vulcanizing (RTV) sealants became available outside the factory. Dr. Waldo Semon originally had developed PVC for B.F. Goodrich in the 1920s, while trying to create an adhesive to bond rubber to metal. (He developed many adhesives as well.)
- [ ] In 1960, UL developed its smoke detector standard.
- [ ] In 1961, the NiCad battery was introduced. That same year, the first integrated circuit was announced.

Drills Here is a brief early history of the drill motors on which tradespersons rely so heavily.

In 1868, years before the electric drill made it out onto construction jobs, George F. Freen, a mechanic for the S.S. White company, invented the first electric dentist's drill.

In the late 1800s, shop tools were belt-driven from one big electric motor, often rated as high as 90 HP. In 1870, William or Wilhelm Fein started building electric motors in Germany to drive individual tools. Robert Bosch started as an assistant to Fein. By the early 1890s, Fein's company was selling lathes and drill presses that incorporated their own motors.

In 1895 two of Fein's mechanics, Wahl and Heeb, created the world's first portable electric drill, though electricians relied on the brace and bit well past World War II. Fein was the first to use a universal AC/DC motor on portable power tools, in 1912.

In 1910, Duncan Black and Alonzo Decker set up a machine shop in Baltimore. In 1916, they started building a drill with a pistol grip, a design based on contract machining work they had done for the Colt Firearms Company.

Publications Should you need to perform your own research to more precisely date a particular installation, you have a few options. One involves looking through old books, most especially old editions of the NEC. The other is leafing through old magazines, which provide the most current snapshots of actual field practice. The *Journal of the Telegraph*, arguably the very first U.S. electrical trade magazine, was founded in 1867. The following list should help you trace back two of the major ones that are still around.

EC&M's current owner, Intertec Publishing, Inc., is not the magazine's first or second owner. Its predecessor, *The Electragist*, was first published, by NECA, in 1901. NECA's present organ, *Electrical Contractor*, got that name only in 1971. As of 1917, their publication was called *National Electrical Contractor*. In 1939 they began publishing *Qualified Contractor*, which kept that name for 32 years. *IEC Quarterly* began publication in 1980, so that, alas, will not take you very far back.

Eustace Soares's book, *Grounding and the NEC*, was first published as an article in the November 1959 *EC&M*. (Soares was the outspoken field representative for the Pringles line of bolted safety switches.)

Organizations and companies Just to satisfy stray curiosity, here are the dates associated with the founding of various names in the trade. Many company histories are given in the references listed towards the end of Chapter 14, "Resources."

☐ Klein tools was founded in 1857.

☐ Waldo C. Bryant patented a push-pull switch in 1888.

☐ Minerallac was founded in 1894, selling insulating lacquer and varnish. (Now you know where it got the name.) Conduit straps (and meters, and voltage sniffers, and pulling compound) came a few years later.

☐ Leviton was founded in 1906 as a manufacturer of mantles for gas street lights.

☐ Amprobe was not founded until 1947.

☐ The IBEW was founded in 1893.

☐ NECA was founded in 1901.

☐ The Illuminating Engineering Society was founded in 1906.

☐ NFPA's electrical section was broken into 10 standing committees in 1919. The NFPA correlating committee was formed in 1930. The NFPA reorganized its Electrical Section in 1949.

☐ The Edison Electric Institute was founded in 1933.

☐ The Electric Power Research Institute was founded in the 1970s.

☐ The IEEE was created, by merger of two predecessor organizations, in 1963.

☐ UL listed its first product in 1913, although the first investigation by its founder, William Merrill, took place in the nineteenth century.

☐ The Canadian Engineering Standards Association was

incorporated in 1919. Its first electrical code was created in 1927. It became CSA in 1944.

- [ ] The Association of Independent Electrical Contractors of America was founded in 1957. The antiunion, or at least nonunion, group changed its name to Independent Electrical Contractors, Inc. in 1980.

Great Britain

Each category up till now, if not of some use in dating American installations, at least touched on the history of electricity on this continent. This final section is purely of interest in drawing comparisons: it lists some dates in British wiring history that go back quite a bit earlier than most of the U.S. information.

- [ ] Early in the nineteenth century, electric bells (e.g., call buttons, door bells, and then a little later, annunciators) were used.
- [ ] In 1840, rubber was first used to insulate telegraph cables.
- [ ] In 1845, lead-sheathed cables were first used.
- [ ] In 1847, Gutta Percha Insulation was tried and found satisfactory for insulating telegraph cables under water. It was found to deteriorate when not submersed.

295

- [ ] In 1874, the Phoenix Insurance Company employed Musgrave Heaphy. He developed the "Phoenix Rules," the very first guidelines for safe electrical installations. In 1892 an American magazine, *Insurance*, mentioned the problem of electrical fires in the United States. Such fires were rare in Great Britain; this was attributed to the Phoenix Rules. (By then, American electricians had for some time had rules for safe electrical installations, including inspection, published by Henry Morton. Perhaps the latter were not widely adopted or enforced, compared to England's Phoenix Rules.)
- [ ] From 1880 on, until supply at over 200 volts began to be employed, concentric conductors were used.
- [ ] Starting about 1898, conduit came to be widely used.
- [ ] Still, till about 1914, in England (in parts of Scotland, till about 1930) wood casings were used to separate conductors, because the wire insulation was not trusted.
- [ ] After World War II, the steel shortage in Britain led to the use of aluminum and plastic conduits, flexible armored cable, and T.R.S. cable (roughly equivalent in design to our UF cable).

Resources

THIS CHAPTER COVERS ORGANIZATIONS, BOOKS TO WHICH you can turn for assistance or education, sources for materials, and tools that will make your jobs easier. It also contains references to the handful of legal cases mentioned in earlier chapters, and a few Web sites. Listing here means that an outfit or a tool may be useful, but it is no guarantee. Feedback on your experiences is very welcome and will be incorporated into future editions of this book.

Fixture parts

The following are companies that as of late 1997 stocked replacement parts for old lighting fixtures:

Midwest Lamp Parts Company
3534 North Spaulding Avenue
Chicago, IL 60618
377-539-0628

B&P Lamp Supply, Inc.
McMinnville, TN 37110
615-473-3016

W.N. de Sherbinin Products, Inc.
POB 63
Hawleyville, CT 06440
203-791-0494

Tools

The following tools have been found helpful for use in old work, some of them particularly for use in work on old buildings. The particular tool described is not necessarily the only one that performs a task. If no source is listed for a tool, it supposedly is available through your local electrical distributor.

Hand-held router-type instruments

The Rotozip™ is used to make cutouts in walls or ceilings around electrical boxes. Unfortunately, if used carelessly, it can cut right through plastic boxes. The same caution applies to any small router or Dremel™-type tool.

Armored cable strippers

"A can opener" for armored cable, such as the Rotostrip™, is much less likely to nick conductor insulation than is a hacksaw. (This is no guarantee.) While the handle of such a tool is designed to be rotated 360°, it can be wiggled like the handle of any mechanical can-opener, ratcheting. There is still the question of whether you can fit it into the access you have created to cut back the cable armor. It is better for use out in the open, but then so is a hacksaw.

Reciprocating saws, such as the Sawzall™ and Tiger Saw™, when handled with plenty of care, can be used to cut cable armor as well as wallboard, paneling, frozen screws, and more. A reciprocating saw can be used to cut a box away from its mounting bracket or nails so that you can pull the box out of the wall. Using one to cut away the armor of armored cable, as opposed to simply chopping the cable, takes care. It is easy to nick conductor insulation. For armor removal, a hacksaw or specialized tool is safer, being easier to control.

Hacksaw blade holders fit where a hacksaw will not and are far easier and safer to use than a hacksaw blade held free in your fingers, or wrapped with tape for slightly safer handling. When you have so little cutting to do that a reciprocating saw would be overkill, or when one would be too cumbersome, a blade holder may serve nicely. Bell hanger bits and Kett™ drills have been discussed elsewhere. Shortened drill bits can also be useful for drilling at right angles in narrow spaces. At least one company sells $4\frac{1}{4}$" long spade bits—but many cut them down from standard lengths.

Testers

A receptacle tension tester allows you to definitively answer the question, "Is this 120-volt receptacle too worn-out, too loose to make reliable low-resistance contact with the prongs of a cord connector?" Usually, if you even have to ask the question, the answer is, "Yes." Sometimes, though, checking a receptacle that felt okay, you will be surprised to find when you use such a tester that one prong is not held with adequate force. This tool lets you prove

to the customer—or to the electrician, if you are an inspector—
the need to change a receptacle. If the receptacle does not grip
each prong of the tester with at least 4 ounces of tension, the receptacle fails.

An outlet tester that includes a load simulator tells you when there
is more than 5% voltage drop on a 15-ampere (or 10, or 20, depending on the model) load. A high-resistance connection has far greater
likelihood of being the reason for a "no-pass" reading than too long
a run. That interpretation is especially likely to be correct when
other loads on the circuit have been turned off for the test. The
Philadelphia Housing Development Corporation requires that insulation contractors who are planning to blow loose fill into old attic
crawl spaces confirm that wiring through those spaces shows less
than a 10% voltage drop with a 15 ampere simulated load. Before
they established this criterion, their buildings suffered a handful of

■ **14-1** *Tension tester*

fires associated with hot wires in blown insulation. In two years subsequent to its adoption, there were none. How about polarity checking? How about fault simulation? Receptacle testers that incorporate GFCI testers can be useful, but a solenoidal voltmeter such as a Wiggy™ can do all that. It even should reliably trip a properly-calibrated Class A GFCI.

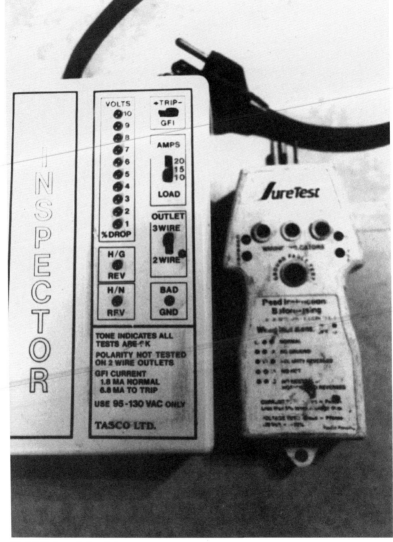

■ **14-2** *Load simulators.*

A megohmmeter will give you some clues as to the condition of insulation and the presence of shorts. Before using one, you should disconnect anything more complex than snap switches and receptacles. A Megger®—or a thumper—could damage any electronic equipment, any equipment containing capacitors, anything with printed circuits. It even might injure lamp filaments.

A circuit tracer is a fancy variant on a current sniffer. The term "current sniffer" refers to a noncontact device that detects the presence of power, even inside a wall. A circuit tracer allows you to insert a unique signal into your wiring by plugging a sender into an outlet. The handheld, noncontact receiver then identifies that signal, so you can follow the circuit through the walls to outlets and junction boxes—even buried ones.

Calling the following type of device a tester may be loose usage. A radio or other 120-volt noisemaker can be invaluable, however you categorize it. This is especially useful as a "helper" to tell you from several rooms or even floors away, "The circuit's on," freeing up any flesh-and-blood assistants for other tasks.

Case history

The following example shows the value of having the right tools for a troubleshooting job. In this case, these tools were an inductive ammeter and a load simulator for checking voltage drop.

A customer complained of the cold in her house. For what she considered good reasons, she was not using the furnace, in Northern Virginia in mid-December, but was relying on portable electric heat. This was in a house where the upstairs was served by one 15-ampere circuit.

The reason the customer shivered was that the circuit breaker for the heater she carried around to wherever she was working kept tripping. The reason the electrician made it a priority call is that the customer said that at first the problem occurred when the heater was plugged into one particular outlet, but not others; then it happened at a second outlet, and then at all outlets on that circuit. She said that at first it would happen only after the heater ran for a long period, then later it began to happen much more quickly. It seemed possible that a bad connection had deteriorated, creating the risk of a fire.

The first thing the electrician looked for was damage at the outlet that initially caused a problem. That was guesswork, albeit reasonable guesswork, rather than thinking things through. Strike one: the wires and connections at the outlet looked, felt, and smelled okay.

The heater showed no damage. Then he went to an orderly approach. This meant first visiting the loadcenter in order to come up with a decent guess as to which of the affected outlets was furthest downstream, based on the house's geography. While at the loadcenter, he took a quick look at the breaker, the hot wire, and the return conductor associated with the circuit. They looked fine, and the terminal screws were snug.

Now he enlisted the first of his powerful tools. He went back upstairs to what appeared to be the end of the run, which also happened to be an outlet associated with the tripping. Using a load simulator, he learned that it did not show an untoward degree of voltage drop. This indicated several things, assuming that the tester was working right:

☐ There were no bad connections between outlet devices and conductors;

☐ There were no bad splices;

☐ The receptacle into which the tester was plugged had enough spring pressure against its prongs;

☐ And the connections of the branch circuit back in the loadcenter were adequate.

True, connections change as they heat up. Still, for a first model of the system, the indication was that all these were good.

Next he snapped an inductive ammeter around the hot conductor back in the loadcenter. It showed a load of about 1 to $1\frac{1}{2}$ amperes being drawn, between the upstairs lights, clocks, and a laptop computer charger. Then the 1,000-watt heater was added. The reading shot up to 15 amperes before the breaker tripped. He duly informed the customer of two things. First, he explained that her unlisted, so-called 1,000-watt heater was actually more like a 1,700-watt, maybe 2,000 watt, heater. So maybe the custom heater designer she had patronized, because he appreciated her beliefs about indoor pollution, was not such a great person with whom to do business. Second, he explained that her breaker appeared to be dysfunctional. This time, he delved further, disconnecting the wires, and pulling the breaker out of the panel. Indeed, there was no distortion and no discoloration, and the wires looked fine. But the unit smelled of burned...something. Plastic, perhaps.

The point of this story is that the troubleshooting process could have taken far, far longer without these tools. Sure, the electrician might have made a lucky guess. A thoroughly ignorant per-

303

■ **14-3** *Edison-base adapter. These used to be far more common when, as in this 1918 illustration of an obsolete version, there were very few wall receptacles—and even some of those were Edison-base outlets.*

son might have automatically replaced the breaker because it tripped, based on the treacherous reasoning that messengers bearing bad news deserve to be shot. Of course, such a person might have replaced it with a 20 or worse. Another troubleshooter might have insisted that the problem was that it is foolish to use any heater on a 15-ampere circuit serving additional loads and that the customer needed another circuit or two installed. Unfortunately, if another circuit were added, some danger would still remain. A bad breaker is not going to heal itself. Also, in this instance, the customer would not have considered additional wiring, for certain idiosyncratic reasons associated with health concerns. Finally, she needed to learn that the heater was at least in part responsible—and before being sold new circuiting.

Adapters

On to more tools. A receptacle adapter, for screwing into medium-base lampholders, can be handy. It provides a convenient way to power a drop cord in a room that lacks working receptacles. Furthermore, it enables you to use your 120 volt noisemaker to tell you when you have killed a lighting outlet.

Three-prong adapters, or "cheaters," have two uses. First, they can allow you to run a grounded tool safely for a few minutes without changing a receptacle, presuming that the outlet box itself is grounded. Second, they allow you to plug in testers set up for three-prong operation.

GFCI extension cords and receptacle adapters are very important. (Take note of the recalled ones mentioned in Chapter 17, "Recalled equipment"!) It is easy for an incautious worker to get fried while working on old houses, These tools will protect you as you work to protect others. The GFCI receptacle adapters that you should seek have retractable grounding prongs for use with old, two-prong receptacles. This makes a more secure connection than you would obtain by using a three-prong adapter in combination with a GFCI tester. On the other hand, the grounding that can be provided by that three-prong adapter makes your GFCI doubly safe when used at a two-prong receptacle with a grounded yoke. Note that a super-strict reading of NEC Section 305-6 means that you and your employees must always use GFCI protection for your hand-held power tools when doing electrical work, even repairs. The reason is that while you are working on the building, it can be classified as a construction site, with its wiring considered, by a stretch, "other than permanent."

Fishing rods

There are at least two brands of slightly-flexible, extensible, rods, with grabbers, used to push or to reach for cables in walls and ceilings. You certainly can come up with kludges to use for that purpose, but the right tool is not terribly expensive, holds up better, and often is not only more efficient but safer.

Personal protective equipment

Old buildings can pose all sort of dangers. Just because you are not working under cranes and riveters does not mean that you can ignore the need for safety equipment. Here are some simple examples. If you are inside an old house, perhaps you do not need a hard hat to protect you from bricks dropped 20 stories. But you still can bean yourself solidly on a low beam or gash your scalp on the underside of a cabinet. Perhaps no one is mixing concrete near your face, or gluing plumbing pipe in the same room, but on the other hand, you may be crawling through rockwool; that could be at least as harmful to your lungs. Chapter 17, "Hazards and benefits," further underscores the importance of appropriate lung protection, hearing protection, eye protection, head and foot protection, and skin protection. It has to be fitted to the worker, who needs training in its inspection and use. And it has to match the specific hazards of the particular work environment. All this may amount to very little fuss on many jobs, but it cannot be ignored just because you "are not on a construction job."

Materials

Certain materials are of particular value in doing old work, especially work on older buildings.

Generic items

Long screws such as flat-, round-, and pan-heads 6-32 and 8-32 by $1\frac{1}{2}$" and even by 2" and 3" will let you fish easily through the ear of a device such as a switch or receptacle, plus a box extension, to find a box illegally recessed deep in a wall or ceiling. Once the screw is started in the recessed box, it often can be used to pull the box forward.

Box extenders, plastic and metal, are readily available. Even if designed for use with single-gang boxes, with care they can be used with too-shallow single-gang device rings. Moreover, where you need to bring a multigang box forward, the AHJ often will accept your modifying box extenders rather than fabricating an extender from sheet metal.

Caulking compound, silicone usually, or silicone-fortified latex for water cleanup, is also known as RTV or Room-Temperature Vulcanizing compound. It serves as a glue and as a sealant, but unlike duct sealant it loses pretty much all plasticity as it sets. Oil-based and even straight latex caulks have too short a life span to be employed in permanent installations. Construction adhesives, which are not all that different from RTV, could prove mighty handy in securing equipment in odd locations and to impenetrable surfaces. Unfortunately, because these are not a traditional type of hardware like nails and screws, and because they have not been specifically investigated for your purposes, if you try to rely on them you are liable to meet inspector resistance and risk getting your installations turned down. On the other hand, you are less likely to get into trouble if you simply use adhesives to supplement screws and anchors.

Some industrial and commercial settings, some historic buildings, and some other special locations will be quite difficult to light, or at least to relamp. Even removing lenses or positioning scaffolds to gain access for relamping can disrupt work—or delicate finishes—enough to give some customers conniptions. In these cases, you have options. One is the use of light guides or fiber-optic lighting, so that relamping can take place in an innocuous location. A plus is that light guides and fiber-optic cable will not transmit heat or, indeed, anything but the kind and amount of radiation that is desired. A second option is induction lighting, which, because the lamp is a completely sealed system, can offer lamp life of a decade. Your favorite supply house may be able to obtain either system for you, though neither is going to be a stock item. A third high-efficiency, long-life, low-heat option likely to become available is LED lighting. With the 1997 development of true-blue LEDs, white LED light should be possible.

Sources for specialty items

Suppose you have 60°C wiring, and you want to install a fixture that specifies 90° wiring? There is a distant cousin of loom that will do the trick. A braided fiberglass insulating sleeving will slip over either a bare conductor or existing insulation and raise its temperature rating through the roof. It is manufactured by Hilec, Inc., under UL category UZIQ2, "Component Sleeving, Flame-Retardant," and sold in inexpensive 100-foot lengths sized by the AWG of the bare conductor it will fit around snugly. With a 1200° (F, but who cares?) temperature rating and a 250 volt dielectric strength, it could come in quite handy. Hilec can be reached at POB 157, Eleven Railroad Avenue, Arcade, NY 14009. Their telephone number is 716-492-2212, and their fax number is 800-450-8193.

Sources that advertise obsolete breakers (as well as motor controls and other equipment) include The Electrical Outlet, 800-227-5731; and Romac, 805-323-0896.

If you are upgrading an existing commercial or industrial site and now are required to add emergency lighting, one option is to install battery packs in existing fixtures. Bodine (800-223-5728), for one, manufactures supplementary ballasts for both compact and tubular fluorescents that include NiCads and inverters in the package. The adapters are installed in parallel with the regular ballasts.

If you need to add exit signs, one option to consider where there is no wiring in place is to use self-luminous signs. These are powered by sealed tubes of slightly radioactive tritium (heavy hydrogen) gas, but, like smoke detectors (which contain americium), are Listed and approved because the radioactive material is not considered a danger. Two manufacturers of these units, which take no wiring and are good for ten years or more before renewal, are Self-Powered Lighting, Inc., 914-592-8230, and Brandhurst, Inc., 914-878-2033.

Researchers

If you have questions about historic preservation or landmark buildings, there are any number of magazines and newsletters to consult—regional, national, and international. Most of them can be found, along with loads of other reference material, at the library of the National Trust for Historic Preservation. The librarians there can be quite helpful, too. They are located at the McKeldin Library, University of Maryland at College Park, College Park, Maryland, 20742. Their phone number is 301-405-6320. Their e-mail address is <NT_Library@umail.umd.edu>. Their web site is //www.itd.umd.edu/UMS/UMCP/NTL/ntl.html.

Other experts can be found at the University's graduate program in Historic Preservation. Its phone number is 301-405-6309. Your local university might offer similar resources.

Another resource is Building Conservation International, a not-for-profit educational organization that can be reached at 215-568-0923. Its founder, while not a hands-on electrician, is a second-generation electrical contractor and a museum lighting specialist.

Legal cases

A number of legal cases have been mentioned in the course of this book. If you wish to look them up for yourself, the following citations may help.

☐ The case over packing the NFPA to keep RNMC out of the NEC was Allied Tube and Conduit Corp vs Indian Head, Inc. (Carlon), #87-157. It was argued in front of the U.S. Supreme Court 2-24-88, and decided 6-13-88.

☐ This is not a court case, but the following citation is mentioned in Chapter 11, "Commercial Issues." Union Cheese Company's factory in Sugarcreek, Ohio, is the one that was inspected in September 1994 by the Food and Drug Administration and found to be very unsanitary.

☐ Bridgestone Tire and Rubber fought their OSHA citations resulting from employee deaths alleged to have resulted from inadequate observance of lockout/tagout requirements. As of this writing, the cases had not reached the courts. An independent judge reviewed the citations and confirmed the guilty finding, though reducing the level of citations and fines. Appeals should stretch at least into 1998 before even leaving the OSHA Review Commission for Federal Court.

☐ Finally, the case that established that "Let the buyer beware" does not generally apply to the relations between tradespersons and their customers is Vanderschrier vs Aaron, a 1957 Ohio state appeals decision. The formal citation is 103 Ohio App. 340, with the parallel citation 140 N.E. 2d 819 (meaning that to find it you would look in Volume 103 of *Ohio Public Decisions*, starting on page 340—or in Volume 140 of *Northeast Decisions*, 2nd series, starting on page 819.)

Publications

Most of the following publications served at least in some small way as sources for Chapter 13, "Dating and History," and Chapter 10, "Historic Buildings." A few are just curiosities.

Books

This section will start out with books, or at least monographs (single documents halfway to being books). Any that no longer are in print may be obtainable from a library.

Electricians' handbooks

Some readers may not be experienced electricians. For those newer to the field, here are standard resources that many or most electricians use.

The *National Electrical Code* (NEC) is obtainable at any supply house, or directly from the National Fire Protection Association, the NFPA, 800-344-3555. It is your bible; no book on electrical work should be used without having the NEC alongside for reference.

The *NFPA Inspection Manual* has one short chapter on inspecting electrical systems.

NPFA 73, informally known as the "reinspection code," offers few answers to the sort of judgment calls that face an inspector or repairer, but does at least offer general guidance. Quite a number of people have expressed frustration with the lack of specificity in at least the 1994 first edition.

The NFPA is preparing their own, reportedly massive, *Glossary of Terms*. While it will not be devoted to electrical terms, it too may be of value. According to the February, 1997 *NFPA Codes and Standards News*, the various NFPA Codes included about 7800 definitions as of Fall, 1996; this document will correlate and consolidate them.

The Institute of Electrical and Electronics Engineers, Inc. (IEEE) publishes a number of design books that, while not specific to old wiring, do come highly recommended. These are two.

The formal title of the IEEE green book is *Recommended Practice for Grounding Industrial and Commercial Power Systems*.

The formal title of the IEEE gray book is *Recommended Practice for Electric Power Systems in Commercial Buildings*.

Before leaving the IEEE, the new *IEEE Standard Dictionary of Electrical and Electronics Terms* (including abstracts of all current IEEE standards) may well serve as a starting point from which you can identify other resources. "Triplen," for example, is defined there, but in few other dictionaries.

Contact the IEEE Service Center, 445 Hoes Lane, POB 1331, Piscataway, NJ 08855. Their phone number is 800-678-4333. Their email address is <member.services@ieee.org>

The International Conference of Building Officials (ICBO), publishers of the *Uniform Code for Building Conservation*, is at 5360 Workman Mill Drive, Whittier, CA 90601-2298. Their phone number is 310-699-0541.

UL product directories do not substitute for product standards, but do give a little information about what uses products are designed for. They also tell you what manufacturers make a particular category of product you might be seeking. You may find their green book, the *Electrical Construction Materials Directory*, and their orange book, *Electrical Appliance and Utilization Equipment Directory*, particularly useful. These annual directories can be ordered from Underwriters Laboratories, Inc., 333 Pfingsten Road, Northbrook, IL 60062. Their phone number is 312-272-8800.

The American Electricians' Handbook, by Croft and Summers, is perhaps the most exhaustive, if not the most accessible, guide to electrical installation.

Practical Electrical Wiring, by Richter and Schwan, is probably the most-accessible, best-written introduction to wiring for relative novices. You certainly need to master most of its contents before getting full value out of the book you are now reading. Incidentally, a related volume, *Wiring Simplified*, does not earn quite the same endorsement. Still regularly updated by the erudite Schwan, it is well-written, accurate, and accessible. But it is slim: despite the fact that *The Essential Whole Earth Catalog* review calls it comprehensive, it is not.

The Soares Book on Grounding, by J. Philip Simmons (because it is now published by IAEI), is a valuable resource to people who have difficulty understanding NEC Article 250. The sixth edition is a bit clumsily written, and only moderately well illustrated. Even so, like Earl Roberts's slim book, *Overcurrents and Undercurrents: All about GFCIs*, it has definite value. Soares's 10-page appendix gives information on the history of grounding practices that you will not readily find anywhere else.

John E. Traister's *Illustrated Dictionary for Electrical Workers* is not exhaustive, but you may find it useful. It does, however, contain assertions that some would question.

The very knowledgeable Joseph A. Tedesco has worked on a number of Code-related books. In addition, he self-published a small book in 1990 called *Basic Checklist for Building Electrical Inspections*. It may be hard to find.

The National Electrical Manufacturers' Association, Inc. (NEMA) publish *Guidelines for Handling Water-Damaged Electrical Equipment*, which you should take under advisement if you have occasion to deal with old systems that have suffered from flooding.

As a transition into purely historical pieces, here is another book that might be worth looking at. *Mechanical and Electrical Systems for Historic Buildings* has a 16-page chapter devoted to electrical systems, plus other scattered bits of interest to electricians. It is written by Gersil N. Kay, the founder of Building Conservation International, which is mentioned above. She knows what resources are available for historically accurate restoration. For example, she mentions that Otis Elevator will sell parts for any equipment they manufactured from 1852 forward. Aside from the useful tidbits, you may not be impressed with the way the book turned out; neither is the author. To her mind, its greatest strength is the 100-plus pages of diagrams showing the hollow spaces in the walls of antique buildings, which can be used for fishing.

Histories

The following books range from straight chronologies to corporate puff pieces to curiosities.

The armored cable people In 1984, AFC/Nortek published a book called *Electric Wiring and Lighting in Historic American Buildings*. Out of print for some time, you may be able to find it in engineering libraries. Co-authored with a consulting firm, The Preservation Partnership, it both offers a considerable amount of history and provides recommendations for repair work. It is highly worth reading. One of the many valuable points its authors make is that some insulating materials were appropriately applied in dry locations, but deteriorated later as the buildings aged and moisture penetrated. The same, alas, can be said of materials other than insulation. See Illustration 6-1 for an example.

AFC makes the valid point that for fishing in historic buildings, type AC or MC cable is less subject to damage than type NM. (That is also true of Type MI, but the latter is less flexible than armored cable, and relatively expensive.) They also argue that because it is more rigid, armor can be pushed through obstacles more easily, and they claim that it literally can be screwed through some obstructions.

Luminaires The next book could serve as a practical reference if you are restoring historic buildings. Nadja Maril (POB 6180, Annapolis, MD 21401) is a lighting consultant and antique fixture dealer. Her book, *American Lighting 1840-1940* (Schiffer Publisher, 1995) shows the fixtures associated with various periods, including appropriate shades.

You get a little history and a lot of pictures, pretty pictures. Unfortunately, as a moment's thought makes obvious, a book covering that long time span is not exclusively devoted to electrical equipment. There are perhaps 40 pages that talk about and show electric or electrified chandeliers. The rest is fixtures burning whale oil, lard, gas, or gasoline, plus table lamps and shades.

A frolic While the book you are reading is written primarily for professionals who want practical information showing how to safely and legally work on old wiring, you have encountered numerous illustrative anecdotes and a little pure diversion hearkening back to the way things were. Some reminiscence is not out of place in a reference book; for more, try autobiography. *Sparks: Memoirs of a Tradesman* is 250 pages of unadorned, gabby, old-time electrician. Norman Koch was not a polished writer, but this 1984 Vantage Press book tells it like it was from the middle of the twentieth century up through the 1970s. An author-subsidized publication, it suffers from minimal editing and sketchy proofreading; it may be offensive in spots. But it is fun.

Just the dates, Ma'am In 1946, NEMA published a book called *A Chronological History of Electrical Development from 600 B.C.* It is interesting, but will be of limited value in dating systems and components you encounter.

Here are some tastes from that tome. The first commercially successful electric motor was invented by Thomas Davenport in 1837. The first hydroelectric plant lit incandescent lamps in Appleton, Wisconsin in 1882. Fired up twenty-plus years after vulcanized rubber insulation was invented, this installation still used bare wire. The first electric power used in homes was battery power, running sewing machines in 1886. What about the 600 B.C. in the title? That was when the Greek, Thales, first documented static electricity.

Standards savvy *This Inventive Century*, by Norm Bezane, is a UL publication stuffed with curiosities. In 1893, electrical inspector William Merrill was called on by the City of Chicago to do the deed that led to his forming UL (initially, the Underwriters' Electrical Bureau) the following year. The City Fathers needed him to evaluate electricity's responsibility for the fires plaguing the Palace of Electricity at their "World's Fair," the Columbian Exposition. Alas, the volume does not report Merrill's findings.

Despite omissions, the mini-book is fun. Thirty-one of its 57 pages are devoted to pictures and paragraphs celebrating UL's introduction of Listing for then-new products. Products-plus, actually. On page 38, you learn that not only did UL certify airplanes—and cars, for a while—but also pilots. You probably knew that they test extinguishers to make sure that they will put out fires. You may not realize that they also test safes and bank vaults, to make sure that burglars will be put out by being kept out.

After the pretty pictures and fun blurbs comes a parochial UL chronology. Readers are on their own if they want to see how UL's landmarks line up with respect to other events in the history of this industry. Mainly you learn when UL first investigated various categories of product or opened new branches. UL started simply inspecting and Listing products (and, presumably, "Recognizing" components) in 1913. Until then UL did "Approve," not merely "List," products that passed its tests.

In the late 1960s, Underwriters Laboratories, Inc. also published a pamphlet, called *A Practical Miracle*, extolling a decrease in electrocutions despite the increased use of electricity. It includes a number of statistics.

An even curioser curiosity *The Electric Chair, Hot Seat of Controversy* was published in 1992 by Forces, Inc. (31 W. 350 Diehl Road, Naperville, IL 60563). Besides giving a peculiar and amusing look at a piece of electrical history, it highlights the greed and infighting that developed between the forces supporting Edison and Westinghouse, or DC and AC, forces in the early years. A historical curiosity, it includes the list of 18 words that were developed in the late nineteenth century for what people now call electrocution. They are (and no, the word forms are not parallel): electromort; thanelectrize; thanatelectrize; thanatelectresis; electrophon; electricise; electrotony; eletrophony; electroctasy; electricide; electropoenize; electrothenese; electroed; electrostrike; fulmenvoltacuss; and, finally, electrocution. Ouch.

Edison's heirs Another source of fascinating pictures and other antiquarian information is G.E.'s Elfun Society of Schenectady, N.Y. In creating the Edison Illuminating Company and then G.E., Edison and Steinmetz made great contributions to the development of the industry. The Society has information from the very beginning of wiring up until, barely, World War II. And the stories can be fascinating. For instance, their John Anderson is the person who described "undertakers' wire." They do charge for research and for duplication of materials from their archives.

Volumes I and II of "The General Electric Story" are: *The Edison Era, 1876–1892* and *The Steinmetz Era, 1892–1923*.(Volumes III and IV are material that only a historian of business management could love.)

Both books I and II are abundantly illustrated with portraits and (monochrome) period photographs. There are useful, albeit not annotated, bibliographies. The Edison volume has a list of historic sites and a selection of pungent quotations; the Steinmetz, a page of epitaphs. What about the stories? These are clearly chronicle rather than history. Both are much more history of the company and its antecedents than history of electricity, or biography.

Lamps for a Brighter America: A History of the General Electric Lamp Business by Paul W. Keating (N.Y.: McGraw-Hill, 1954) is fun, well-written, somewhat pious, and out of print.

The folks who produced the first noninterchangeable breaker G.E. was far from the only manufacturer to arrange to have its history published. *Square D Company Yesterday and Today* is an elegant volume that was published in 1988. Founded in 1902 as the McBride Manufacturing Company, the outfit underwent many name changes and mergers and acquisitions. What not everyone knows is that its business changed radically. In 1905, it introduced an enclosed fuse, and in 1908 changed its name to Detroit Fuse and Manufacturing Company. It then secured the rights to manufacture an enclosed switch developed in Britain. This led to its main claim to early fame—the safety switch with the square D on the cover.

In 1917 it sold its fuse business and, bowing to popular usage, changed its name to Square D. It was not until 1929 that it started producing Westinghouse circuit breakers under license. In 1951, it manufactured the first panelboard using plug-in breakers, and in 1955 the QO breaker designed to be noninterchangeable with other brands. The "visi-trip" feature was not introduced, though, until 1967. That is the sort of thing you will get from this volume, along with information on Square D's finances and management.

A handful of other histories *Fifty Years of Electricity: The Memories of an Electrical Engineer*, London and N.Y.: The Wireless Press, Inc., 1921 by Sir John Fleming, the man who developed the vacuum tube, is British more than it is American in focus. It is well written, with technical details and decent plates, starting with the period before the advent of telegraphy.

The next volume is highly recommended to history buffs: *The Electrical Manufacturers 1875—1900*, by Harold C. Passer (Cambridge: Harvard University Press, 1953). The subtitle is *A Study in Competition, Entrepreneurship, Technical change, and Growth*. The writer is a former Navy electronics specialist and a member of the Harvard research center for entrepreneurial history. The volume gives details of business/money/management factors but does not neglect inventions and technical improvements. The bibliography is excellent. The focus is mainly on the U.S., with bare nods to European developments. It extensively covers lighting, generating and transmission, motors, and railways.

Percy Dunsheath's *A History of Electrical Engineering* (London: Faber and Faber, 1962) offers a collection of personal biographies, somehow more intimate than those that G.E. sponsored; detailed technical description from someone solidly in the field, a là Dr. Fleming; and a longer perspective than any of the other volumes, starting with a Chinese magnetic compass of 2637 B.C.E. It is delightful to read a thoroughly British perspective on the history of electrical development. It includes passing mentions of what was happening in America and a bit more about developments in Germany and France.

Finally, like many organizations, the International Brotherhood of Electrical Workers has published its own history, in a couple of versions. *The Electrical Workers' Story* is a 29-page booklet containing 10 pages of history, including some dramatic pictures. *International Brotherhood of Electrical Workers History and Structure* is about 40 pages, of which 19 are history. Although thoroughly focused on the organization, they do provide a worthwhile perspective.

Those are all the books and booklets there is room to talk about here. There are of course hundreds or thousands more for you to discover, if you want to explore the background of your trade.

Articles

Before closing with a few how-to pieces, here are some articles that will provide lots more information about electrical history.

The May, 1991 issue of *EC&M* was devoted to a history of electricity. Not only did it go way back to nod at Chinese and Greek experiments with amber and lodestone, it also gave capsule histories of a goodly number of companies that were and remain significant players in the electrical industry.

Their July, 1987 issue had a shorter, somewhat similar feature, with lots and lots of pictures of early versions of electrical products.

The March, 1985 issue of *Electrical Contractor* had an article on the Burndy Library in Norwalk, Connecticut, which houses many thousands of books in various languages on the history of science, particularly the history of experimentation on electricity and magnetism.

The June, 1988 issue of *Electrical Systems Design* was devoted from beginning to end to a history of electricity, including inventors and companies.

The November, 1990 issue of *IEEE Spectrum* had an article, starting on Page 120, explaining the rationale behind 25 Hertz power.

If you are a P.E., an article called "Electro-forensic Engineering" might be of interest if you want to be consulted as an expert witness. It originally appeared in *IEEE Industry Applications* magazine, 3(2), 46-53, March-April 1997.

The January, 1989 issue of *Electrical Construction Technology* has historical material, including an article by the author of this book called "A History of Wiring," based in part on a number of the sources referenced in this chapter.

For fun, if you have a major reference library handy, you might want to look at the November–December 1889 issues of *The North Atlantic Review*, which feature Edison and Westinghouse arguing about AC versus DC.

John Fidler's articles in the June and July 1988 issues of the British magazine *Traditional Homes* offer useful perspective, including solutions to the problems of upgrading electrical systems under U.K. rules. His suggestion that floor outlets be used to prevent marring the beauty of walls, mentioned in Chapter 10, "Historic Buildings," in the section called "Satisfying customers' aesthetic requirements," is found in the June 1988 issue, 4 (9) pp 102-106.

So much for history. Here are just a few more resources. Amory Lovins and William Browning have an article called "Negawatts for Buildings" in the July 1992 issue of *Urban Land*. It is a useful summary of what you have been reading in industry magazines about reducing power usage. Lovins's organization, the Rocky Mountain Institute, offers plenty more of the same, including a newsletter, position papers, and consulting. Their address is 1739 Snowmass Creek Road, Snowmass, CO 81654-9199.

Handouts

The CPSC's Document #4701 is the "Safety for Older Consumer Home Safety Checklist." You as a professional do not need it to ensure the safety of your customers' electrical systems, especially after studying this book. Still, it is an educational tool that is worth providing or at least recommending to customers. It covers far more issues than electricity and lighting.

The CPSC's Document #4513 is the "Home Electrical Safety Audit: Room by Room Checklist." A good educational tool overall, it conceivably could have an unfortunate affect on some people who read statements such as "If you do not have a 3-hole outlet, use an adapter to connect the heater's 3-prong plug. Make sure the adapter ground wire or tab is attached to the outlet." The main problem with that advice is that while, more often than not, following it will greatly improve matters, without testing to confirm presence of a ground it will in some cases lead to a false sense of security—and increased hazard. A consumer education package that includes this checklist plus an impressive video is available, free, from "Home Wiring," U.S. Consumer Product Safety Commission, Washington, DC 20207.

Web sites

The multipage article, "Reducing the fire hazard in aluminum-wired homes," by Jesse Aronstein, Ph.D., P.E., might be found on the World Wide Web at <www1.mhv.net/%7Edfriedman/aluminum/alreduce.htm>. Friedman, incidentally, is also the author of the IEEE monograph mentioned in Chapter 6, which found corrosion to be a common problem in residential electrical panels. IEEE publications are available through IEEE Customer Service, 800-678-IEEE. If Aronstein's paper is no longer available at Friedman's site, it may be necessary to email Aronstein at <Protune@AOL.COM> for a new location.

The CPSC's site is <www.cpsc.gov>. This is where you can find up-to-date information on product recalls. It also has the text of the consumer safety articles described above.

Additional useful information may be found at other individuals' Web sites, including those of Joe Tedesco: <http://pw2.netcom.com/~joetede/index.html"> and Redwood Kardon: <http://www.codecheck.com>

Finally, information on the Scott-T transformer, used to derive two-phase power from three-phase, may be available at <http://www.LIBRARY.KMITNB.AC.TH>

Training, referral, and consultation

For training in Copalum installation or referral to a colleague in your area who already installs that system, contact AMP Incorporated. Their number is 800-522-6752.

For information on Elan low-voltage controls, call 800-622-ELAN.

For information on Smart House Limited Partnership, Inc., call 301-249-6000.

Product certification

If you have problems because a piece of equipment is neither acceptable as is to the AHJ nor grandfathered, you may need to get it field-labeled. If the representative of a NRTL certifies it, you are set.

You have a growing number of choices. As of late 1996, the 14 NRTLs included the following companies which have some interest in field investigation. All are certified to offer the service; not all necessarily are set up for or actively pursuing such business.

Underwriters' Laboratories Inc. 847-272-8800

MET Laboratories, Inc. 800-638-6057

ITS 800-345-3851

Electro-Test, Inc. 510-824-0330

TUV Rheinland of North America, Inc. 203-426-0888

United States Testing Company, Inc./California Div. 909-624-1244

Communication

AS AN APPRENTICE, YOU MAY NOT HAVE BEEN EXPLICITLY taught how to deal with customers. You can be a perfect electrician from tip to toe, get along fine with suppliers, employees, plan reviewers, and inspectors, and still fail in business if you are unable to maintain cordial yet professional interaction with customers. Effective communication bridges the gap between your understanding and your customers' ideas. It takes two parties— but you are the one who has to do the work if you want the job to go smoothly and hope to get referrals. You are commonly dealing with ignorance, sometimes with suspicion, and on occasion even with muddle-headed arrogance.

Customers with old wiring tend to be more involved

Before you can tackle the special problems and opportunities presented by old wiring, you have to land and keep the work. That is true with any job, but two factors are more prominent in old work than in new. First, old buildings more often are in use while you work. Second, you are more likely to uncover unanticipated problems as you work on their wiring: problems that require you to turn to your customers for decisions unless you want to wing it. This makes effective communication with customers an essential skill.

Attitude

Customers vary all over the map, from those who act helpless— "How bad is it, Doctor? Whatever shall I do?" to those with attitudes: dismissive—"Just fix it and give me the bill" or, worse, antagonistic—"I can't believe it's such a big deal—it was working till just recently!"

There is little you can do to force change in others' attitudes. Your best bet is simply to act in a professional manner. That does not mean standing aloof and ignoring their concerns. Because you know what you are doing, and because customers rarely share your competencies, there is a place for education.

Language

Your first communication problem stems from the fact that customers and electricians have different backgrounds and use different language. This is not an issue that comes up only when dealing with old buildings, but it is particularly acute when doing old work compared to working on new installations. The problem is present on every job, except for those where you are dealing with a competent expert who prepares blueprints and specifications and subsequently approves your punch list. When you take on old residential and small commercial repairs, you normally are not dealing with architects, engineers, and general contractors. You are not working from written specifications, which help level the playing field.

Here is a blunt statement: an expert usually has the better, meaning the more useful, language—whether he or she is a scholar or a salt of the earth. As an expert in wiring, your terms are more precise, therefore clearer, and therefore more suitable for describing what you are trying to get across. This fact points to two wise (respectful) and two foolish (arrogant) options for making a go of this business.

To start, here are some smart options for connecting with customers. If your customers are open to the possibility, experiment with using the proper language as you talk with them, helping them to understand and even learn the correct terms for what you are doing. If customers are not up for that, you have the second option: talk their talk as best you can. Even, and maybe especially, in that case you need to make sure you are understood—that there is a meeting of the minds. If necessary, show them pictures or actual devices and equipment, and walk them over to the physical locations being discussed.

There is also a middle ground. "Conduit" does not necessarily mean anything to a layperson. If you say "conduit, which is what electricians call their special kind of pipe," the likelihood of confusion is much less. When you say, "receptacle," context makes it pretty clear that you are not talking about a trash receptacle. Holding up an example works well, too. "You see, this three-way switch has this green ground screw and three metallic-colored screws on the sides." And hand it to him or her to look over. Pointing at something is not nearly as reliable as walking over, because people sometimes misidentify what it is you are pointing at. (Gee, you knew what you had in mind!)

So much for the two wise options. It is foolish to operate on any basis other than a meeting of the minds. Even when you have signed, very detailed, contracts, if the customer and you do not share the same picture of what you are going to do, this imperfect understanding often will result in grief for all parties. Sure, sometimes customers are sufficiently cowed to pay up. Sometimes they are grateful just to get any work done. But if you want repeat business, if you want referrals, if you want to reduce the chances of being badmouthed, try for something better than that. Repairs on old buildings can mean a lot of referrals and repeat business.

What interferes with a meeting of the minds? One example of foolish behavior is to listen to customers' plaints and say, "Yeah, yeah, I'll take care of it" without making sure you understood what problem they are experiencing or what kind of service they are asking for—not to mention letting them know what you might have to do. Another example is to slap them silly with jargon. Sure, there sometimes is a need to use technical language, terms that are not part of everyone's general vocabulary or that are used by electricians in a special way. But always check with even highly educated customers to find out whether they follow your meaning. Really check. Just asking, "You follow?" sometimes gets a "yes" when it should not. Otherwise, your so-precise language is a weapon, not a communication tool. Oh, it communicates—but what it is communicating is something like "I'm the expert, and I can't be bothered explaining myself to the likes of you."

Most problems with customers have nothing to do with malice but result from the fact that everyone relies upon prior knowledge, and on context, for understanding. If you say, "loadcenter" or "panelboard" while gesturing towards a customer's panel, he or she probably will get your meaning. Over the phone, without the context of your visual cue, he or she might not understand—and might not think to ask. To an electrician, "fusebox" means a panel or cutout box containing fuses. To many customers, even those with modern houses, it means, "Where I can turn the power off." You operate from different assumptions, different knowledge bases.

It is time to get down to cases. What is there for a customer to misunderstand? Try some common terms:

□ cord connector;

□ ground fault (not to mention getting an uneducated person's mouth and mind around, "ground fault circuit interrupter");

□ line-to-line;

□ multiwire.

These are just a start. How many of the items in NEC Article 100 are common-usage English? How many of the items in the NEC index? In the index to this book? You need to make sure that your audience understands what you mean every time you use such terms.

The other end of the transaction is just as bad. What do you suppose it means when someone says, "I think my switch has a short in it"?

The meaning of a statement depends on the speaker. If the speaker is an electrician, even a retired or disabled electrician, you probably can take the information at face value—not necessarily indicating that a switch has bridged contacts, but at least that that is what the speaker believes. Not so with the average customer, or even carpenter. "Switch," in lay terms, can mean what you would call a switch, what you would call a circuit breaker, and even what you would call a receptacle! Ouch. Similarly, "has a short" most commonly means "is behaving some way I don't like."

This is far from the end to parsing—eagle-eying and pulling apart—that sentence. What does "the switch has a short in it" mean? Usually it means "the switch is what I blame for the problem." However, "in it" can also merely mean "seems somehow associated with the circuit I think is involved, though I have no idea whether it's the particular part that's gone bad—nor do I have any valid reason to believe so, now that I think about it."

Table 15-1 does a brief job of untangling some of the loose language you may hear from laypersons. In Appendix 3, "Electrical Myths," you will find a slew of laypersons' misconceptions about electricity and electrical safety. Some of them are important to recognize because they affect customers' sense of the work you need to do. Others are simply there to reveal the gap between your understanding and your customers'. That appendix directly complements this chapter. Whenever you have a chance to educate customers about electrical safety, you do them a service. For instance, always mention overlamping when you observe it, even though it is not your problem.

322

Laypersons' language	Technically precise Language
Box	Panelboard; meterbase
Bulb	Lamp, including fluorescent tube; globe
Cable	Cable; coax; extension cord; video interface; bell wire; cordset
Conduit	Conduit, cable, BX
Doorbell	Bell, buzzer, chime, annunciator, pushbutton, intercom, bell transformer
Electrician	Electrician; estimator; handyman; apprentice; inspector; plumber; HVAC installer; plan reviewer; engineer
Estimate	Estimate; bid; troubleshoot; prepare specifications; design
Flush	Flush, surface
Fuse	Fuse; "minibreaker"; starter; circuit breaker
Insulation	Insulation; sheath; armor; raceway
Move	Move; remove; extend; relocate
Outlet	Outlet; switch; junction box; receptacle; phone jack
Pipe	Pipe (plumbing); conduit; tubing (EMT, flex, or "Smurf™"); cable; surface raceway
Plug	Cord connector; receptacle; switch; fuse; starter; k.o. seal
Short	Short; open; overvoltage; malfunction
Starter	Starter; ballast; disconnect; contactor
Switch	Switch; receptacle; circuit breaker
Three-way switch	Three-way lamp socket
Two-phase	Single phase, 240 or 208 volt; much more rarely, three-phase
Two-way	Three-way
Wire (noun)	Wire; conductor; cable
Wire cover(ing)	Surface raceway; insulation; sheath

323

Getting past language barriers

Customers' muddled understandings can incline an electrician to dismiss presenting complaints with a "Yeah, yeah." Even if muttered silently under your breath, this is a short-sighted response. The sensible response is to ask questions. Phrasing, though, can make a big difference. "Does 'has a short' mean it killed the circuit?" This may not be a good way to ask the question. How is a layperson to know whether a circuit is killed? Often they have only the muddiest idea of what constitutes a circuit! Try again: "When you say it may have a short, do you mean you had to change the fuse or reset the circuit breaker?" "No, never had to do that. Why?" Now you have some real information. On to your next question: "Did it stop working? Did it spark?

If your customer is articulate, there is an even better way to proceed than playing "Twenty Questions." Ask for a behavioral description: "What have you observed and when?"

An answer in behavioral terms, as mentioned in Chapter 2, "Troubleshooting," might be, "the light dims sometimes." It could be, "The light just went out recently, and it couldn't have been the bulb because it wasn't an old bulb." The latter throws in a little amateur diagnosis along with the behavioral description—but at least it adds more information. If the customer tells you that one fluorescent lamp keeps flickering but never quite lights, while the other is just fine, you can begin to suspect an old fluorescent fixture, with two independent lamps, fired by separate starters. (See Appendix 1, "Fluorescents," for more on that.)

Getting things straight from the start

From your very first contact, miscommunications bear consequences. Consider this scenario: a new customer calls and says, "I need you to change my switch; the handle is broken." Your service truck arrives, and you learn that "switch" meant "circuit breaker," and that "broken" meant "in trip-free mode." This could be annoying if you unsuspectingly took the customer at his or her word and slotted the call in as a quick stop between other jobs, not allotting sufficient time for extensive troubleshooting. If the difficulty stems from a defective breaker, there is still no major problem if the breaker is, say, a QO 115 or a BR120. But what if it is a Zinsco, or some other brand you do not normally carry?

A way to minimize that risk is to spend a little time on the phone confirming the nature of the problem. "You mean a piece has broken off the handle, or it's loose, or what?" If the answer is "No, when we plugged in the heater, power went off, and the handle is loose, different from the others in the box" you have been pretty well clued to the likelihood that they are not talking about a snap switch. "You mean the box where all the electricity in your house is controlled?" "Yes...I guess so." "Well, look on the cover, and see if you can find a brand name. It may be inside, on a piece of paper pasted inside the door." Alternately, if you ask, "What does the switch control?" and the answer is not, "A ceiling light" but "The light in the hall, the TV, and we think the bathroom," again you have saved yourself considerable frustration—you can be pretty sure they mean "breaker" when they say "switch."

Here is another communications challenge. You are called to fix a lights-out complaint. Your truck makes it to the job when promised, and you spend a goodly chunk of time troubleshooting. This turns up a burned-out fuseholder. It looks as though it is time to upgrade the service—in any event, the existing service equipment is shot. Pigtailing the wire from the burned-out fuseholder onto another circuit is a temporary fix.

Then comes trouble. You give the customers an estimate. They tell you they will think it over; for the moment, they will live with the temporary fix you have lashed up. "Thanks for coming by," they say, "we'll be in touch." When you hand them the bill, they act outraged. "You haven't fixed the problem. Why should we pay for what you tell us is a lash-up, and for giving you a chance to submit a bid?"

Unless the customer is out-and-out crooked, this problem is even easier to minimize than the previous one, that of the misnamed switch. If customers call you in and fail to ask how you charge, their dissatisfaction with your bill is their look-out, so long as what you charge is in keeping with your standard rates. Still, if you want to minimize such difficulty, it is your responsibility to communicate your terms of doing business. When setting up the appointment, it takes only a minute or two to explain charges and payment terms.

Here is an example of what looked at first like a customer trying to stiff her electrician. It turned out just fine, precisely due to careful, professional communication. The customer, a sharp widow lady, griped at her bill and complained that the people who referred her to the electrician had mentioned a much lower hourly rate (one that was six years out of date). Fortunately, the contractor had both described his current charges over the phone when setting up the appointment and mailed her a copy of his printed "Terms of Engagement" (a document that he found worthwhile to write out some years earlier). Prompted with those facts, she acknowledged their validity, and her grumbles turned out to be nothing more than that—passing grumbles. One reason they got on fine subsequently probably was that the contractor had nothing to be defensive about, because he had been up-front about all charges. So her complaints had neither infuriated nor threatened him.

Estimates are a fact of life. If you do have occasion to give estimates, you may be concerned about a paradox. On the one hand, if you provide insufficient information about how you plan to solve the customer's problems, you could lose the work to someone who will do a far lesser job—at a lower price. On the other hand, you may well be reluctant to essentially draw up specifications that the customer could use to work up a contract with your competition.

What works for some is to give a written estimate with sparse detail, and fill in the details verbally—at least until the customer sits down to sign a contract. For others, estimates are consultation—and it is not appropriate to give away consultation without billing for the time.

Bearing bad news

People may get along like a charm until trouble arises. Remember the customer with a combative attitude, the one who says, "I can't believe it's such a big deal—it was working till just recently!"? Many a time, electricians have wondered how a fixture managed to conduct power to and from lamps as long as it did. The answer usually is that the break in the wire did not matter much, in terms of function versus nonfunction, because the broken ends were still in contact. Mind you, it was a high-resistance contact and could have caused a fire, but by golly the light lit!

The reason for emphasizing that such a fixture functions goes right to the issue of customers' understandings. Some are so fearful of electricity that they will accept your every word as gospel—not necessarily in the etymologically pure sense of good tidings, but in the sense of a not-to-be questioned message. Some others are so indifferent to safety matters, or so suspicious of tradespersons, that any warnings about looming risk will be dismissed out of hand. Among the rest, though, when you find something deadly and fix it, it is important to say, "I may have just saved you from a fire." Drive home the idea that mere function does not prove safety.

Tooting your own horn is fine and has its place, but as a conscientious professional there is a much more important reason to speak up. It is this simple: if one fixture is shot, what are the odds that the rest of the wiring is up to snuff? If those odds are slim, it is your duty to alert your customer to the risk.

What your customer does with your unhappy information is up to him or her. Chapter 6, "Safety Surveys: Looking at the Big Picture," describes a comprehensive safety survey. This is an option you can offer. Other options are addressed in Chapters 5, "Accept, Adapt, or Uproot" and 7, "Setting Limits and Avoiding Snares."

Avoiding information overload

There is yet another important aspect to communicating with customers. Respectful communication means recognizing how much detail your customer both does and does not want. He or she may not wish to hear everything about what you find and what you do. Respect that preference, but, if for no other reason than to cover your own liability, note everything that may be of concern. Note it in writing, perhaps on your invoice, even when you choose not to speak up right at the time you encounter it. You may be tempted to figure, "They can't afford to have me correct this, based on the decisions they've made so far," and simply bypass problems that do not strike you as absolutely urgent. This train of thought and action usurps the customer's responsibility. And making somebody else's decisions for them is not consistent with professionalism.

Incidentally, nothing but good ultimately has come of saying, "It seems to be working now, but I can't assure you that I've fixed it." Sometimes problems will reverse themselves without telling you why. You know that they can easily reverse themselves back again once you have stepped out of the door. Sometimes, you may correct one clear problem but be less than certain that what you have cleared up is responsible—or at least solely responsible—for the symptoms that caused the customer to call you.

"Nothing but good has ultimately come of it" does not mean that your "I'm not sure" will please customers. Certainty is nice, but only fools demand it. And if someone demands certainty, you can offer it...for a price: rip out the old and install everything new. That job you can guarantee.

To further explore the confusions that arise out of laypersons' misunderstandings, you can turn to Appendix 3, "Electrical Myths."

Recalled Equipment

16

EQUIPMENT MANUFACTURERS MAKE MISTAKES. WHETHER you are called onto a site to perform a consultation such as a safety survey, or simply to fix something, alerting your customers to the hazards presented by recalled products benefits them and costs you nothing. The United States Consumer Product Safety Commission publicizes product recalls, and a large part of the following list—which is not current, and is not guaranteed to be exhaustive—came from their World Wide Web site, which is listed in Chapter 14, "Resources."

Identifying equipment as being on this list is not the same thing as locating defective equipment. In some cases, a particular sample of product subject to the recall program will not be defective: the manufacturer will need to check it to find out. In other cases, after the manufacturing defect was discovered, a company produced corrected equipment with the same model number and outward appearance as the equipment subject to the recall program. Regardless of those possibilities, saying, "I think this model you have installed was subject to a recall; we'd better check" certainly is not crying wolf.

Since 1973, the government has served as a clearinghouse for information about flawed products. Should you ever run into defective products that might deserve to be recalled, get in touch with the CPSC at 800-638-2772. This is also the number to call to find manufacturers' telephone numbers. CPSC news release numbers have been provided to help you identify the recalls when communicating with CPSC personnel.

Consumer goods

The lists in this chapter are largely restricted, of necessity, to hard-wired items. There have been many recalls of products which, while electrical, are of little concern. You simply are not involved. Take appliances, such as toasters or juicers. They are none of your business, except for the need to provide appropriate, safe circuits and outlets for them. Holiday light strings and portable lamps are two more products whose manufacturers have several times discovered defects and issued recalls. Neither involves an electrician or is part of a structure's permanent wiring. Moreover, UL's standard for light strings changed in 1997. Any produced under the old standard, even if not defective, are also not as rugged and safe as the latest versions. Halogen torchiere floor lamps, especially those sold with 500-watt lamps, have been associated with some fires. Again, these are not directly up your alley. Heat tapes are almost the same—plug-in items that are notorious for deteriorating and causing fires. The risk might be worth mentioning to a customer if you feel chatty, but the replacement of damaged heat trace units is not a problem needing an electrician. Fluorescent lamps are an intermediate issue, in that a mismatch between ballast and lamp is something about which you might want to alert customers, something that you might be called upon to diagnose, or that you might be involved in correcting.

Your own lookout

Some consumer goods are items that electricians might well use, so it is tempting to include them in this chapter. For instance, many portable lamps and spotlights have been the subjects of recall campaigns. Most commonly, a metal lamp guard has been inadequately grounded, or the plastic covering a drop lamp's convenience receptacle has permitted users to contact live parts. (Example: Release #97-073, SmartLite and RiteLite twin-tube fluorescent work lights, which can shock.) There also are hand tools which have been recalled, and which are as likely to injure you as any other tradesperson. (Example: Release #97-084 regarding Hitachi cordless drill battery packs, which can overheat during recharging.) Unfortunately, it is necessary to limit what is reprinted here. The only type of electricians' equipment discussed here is defective GFCI cords and adapters. Because these are devices you may rely upon for safety in particularly hazardous situations, they are worth an extra alert. So we start there.

Woods Wire

The Woods Wire brand Model 1651 GFCI adapter could not be relied upon to trip in response to a ground fault. Contact Woods Wire, 510 Third Avenue SW, Carmel, IN 46032.

Columbia Wire and Cable

The CPSC's release #94-079 concerns 39,600 so-called GFCI extension cords sold in 1992. The 12-foot-long white cords actually contain IDCIs—and disconnected ones to boot—rather than GFCIs.

Most have been recalled, but 14,000 are still out. The bottoms of the plugs on these "GF1812 Ground Fault Interrupter" units say, "Cat. No. 6575, E-96425." The receptacles are labeled, "JC."

Hard-wired equipment

When equipment is hard-wired, it definitely falls into your territory. You may or may not take it on yourself to warn customers against overlamping. An example of something about which any responsible electrician will speak up is a defective lampholder, whether or not he or she actually has worked on the fixture. The following recall campaigns concern such hard-wired equipment.

Lighting

Lighting is a good place to start because so many fires seem to originate at light fixtures.

Lampholders

The earliest CPSC notice recalling hard-wired equipment that will be mentioned here was a July, 1974 announcement about weatherproof floodlight lampholders. Approximately 190,000 Bell/Light lampholders sold since November, 1973 were recalled because of shock hazard.

Not all of the units were for hard-wired installation. The word, "BELL" is cast on two sides, and the porcelain socket is secured by a fluted screw through the back of the chrome, gray, or green lamp holder. Bell Electric Company may be out of business.

Shocking fluorescents

The CPSC's release #94-002 concerns two brands of outdoor fluorescent lights that were made in Korea. Well over 9,000 of the units were sold here in 1991–1992, with indoor ballasts and without grounding. Any "Pay-N-Pak Deck Lite Fixture" you find should be disconnected and trashed. For further information, contact Pay-N-Pak Home Center, 116 Morain St., Kennewick, WA 99336-2908. Phone 509-783-4171. Any "Builder's Square Wall Bracket Light," SKU 9300117, is equally dangerous. The latter model can be identified by its black aluminum body, clear lens, and a tag saying "UL Issue #696089."

Chandeliers

In late 1996, Underwriters Laboratories Inc. warned purchasers of Good Tidings, Inc. "Electric Candoliers" that some units had been badly assembled.

They could short internally when a lamp is installed, creating a fire hazard. The units are marked with the trademark, "Toyo" and the number "E55068." The manufacturer may be out of business.

Wiremold™ covers

The CPSC's release #91-35 concerns 300,000 cover plates for Wiremold™'s "On-Wall" plastic raceways sold between 1986 and 1989. The covers can deteriorate and crack. Models affected include #840 and NNM13 (switches), #843G and NM12 (duplex receptacles), and #NM100 (outlet kit).

Switch

332

The CPSC's release #96-108 concerns about 12,000 "TWI LITE Illuminated Switches" sold in 1995 and 1996. The item numbers on the packaging start with the letters "AWL." If they are in place and working, they should be okay; the problem was an incorrect installation schematic. Installed in accordance with the original instructions, they produce a bolted short. If a customer has been saving one, uninstalled, for your visit, use the corrected instructions obtainable by calling 800-220-9294.

Surge protection

There have been at least two recalls of surge arresters.

Trilliant™

The CPSC's release #95-168 concerned about 4200 secondary surge arresters that could cause fires. Used only in the Trilliant SDT Series 1 Home Power systems, this version of their "Surge-breaker" is Model #SDT1175SB. If you find one, call 800-666-7557.

Tranquell™

On July 30, 1990, the General Electric Company's Power Delivery and Control division warned that certain of their products could seriously endanger people. Their "9L15" E and F-series 650-volt Tranquell Secondary Surge Arresters sometimes fail catastrophically—on some occasions propelling parts violently up to 100 feet away. Sold since 1983, they can start fires. If you do come across any of these on the job and are not authorized to remove them, keep everyone well away on any occasion when they are deenergized and then reenergized, unless each pole has been protected with a current-limiting fuse. While they can be protected with current-limiting fuses, doing so also reduces their efficiency as surge arresters. Therefore, a prudent electrician would replace them.

Circuit breakers

In November, 1988, CPSC release #88-95 announced the need for replacement of 9000 Challenger type HAGF circuit breakers, single-pole 15- and 20-ampere GFCIs manufactured in early 1988. The bad units may provide short circuit and overcurrent protection but fail to detect Class A ground faults. The indication that you have found an affected unit is that the ampacity is printed on the handle in white letters, rather than black, and there is no label with white lettering next to the word "TEST" on the side.

Loadcenters

In November, 1988, CPSC release #88-91 announced Siemens's offer to inspect and, as necessary, correct ITE loadcenters whose lugs had not been inspected. The date codes on the labels of affected panels are within these ranges: H2586-H3186; J0186-J3086; and K0186-K2486. If the labels are not stamped "Inspected Lugs," and there is no "Inspected Lugs" sticker, Siemens should be contacted.

Electric heaters

There have been a number of electric heater recalls.

Dayton ceiling units

CPSC release #95-105 concerns 10-inch-diameter heaters whole-saled by Grainger's in 1993 and 1994. Just 661 of the Dayton Model 4E154 Electric Ceiling Heaters are potentially dangerous. The affected units have the following date codes: A93, B93, C93, D93, E93, F93, H93, J93, K93, L93, M93, A94, B94, C94, or D94. The manufacturer was Fasco.

Nutone

In early 1987, Nutone, Inc., Cincinnati, Ohio, recalled its ceiling heater fans, Models 9905, 9960, and 9965. Some of the units could ignite their grills if the fans stall.

Square D and Nelco

From 1970–1986, a million defective electric heaters were sold, including both wall and baseboard types. They can overheat, and are alleged to have killed people. Their labels say Square D or Nelco. The wall heaters are type "TW," like the wire insulation. The baseboard heaters are built-in tan units, 2 to 10 feet long.

The baseboard heaters have model numbers beginning with the two numbers "18," the two numbers "25," or the two numbers "30," followed by four more numbers and then either the letters "SER A" or the letters "SER B." If you suspect you have come across one, call 800-666-7557.

Berko Electric

Marley Electric Heating also sold some defective tan baseboard heaters, 2 to 10 feet long. The affected units have identification labels inside the heaters on the right hand side, just below the silvery (or perhaps scorched) air deflectors. The identification labels have model numbers of the form "MBBe-fgh-M," where the "e,f,g,h" could be any numbers. Just to the right of the labels on the affected units, there is a date between January 1980 and March 1987 If you come across one, call 800-545-8306.

Smoke detectors

Here are four recall campaigns relevant to smoke detectors. The first three concern the detectors themselves, which may fail to sound. The fourth concerns control panels that monitor their function. The supervisory warning system may fail.

Substantial hazard

In November 1978, the CPSC issued report 78-89 concerning tens of thousands of smoke detectors that not only could fail to detect, but that could themselves spontaneously catch fire. Pittway Corporation, Northbrook, IL, doing business as BRK Electronics but selling detectors under many other names, paid a substantial fine.

The hard-wired version of the detectors was sold under the following names and model numbers:

BRK SS749AC

BRK SS749ACS

AMF (American Machine and Foundry) 2000AC

ITE (ITE Imperial Corp.) ITO1-AC

Sears 9-57049

If you should still run across one of them, call BRK Electronics at 800-392-1395 or write to them at 780 McClure Avenue, Aurora, IL 60507.

Bad horns

In September, 1992, The Consumer Product Safety Commission issued alert #92-130, recalling perhaps 3.5 million hard-wired smoke detectors. Sold under a number of brand names and models, they were manufactured by BRK between October 1987 and March 1992.

- Under the BRK name, they include models 1839I, 1839 WI-M, 1839WI-12, 1839I12R, 2839I, 2839WI, and 2839TH.
- Under the First Alert name, they are model SA1839WI.
- Under the Family Gard name, they are models FG1839I and FG1839IHD.

Insensitivity

In July 1977, the CPSC issued release #77-81 concerning 32,000 possibly defective smoke detectors manufactured by Master Lock Company, 2600 N. 32nd St., Milwaukee, WI 53210, in late 1975. The units may or may not operate under test, but their calibration cannot be relied upon. Their serial numbers fall within the following ranges: 1 to 10,000; 17,735 to 20,000; and 2,000,001 to 2,020,000.

Defective control panels

The CPSC's release #96-053 concerns 4,700 Moose-brand security control panels sold to professional security firms in 1994 and 1995 for use with two-wire smoke detectors. They include Model Z1200 with date codes 0594 through 4795; Model Z1250 with date codes 3195 through 4795; and Model D3000 with date codes 4894 through 4595. They were manufactured by Sentrol, 800-547-2556.

Paddle fans

Into some lives fans fall. For this reason, there have been several recalls.

Kiddy fans

The CPSC's release #88-44 concerns 1,500 defective 48" white, four-blade, reversible paddle fans with pictures of Snoopy or of a Smurf. Sold between 1984 and 1985 by Linsley Home Decorating Centers of Miami Lakes, Florida, their mounting brackets and balls need replacement. Some fans have fallen.

Antica fans

Two companies sold a number of paddle fans with defective metal blade brackets during 1984.

Columbia 52" fans. Fantastic Fans, Inc. sold them as Antica Columbia 52" "Hugger Ceiling Fans." All known customers were offered replacement blade brackets. This does not mean that they succeeded in reaching all purchasers.

Kenroy Casbah Kenroy sold them as their Casbah Flushfan, Model 3052, 52" ceiling fan. The brackets are marked, "Antica" on the undersides.

Casablanca

The CPSC's release #94-022 concerns 3,264,000 Casablanca paddle fans manufactured between January 1981 and September 1993, which were found to have unreliable ball-and-canopy mountings. To arrange for a retrofit kit, call 800-390-3131. The nameplate says, "Casablanca," and the second letter of the serial number is A, B, C, O, P, R, S, T, U, V, W, X, or Y.

NuTone

The CPSC's release #92-59, in March 1992 warned of 310,000 fans manufactured in 1983 and 1984, whose die-cast mounting brackets may crack. Included are these 36-inch and 52-inch models: Veranda, Hacienda, Sea Island, Decorator, and Slimline.

Pool and spa motors

The CPSC reported in January 1987 that Century Electric, Inc. sold something under 2,000 motors, $\frac{1}{6}$ through 3 HP, with inadequately insulated wires. The motor codes were 889 or 8810, and they contained a white 18-AWG lead marked "Xlink Wire, Inc., AWM Style 3173 125 C600V." Century wants people to look for the six-digit part number on the motor name-plate and call them to verify whether the leads need replacement.

Well pumps

The CPSC's release #95-114 concerned about 250 severely hazardous, ungrounded well pumps manufactured by Monarch Industries, Inc. and sold in 1993 and 1994 to distributors or contractors in Missouri, Minnesota, Michigan, and Wisconsin. (Actual sale to end-users, and installation, could be more more recent.)

The following model numbers are involved: JKC-1, JKC-20, JKC-S2, HJKC-S2/JR44HS, JKC-S3, JKC-S4, JKC1/JR-15S, JKCS2/JR-15S,

JKS-1, JKS-20, JKS-30, JKS-40, JKS-S2, JKS-S3, JKS-S4, JKS1/JR15S, JKSS2/JR15S, RLC-1, RLC-2, RLC-2/RLI14H, RLC-2MM, RLC-3, RLCI-RL4H, RLC2/RL4H, and RLS-2. The following serial numbers are affected: 5093, 5193, 0194, 0294, 0394, 0494, 0594, 0694, 0794, 0894, 0994, 1094, 1194, 1294, and 1394.

Sprinkler controller

Toro Vision II series irrigation controllers in metal cabinets may not be grounded. This could result in electrocution. Affected cabinets have the model numbers 189-66-01, 189-96-01, and 189-06-01. They are painted white, as opposed to gray or tan, have a red "Toro" logo, and have a key lock on the door. A "Vision II Series" paper label was affixed inside the door. Toro provides a metal cabinet modification kit, 91-01034, to ground the cabinets. The company's number for information on the program to rework affected controllers is 800-664-4740. The address is Toro Irrigation Division, POB 489, Riverside, CA 92502.

Sump pumps

In April, 1978, CPSC release #78-25, concerning Compliance Case #780062, warned that pedestal pumps manufactured under many names by Wayne Home Equipment, Inc. could shock. The potentially dangerous units have black plastic float switches; those with metal float switches are said to be safe.

The alert concerns models SPV-500, BPV-500, and PSP-330. They were sold under the following brand names: Agway, American Jet Stream Pump, Hydro-Lux, Menards, Super Flo, True Value Sump Pump, Wayne, and Wheatbelt.

Furnace igniters

In early 1987 Honeywell, Inc., Golden Valley, Minnesota, reported a problem to manufacturers to whom they had supplied their S87 A&B Electronic Ignition Controls for LP furnaces. The S87 units manufactured prior to October 1982 could deliver gas even if the igniter was not providing a spark—and some would not spark reliably.

Attic ventilators

CPSC announcement 82-20 concerns compliance case RP820036, a voluntary recall of 11,500 Emerson Electric Company power roof ventilators sold through Sears in 1980 and 1981 that could cause fires. The ventilators have a gray vinyl outer hood, and the ID plate on the back of the motor has Model No. 758.648360.

Permanently installed stereos

CPSC recall #96-083 resulted in a settlement agreement and order (release #97-068) regarding NuTone, Inc.'s ST-1000 wall-mounted stereo system, which was sold by electrical distributors, among others. A dozen units overheated between 1993 and 1995, and NuTone paid a $110,000.00 civil penalty for tardy reporting.

Being a good neighbor

This chapter is about going an extra step for your customers, and about not getting tangled in problems resulting from the installation or reinstallation of bad equipment. Only you can decide how much energy to devote to vigilance regarding recalls. As pointed out in the opening discussion, this list is incomplete—it has to be, if only because of new recall campaigns. And it contains very few addresses and phone numbers, because even the CPSC does not keep up with the manufacturers of recalled products to see if they have gone out of business.

You are not a stand-in for the CPSC. If you do choose to keep informed about recalls, you can provide added value to your customers at the cost of small effort. Many, many holiday light sets have put the "Ecch"s in "X"-mas by starting fires due to their shoddy construction. Many, many indoor and outdoor extension cords, "power strips" (including some with surge protectors), and portable lamps have been manufactured with undersized conductors and flammable insulation. For a small sampling, read the CPSC's releases #96-117, 96-135, 97-008, 97-018, 97-079, 97-034, 97-093, 97-094(the latter six being undersized cords), 97-066 (Chinese cheaters falsely UL-labeled), 97-109, 97-119, 97-131, and 97-132.

The recalled item need not tie into the job for which you yourself have been hired. Release #96-181 concerns bad plug-in "Gas Sentry" carbon monoxide detectors, and 96-184 concerns bad "Ceramic Furnace" portable heaters. Release 91-108, concerning De Longhi radiation-type portable electric heaters was on its third phone number, 800-322-3848, as of late 1997. But you can go even further. If, for example, you notice that a customer's gas stove or dryer is connected with corrugated brass flex lines, rather than stainless steel or plastic-coated brass lines, it is no more than thoughtful to say, "Hey, there's a warning out—a couple of hundred of these connectors have sprung leaks. You might want to talk to a plumber."

17

Hazards and Benefits

This chapter begins with a nod to the risks associated with any construction, proceeds to glance at those peculiar to electrical work, and quickly moves on to the hazards of working on old wiring and in old buildings. Then it suggests reasonable responses to those dangers, and closes on a positive note: a look at how understanding old wiring is of benefit to almost any electrician, whether or not he or she is planning to specialize in such work.

Dangers

On to the hazards. Some laypersons are petrified by the idea of touching anything electrical; that attitude is extreme. But just as the training of a utility line worker does not equip him or her to wire a house, what you learn as an apprentice in dealing with new installations does not teach you all it takes to troubleshoot, evaluate, and repair old systems. It is much less safe to work on old wiring, and generally to work on existing residential or small commercial systems, than to work on modern systems that are more likely to have been properly installed and maintained.

Risks inherent in any electrical work

These risks will be broken out and discussed in some detail. Electrical work has inherent dangers. A defective or misapplied tool can cut, burn, bruise, poison, stab, twist, gouge, shock, or blind its user—or anyone else who happens to be around. Similar risks are associated with materials and equipment. Fatigue or carelessness on your part or on the part of others can expose you to electricity at various voltages even if you have tested, locked out, and tagged sources. And all that is true even when you are dealing with new, modern wiring. Because old buildings are structurally less predictable and because old work often means working around more clutter, these same hazards can be magnified.

Special risks inherent in "old work"

Forget old buildings for the moment. Most electricians have done at least some "old work," whether or not they have engaged in renovation or rehabilitation. "Old work," in this case, simply means maintenance, repair, and remodeling of structures that already are wired. Any "old work" is more risky than most new construction in a number of ways. To begin with, when you deal with pre-existing installations that you have not installed and maintained, you are working without specifications—at least without specifications that you can trust before you have checked them.

Also, when you come on a job that has had others working on it before you, there is some reason to fear that what has been done previously may not be as it should. Why has the customer changed electricians: dissatisfaction? If the previous installer did bad work, any inadequacies in that work could pose threats to you or your workers. You also run the risk of being held responsible for problems that were already there, or which were lurking, ready to pounce. If your predecessor did good work, you may be dealing with a customer who finds fault with good work. Ouch! At least that is not unsafe, though it may be unpleasant.

The added problems presented by older wiring

That is one small concern. There are plenty more beyond what you can extrapolate from your experience at old work to the challenges of work on genuinely old buildings. Old buildings pose far, far greater threats than relatively new ones for three reasons.

☐ First, old systems are likely to show some deterioration. Further detail about modes of deterioration—and their hazards—is scattered throughout this book.

☐ Second, the longer a system has been around, the greater the chance that someone along the way worked on it incompetently or carelessly.

☐ Third, electrical doctrine and standards have changed markedly over the years. The methods and materials that have passed out of the National Electrical Code were more dangerous, even when they were state-of-the-art, than those which have replaced them. Those antiquated systems are yet more dangerous now, deterioration aside, simply because they contain elements that are unfamiliar and perhaps unexpected. Much of the book has been devoted to this factor.

A few paragraphs back, when talking about "old work," mention was made of the fact that taking responsibility for electrical work carries a whole additional layer of risk beyond the hands-on physical danger associated with actually performing the work. These dangers are magnified in work on old wiring systems.

Earlier chapters, such as Chapter 7, "Setting Limits and Avoiding Snares," looked more closely at the risk of innocently opening someone else's can of worms only to have a hungry lawyer peck your eyes out. For now, simply consider the fact that a contractor offers assurances, implicit or explicit, in three directions.

☐ First, there are unknowns associated with old wiring repairs that can cause problems for your customers—and you are the point man (or woman) who has to reassure the customers or to bear the bad tidings to them.

☐ Second, there are questions relating to permits, involving issues of repair versus alteration and of grandfathering.

☐ Third, you have some responsibility for the safety not only of yourself and your employees but of inspectors and other trades—and even of those who come on the premises after you are gone.

Danger lurks when you have employees who are not used to the special dangers of work on old wiring, or colleagues in other trades who make unwise assumptions. Parenthetically, the problems created by such unsafe assumptions and practices seem to be a major reason that the NEC has increasingly restricted the option of using an "assured equipment grounding program" as an alternative to GFCI protection.

Someone who knows nothing of construction work may in some ways be a better risk than someone who thinks he or she knows what to expect. Alexander Pope's "A little learning is a dangerous thing" applies to the apprentice who assumes that turning off a light by flicking the switch, or even unscrewing the fuse, means the outlet is dead.

Other dangers of work on older structures

Dangerous assumptions extend beyond the electrical realm. That apprentice could also sustain serious harm by assuming that thermal insulation is going to be a mild irritant rather than a serious carcinogen. The concern also applies to the carpenter who carelessly breaks open a wall, exposing himself or herself, plus yourself and your apprentice, to the danger of a scalding or of a gas explosion.

You are probably familiar with, and you should be used to protecting yourself from, certain hazards of construction work. These include the dangers of falling in holes or over obstructions, the dangers of being struck by objects that other workers drop, the dangers of other trades' stinks and explosions and glares and flyings. You probably know that (according to some reliable estimates) 90% of construction workers have significant hearing loss; and surely you are no stranger to burn and shock hazards.

OSHA—and competent safety workers such as industrial hygienists—will tell you that protectors for the lungs or ears must meet several tests:

☐ be chosen for the specific types of levels of hazard present;

☐ be fitted to the individual,

☐ be part of a program that includes training the users in their inspection and wearing.

A new set of respirator standards, the first new one in twenty years, will begin to be enforced in 1998, with additional modules to follow. Men with beards still cannot be protected by anything other than a powered air purifying respirator, but even those whose faces can accommodate close-fitting partial masks no longer simply can rely upon HEPA breathers.

Some of the additional hazards associated with old buildings develop because systems have aged, while other problems occur because the structures were built in different days, when contractors knew less. This cannot cover all the possibilities, but we will spend a little time discussing three types of toxin. Each is cumulative—like noise, which even when it does not cause ringing in your ears, or make loud noises less painful (another bad sign), may be insidiously cutting down your hearing. If you have to raise your voice to be heard, you need ear plugs or muffs where you are working. Two readily recognizable issues are lead poisoning and asbestosis; the third, exotic lung disease, may be less familiar. Unfortunately, their effects are far more severe than reduced pleasure in music or than the need to say, "Huh? Speak up!"

Lead poisoning

Until about 1950, most paint—especially white or yellow paint—contained lead. Now surely you are not going to chew on paint chips. Still, if someone is foolish enough to remove old paint by sanding, torching, or wire-brushing, you are at risk of inhaling the dust and of carrying it home on your clothing. Lead was also used in glazing putty, so also look out when working around glaziers. Certainly, if they are removing leaded stained glass windows, there is a good chance that deteriorated lead caning (the ribs between the glass pieces) will release rather pure lead dust into the air. If you have infants or small children at home, be especially careful, as lead dust can harm the development of their brains.

Tin roofs, on the other hand, should concern you little. On the off chance that you are unfamiliar with them, terne ("tin") roofs are lead-coated steel. There has been no widely-publicized case of an electrician exposed to free lead simply by virtue of installing a service mast. Presumably, if a terne roof is being torn off or especially if the old paint is being removed (the lead requires periodic painting) while you are working in the area, you should be concerned.

343

What about those old electrical enclosures made on site out of sheet lead? You are very unlikely to encounter one. If you do, it most likely will be in the context of tearing out a deteriorated system. Find out about local restrictions on disposal of heavy metals. If you have much lead to handle, wear cheap gloves and throw them out when you are finished. Only if you have reason to believe that you may be exposed to lead in dust form do you need to be concerned about a respirator, and in that case it needs to be OSHA-approved for that specific type of protection. So-called "paint respirators" are no longer made. But the new NIOSH-approved ones serve the same puposes even better, and sometimes cheaper.

Pneumonia and worse

There probably is greater reason to be worried about catching some fell disease from bird or rodent droppings than about lead poisoning. Unsavory though they may be, their distasteful nature is not the concern. There is documented evidence that the respiratory illness histoplasmosis has been contracted by people breathing bird droppings in the course of old-house rehabilitation. Particularly prevalent in the Midwest, this fungus disease can cause quite bad symptoms, ranging from pneumonia on up, in the 1%–5% of infected people who succumb.

Lung cancer

Asbestos, installed in different forms from the 1920s through the 1960s, is quite fearsome. There have been ample reports pooh-poohing overcaution, particularly the overcaution that, perversely, impelled people to disturb intact asbestos in order to get rid of it. Still, asbestos dust can give you cancer.

To be endangered, workers do not have to be actively applying or removing asbestos, or even chopping up the stuff. Just being around it regularly, as it deteriorates like any other material, can do a number on human tissue. While less than 1% of the general population suffers from scarred lungs, 28% of New York City school janitors examined in a late-twentieth-century study showed asbestos scarring, presumably from working around pipes and boilers.

Most of the asbestos you encounter may be benign. Asbestos floor tile, shingle, even pipe wrap will do you no harm so long as it is not flaking or chalking. On the other hand, if it comes apart or is cut up or broken up, it can be dangerous. Asbestos also was used in acoustical ceiling tile, wallboard, and plaster.

Besides actual disease, there is a big, bad issue of liability associated with these risks. Asbestos-related lawsuits more than doubled between 1986 and 1991, and they are not going away. In 1996 a New York court permitted 140 railroad workers who had been exposed to asbestos for three years to sue their employer, even though they had no proof of harm.

Rational response to these concerns

How are you to respond to these issues? Make sure that your general contractor has consulted an industrial hygienist to determine abatement measures and has implemented them. If you are not satisfied, or if it is a very small job and you are on your own, talk to your local Environmental Protection people.

There usually is an answer. When a surface bearing lead-based paint needs to be covered, sometimes the paint can be sealed rather than removed. If it needs removal, a paste-type paint remover may be acceptable. If toxic dusts—lead, asbestos, or other—may be generated, the area can be sealed off and a hazardous dust (High Efficiency Particulate Air or HEPA filter) vacuum employed. Finally, if a wall is too dangerous or too expensive to break into, sometimes there are alternatives. Power can be run the long way around or on the surface. Communications systems can be run wireless.

The benefits of learning about old wiring

Has this scared you away yet? Stick around just a bit longer; you are coming to the good part. There are fools who "know" electricity cannot hurt them, because, after all, they have survived a tingle or two. They need no book. This book is for electricians with some experience and prudence, people thoughtful enough to recognize danger and surmount it.

For perspective, consider a different context. There are plenty of people who never would get on a motorcycle—"how could you risk your life that way?"—and there are also some hot dogs who ride them fearlessly, with no helmets, no protective clothing, and no respect for their exposure to "pavement rash" and worse. An intelligent biker recognizes both risks and benefits and will tell you, "Ride scared."

Take this same advice as you tackle old wiring: "Ride scared." But consider this factor. Almost any electrician will have some occasion to deal with old wiring, whether or not he or she has planned to, whether or not he or she makes a specialty of it. It is good to know what you are up against. And it is better for everyone concerned if some people make a specialty of this work.

It is worthwhile work. The plus side of this work in terms of service and pride is that it is needed. There always will be old buildings whose wiring requires knowledgeable consultation and repair. The plus side in terms of financial security is that evaluating and fixing old systems is a reliable albeit low-profile niche: a type of work that builds on your existing skills and that is hard to foresee disappearing. Old systems last. The longer you are in the electrical business, the less apt you will be to say of any type of wiring, "Oh, I don't think you'll still run across any of that outside of some museum display."

Finally, knowing this work can greatly improve your judgment as you do new wiring. Here is an example. A "damp location" is one that is subject to moisture but not to thorough wetting or to immersion. Do bathrooms count? A strict construction, one that, appropriately, is followed by most inspectors, says that a shower stall, below the shower head, is a wet location, and the area above a wall-mounted shower head may be considered a damp location.

The rest of a bathroom (as opposed to a steam room) is considered a dry location.

This is reasonable, and dry-location fixtures and hardware installed in bathrooms generally work fine. There is a big "but." If you have worked on older bathrooms, especially those that are not wonderfully well ventilated, you are sure to have dealt with rust, frozen screws, and peeling finish. That experience can inform your judgment—make you think again about the moistness classification—as you choose materials to wire new bathrooms.

Closing advice

The work is worth learning, and this book fills a gap in the materials available to teach electricians. Still, this is not a stand-alone "how-to" manual. No book can teach you enough to do the job safely. Supervision and experience are essential parts of training. Moreover, any work you perform on old wiring is a compromise, trading less expenditure of money and mess against less safety. That may be blunt, but that is the straight of it. How can you assure safety in dealing with old wiring? The only solid advice that can be offered without any hedges is this: rewire from the service drop on, rather than touch anything that another has done or begun—whether yesterday or yesteryear.

This was said seriously. Wiring anew, you still may have the lousy luck to receive bad materials and equipment from the factory, or you may fail to catch mistakes by tired or careless workers. But at least you will not be gambling on as many unknowns—gambling along with the customer. So "rip it out" is not a stupid choice, just an expensive one. Should you make that choice, you will still get value from this book, as you follow the recommendations in Chapter 8, "Tearing It Out and Rewiring."

Whatever you choose to do, rely on your professional judgment, not just a book or books. Keep your eyes open and your mind alert. If you think you are getting in over your head, pull back or get help.

Glossary

The following terms and acronyms include many that can be found in a dictionary, a few that can be found in NEC Article 100, "Definitions," and many more that simply are current or former trade usage. If you do not find a term here, by all means look in Article 100. If all else fails, check with a dictionary—or an old-timer.

The purpose of this glossary is to help you understand what you hear, see, and read in this book and on the job. It includes both formal language and slang. Uses and code implications of the terms and the concepts they present also are addressed. Some items are included here mainly to provide a place to discuss concepts that have not been covered elsewhere. The terms were chosen as being possibly unfamiliar to or misused by less-experienced electricians.

Generally understood meanings, even for terms listed in this glossary, rarely are included. For instance, "bug" has many dictionary meanings, including electronic eavesdropping. "Bug" as a splicing device, though, is the only meaning included here.

Guide to use

Initialisms (acronyms and abbreviations) and slang terms are identified as such, as are some parts of speech. In the expansion of acronyms (a term that, for simplicity, will be used to refer to all abbreviations)the letters used in the acronym are capitalized. Do not take this to mean that you should capitalize them in normal usage, except for the first letters of proper nouns, such as National Electrical Contractors Association (NECA). Acronyms are listed without periods between the letters, but you may encounter many of them used either way.

Occasionally, you will find an entry that is not explained directly. Instead, you will be directed to a different entry which means the same thing by the use of the phrase "see entry." Other entries are explained directly, but the phrase, "see also" is tagged onto the definition or explanation, referring you to additional entries that are related.

Word list

3/4 sequence An old service arrangement where the service switch is ahead of the meter, which is ahead of the service fuses.

11B A metal enclosure, $4^{11}\!/_{16}$" square, $1\frac{1}{2}$" deep. The deep version is $2\frac{1}{8}$" deep.

17A A metal enclosure, $3\frac{1}{4}$" octagon by $1\frac{1}{2}$" deep.

110 block Equipment for telephone closet terminations, used to make the transition from horizontal feeds to the twisted pairs going to telephone outlets.

1900 A metal enclosure, 4" square, $1\frac{1}{2}$" deep. The deep version is $2\frac{1}{8}$" deep.

1935 A porcelain block assembly used in old cutout boxes, with two-wire service entering, two fuses, and one two-wire circuit leaving.

1935A The same as a 1935, but with only one fuse, hence a continuous rather than a fused grounded conductor.

2-by Dimensional, or standard, lumber is now $1\frac{5}{8}$" $\times$ $3\frac{5}{8}$" for "full-sized" studs, with similar odd reduced sizes for joists. It used to be fully 2" $\times$ 4", 2" $\times$ 10", and so forth. On the other hand, in former times lumber's finish was not smooth, and its exact measurements used to vary sometimes. If you need to work with or replace old lumber, you may have to shim your wood, nail it offset from its neighbors, or, at worst, have a larger standard size planed down to match, perhaps starting with a modern 4 $\times$ 6, for example, to make something that will match a genuine old-fashioned 2 $\times$ 4.

2199 A porcelain block assembly used in old cutout boxes, with three-wire service entering and paired circuits leaving. All conductors were fused.

2199A The same as a 2199, but with continuous rather than fused grounded conductors.

2587 A porcelain block assembly used in old cutout boxes, with two-wire service entering and paired circuits leaving. All conductors were fused.

2587A The same as a 2587, but with continuous rather than fused grounded conductors.

3-Way A single pole, double throw switch with no center-off position. Its name is based on the British use of "way" to mean path or terminal.

45 EITHER a 45° bend, OR a short length of conduit or tubing prebent to 45°, a "made 45."

8B A 4" octagon box, 1½" deep. The deep version is 2⅛" deep.

90 EITHER a 90° bend, OR, more commonly, a short length of conduit or tubing prebent to 90°, a "made 90."

AC (acronym) EITHER Alternating Current OR Armored Cable.

ACL (acronym) Armored Cable with an internal Lead sheath around the conductor insulation. No paper or cloth wrap was required, just the lead; and it was suitable for use in damp and wet locations, at least indoors.

Acorn clamp (slang) An oval clamp, normally brass, with a setscrew, commonly used to mechanically bond a grounding electrode conductor to a ground rod.

AFCI (acronym) Arcing Fault Circuit Interrupter. Although quite different in design, and protecting against a different danger, it is analogous to the GFCI, in that an AFCI may add considerably to the safety of older buildings where insulation breakdown may result in high-impedance short circuits.

AFF (acronym) Above Finished Floor.

AHJ (acronym) Authority Having Jurisdiction—usually whoever inspects for the government.

AJ (acronym) Ackerman-Johnson masonry anchor; steel with a tapered lead insert, it is set with a special tool. Installed backwards, it does no good at all, as the taper does not bear against its surrounding lead. The only masonry anchors more certain of holding in place are split-steel anchors and through-wall bolts, the latter normally embedded during wall or surface construction.

Al The chemical symbol for aluminum.

Al/Cu A marking indicating that a device of 30 ampere or higher rating may be directly connected to aluminum or aluminum alloy conductors. "Cu" is the chemical symbol for copper. See also Co/Alr.

All-thread (slang) A nipple or rod with running thread (see entry).

Amprobe™ An inductive ammeter, after the well-known manufacturer.

ANSI (acronym) American National Standards Institute. The NEC is recognized as an ANSI standard. One of the implications of this is that even where it has not been adopted into law, it is

349

recognized as representing the accepted standard of practice unless explicitly superseded.

Approved EITHER a product characterization, for definition of which see the NEC, OR, in Scotland, the equivalent of the designation "Master" for an electrician.

Armored cable outlet plate Outlet plate (see entry).

Armstrong (humorous slang) Any tool that relies on muscle power. At one point, electricians used literal horse-power to pull the ropes that drew conductors into conduit!

ASHI (acronym) American Society of Home Inspectors. Members are certified in home inspection; this does not mean that a particular member necessarily knows diddly about wiring. Too often, an ASHI member knows just enough to offer false assurance to laypersons. Other ASHI members do a great job. Since they are not up on the nuances of code, they may be less reluctant than an electrician to simply warn a customer, "This house's wiring is old and no good." For an example of how much their electrical expertise can vary, one home purchaser reported that due to a peculiar situation, she had reports on the same house from two home inspectors. One told her it needed $2,000 of electrical work: clearly a service upgrade; plus one or two other things. The second examined the wiring itself and ballparked needed rewiring at $25,000.

Awl A chunk of pointed steel for starting or simply for making holes. Sometimes an awl has a wooden handle, especially if it is a "scratch awl." An awl can serve some of the same uses as a center punch and often is handier. Do check to make sure that you do not use it to punch materials that are too hard for it and could result in injury.

Attix wire Rubber-covered wire enclosed in loomlike material, used in the early 1900s. Attix wire was a sort of nonmetallic cable. By the 1920s, it was supplanted by armored cable.

Auger A self-pulling drill bit with screw threads, which can be sharpened with a triangular file.

Backfeed EITHER to receive power from what normally is the downstream or outlet side of a device, equipment, or circuit, OR to backwire (see entry). The former may be an acceptable practice or it may be hazardous. A GFCI, for example, when connected with power going to the LOAD terminals (backfed), will not operate correctly.

Back strap A support clip that is secured to a surface before a raceway is put in place, and that than snaps onto the raceway.

Most often, a back clip designed for surface metal or nonmetallic raceway is largely concealed by the raceway.

Backwire To use push-in terminals for connection to a device, a design that had been introduced by 1980. This term normally is not applied to terminals that need to be tightened after the wire is pushed in, but only to those that rely on spring pressure. The former tend to be high quality and make very secure connections. You may encounter circuit breakers, switches, receptacles, and combination switch-receptacles with push-in backwiring as the only means of connection, as well as many devices that incorporate both standard screw terminals and push-in backwiring terminals. Backwiring of receptacles now is permitted only with #14 solid copper conductors. The reason is not that the devices' design has changed, but rather that a consensus has developed which says that backwiring with #12 is unsafe. And this includes devices you may find backwired in older buildings. You have to decide whether to modify such installations. As a compromise, retaining old devices if you wish, you can pigtail #14 leads so as to avoid relying on backwiring as a means of daisy-chaining. While "backfeed" sometimes is used to mean "backwire," in other words "hook up by sticking wires into holes in the back," "backwire" is never used to mean "backfeed," signifying "connect power to the load side."

Ball chain Bead chain (see entry).

Base plug Convenience outlet(see entry).

Bead chain Chain used for fishing down inside walls. It has less chance of getting hung up on obstructions than some other forms of chain and has enough weight to pull itself down past minor resistance.

Bell box (slang) A weatherproof cast box, named after the Bell Company.

Bell hanger bit A long, narrow twist drill bit, most commonly 12" long by $\frac{3}{16}$" or $\frac{1}{4}$" diameter. Near the start of the spiral it has a hole into which a small wire, 22 or 28 gauge or so, can be inserted and secured so that withdrawing the bit pulls the wire back through the material you have drilled. Masonry versions are available.

Bell wire Low-voltage, light-gauge, Class 2 wire, normally solid, suitable for intercoms, doorbells, chimes, buzzers, relays, annunciators and similar uses.

Bender A device for bending pipe or tubing that ensures that the appropriate radius is maintained. Being designed to create a specific radius makes it more convenient but less adaptable than a hickey (see entry).

Bilco door (slang) An angled storm door for a basement or cellar (named after a brand). Entrances with these doors are generally considered exempt from the requirement for light switches at entrances.

Bird cage This term is said to have been used to refer to the situation where the strands of a stranded cable come unwound, or at least loose. Such a configuration may be intentionally created to provide a means to snare another wire and then drag it along once the strands have been tightened. It looks a bit like, but works differently from, a Kellem grip(see entry and illustration). See entry for Mare's Tail.

Bird wire An old conductor type used by line workers, rather than by electricians working under the NEC.

Block EITHER fixture block (see entry), OR rectangular openings in face plates, corresponding to rectangular-design switches and receptacles, currently represented by Leviton's Decora™ and Bryant's Fashion Series 9000™ lines, OR Block and Associates, authors of widely-used tests of professional competence. The latter are used by many states for Master Electrician certification.

BOCA (acronym) Building Officials and Code Administrators International, Inc. Publishers of their own widely-adopted building codes. See also CABO, ICBO, and SBCCI.

Bollard A short post, normally containing a light, as for illuminating an adjacent walkway. This is different from a standard post, which would support a separate light, either on top of it or suspended from it.

Bolted short A connection between the hot and grounded conductors, or between the conductors of two phases or hot legs, having negligible resistance. Splicing such conductors together inadvertently is one way bolted shorts are created. As a means of protecting against the possibility that someone will override the lockout while an electrician is working on a circuit, a bolted short sometimes is created by literally bolting the conductors together.

Bond To connect two or more parts, whether with wires, bonding locknuts, bonding bushings, or other suitable means, to maintain the same electrical potential—usually ground potential. The term "ground" is sometimes used in the field when the Code actually requires a bond. It usually, but not always, works out the same. For instance, you can ground a receptacle, as required, by bonding it to a grounded enclosure. An isolated ground receptacle, however, by definition is not bonded to the enclosure, although it is grounded by connection to a grounding conductor (unless the entire enclosure is isolated from the raceway).

Bonding Hardware designed to ensure positive electrical connection, usually with serrations or a pointy, indenting setscrew.

Box A term frequently used to specifically mean "loadcenter" or "service equipment."

BR (acronym) Bryant Electric, manufacturers.

Brace (-and-bit) A muscle-powered direct-drive hand drill using auger bits, or drills, with wedge-shaped bases. It can be braced against your chest as you operate it, unlike a hand drill(see entry).

Break Interrupt contact. Compare Make (see entry).

Bubble cover A device cover designed to make the device weatherproof in use; for example, a receptacle will be weatherproof even with a cord connector plugged in. These have also been used to provide tamper-resistance. Permission to use them as alternative protection for that purpose in pediatric care facilities was introduced in the 1993 NEC and quickly challenged. The problem is that they are not inherently sufficiently tamper-resistant to defy a curious kid access unless they are locked, any more than are push-in outlet seals; and when locked, they prevent quick disconnection of cords. The 1996 NEC's permission to use such covers as an alternative to tamper-resistant receptacles in pediatric care health facilities may be removed from the 1999 NEC. These covers were introduced in response to the fact that ordinary weatherproof receptacle covers are not weatherproof when in use. When replacing a pre-1990 outdoor receptacle, you need to evaluate whether equipment is likely to be left unattended, plugged in. If so, a bubble cover must go on, regardless of the fact that the outlet is preexisting.

Buck-boost A transformer designed to bridge a relatively small difference between the supplied voltage and that required by utilization equipment. Interestingly, while both "buck" and "boost" commonly mean "to raise," the term, "buck-boost transformer," is used regardless of whether the transformer is applied, for example, to raise 208 volts to 240 or to lower 240 volts to 208.

Bug A splicing device, usually two-piece, usually for larger conductors. Generally, bugs, like layin lugs, allow you to splice without breaking at least one of the conductors—to tap it.

Building wire Conductors listed for permanent wiring in NEC table 310-13, and cable containing them. Not flexible cord, and not, by common usage, Class 2 and 3 wire.

353

Bull (slang) Heavy-or hard-or large-(duty, usage, gage).

Bus A heavy-ampacity form of conductor, often bare, often in the form of tubing or bar(s). In telecom and electronics use, it means the common path. See also Buss.

Bushing EITHER a red hat (see entry), OR a bonding bushing (see "bonding" entry) OR a fitting that screws or presses onto the end of pipe as abrasion protection for the cable or conductors that emerge OR a capped bushing, closing off the end of a threaded round raceway.

Buss EITHER a misspelling of bus (see entry), OR perhaps part of the appellation of a Bussman product. ("Buss" is a good English word—it means kiss.)

BX (slang, but common usage) Armored cable. Originally one brand name. The term was replaced by "Type AC," except when speaking of "BX connectors" in any but the most formal circumstances.

BXL Armored cable with an internal lead sheath. See ACL. Until World War II, Washington, D.C. allowed no use of type NM cable. They permitted armored cable in basements, but no more than 18" of it; beyond that, conduit or BXL was required. (New York City restricted type NM similarly.)

Cable EITHER an insulated conductor, normally heavy-gauge; OR, more commonly, an assembly of multiple conductors, parallel or concentric, insulated, covered, or a combination of insulated and bare or insulated and covered.

Cable clamp letters Cable clamps found in metal boxes are embossed with letters to indicate that the clamps are suitable for the following wiring methods:

☐ A armored cable;

☐ F flexible metal conduit;

☐ N nonmetallic sheathed cable;

☐ T flexible tubing;

☐ MCI metal-clad interlocking cable;

☐ MCS metal-clad cable with a continuous smooth sheath;

☐ MCC metal clad cable with a continuous corrugated sheath.

CABO (acronym) Council of American Building Officials, publishers of codes other than the NEC. See also BOCA, ICBO, and SBCCI.

Cadweld™ A widely-used system of exothermic welding (see entry).

Can A term used EITHER to refer to different types of enclosure, such as cabinets, cutout boxes, and meter bases, OR to lighting fixtures, especially cylindrical ones, and especially, when the fixture has several independent parts, as with track lighting, to the part of the fixture that includes the lampholder.

Canvasite A weatherproof cord used in the 1920s for drops and pendants in damp or wet locations.

Carter System A switching system made illegal as early as the 1920s, but used much later, despite the prohibition. See Chapter 4.

Ceiling pan Pancake box (see entry).

Ceiling panel EITHER a radiant heating panel for resistance heating, embedded in a ceiling, or a rectangular panel in a suspended ceiling.

Center contact The base or bottom contact of a lampholder or fuse holder, as opposed to the shell or side contact. It is the one opposite the center of the bulb in a lamp, and opposite the fusible link in a fuse. Three-way lamps have two base contacts; most commonly, but not always, one is central and one circumferential. Neither of them, whichever the design, is to be connected to the grounded circuit conductor.

Center punch A short piece of pointed steel designed to be hit with a hammer against a metal surface to create a dimple. Making such an indentation before drilling reduces the likelihood that the bit will "walk." While a punch is "the right tool for the job," you may find it less convenient than an awl (see entry).

C-H (acronym) Cutler-Hammer, manufacturers.

Channel EITHER channel iron for fixture support, such as Kindorf™ (see entry) OR a fixture (usually fluorescent) capable of being joined to its fellows in continuous strips. Beware of the use of the latter as raceways for older, 60° rated wire. Ballasts are likely to overheat the insulation.

Channels (slang) Groove-joint pliers, after the Channelock Company.

Chase nipple See under Nipple.

Cheater (slang) EITHER a three-prong adapter, often without the tab (or, formerly, the wire) properly attached to a grounded screw, OR an extension cord used to illegitimately power a three-prong device without grounding—either by breaking off the third prong from a three-wire cord's male connector or simply by misapplying a two-wire extension cord to a grounding-type load. Using either approach not only is far more dangerous than using a two-prong, double-insulated device but also far more dangerous than plugging in many an old-style two-wire piece of equipment.

Chicago three-way (slang) Carter System (see entry).

Circulating current Current unintentionally induced in metal such as conduit, enclosures, or building steel. As with most stray electricity, it can waste energy, cause shock, build up heat or create arcing to cause fire or, in some environments, explosion.

Class A, B conduit British equivalents, roughly, to RMC and EMT respectively.

Class of Service A utility designation including but not limited to nominal voltage, frequency, and phase configuration.

Classified EITHER designated as fitting a particular category of hazardous location, OR a UL (or other NRTL) designation indicating that a product has been investigated, but only for use as a component for subassemblies such as a cord replacement or a luminaire retrofit—compared to complete, stand-alone equipment, which would be Listed instead.

Close nipple See under nipple.

CMP (acronym) Code-Making Panel.

Co/Alr A rating, and marking, indicating that a 15- or 20 ampere device may be directly connected to aluminum or aluminum alloy conductors. In the early 1980s, a manufacturer's representative confided, the industry gradually phased out production of devices carrying the Co/Alr designation because of pending liabilities arising out of wrong installations and the like. "Co" is the chemical symbol for cobalt, but that has nothing to do with this abbreviation for "Copper/Aluminum-rated." See also Al/Cu.

Code For this book, and in most electricians' conversation, unless otherwise indicated, "Code" refers either to the NEC, or else to the variant that has been adopted into law in your area. "Building code" is different, and refers to the rules adopted in your area regarding issues such as ceiling height, corridor width, windows, and fire resistance. See entries under BOCA, SBCCI, and CABO.

Combination inspector An inspector authorized by an Authority Having Jurisdiction to pass or fail work done by many different trades. Normally, this means that the inspector has no work experience at and no professional qualification to perform the work of most of the trades over which he or she is passing judgment. Also known as a "multihat" inspector. See Chapter 12, "Inspection Issues."

Common In three-way switching, the contact that alternately is connected to either of the other two by operation of the switch handle or actuator. Compare "Traveler" (see entry).

Concrete box A box intended for installation in, and support by being surrounded with, concrete. It has tabs allowing it to be nailed to the form, to keep concrete out during the pour.

Conductor Anything normally intended to carry electricity along a circuit or between parts of an appliance or other piece of equipment. Wire is the most common conductor, with busbar a distant second. Metallic raceways, such as conduit and cable armor, are commonly intended to serve as grounding conductors, meaning that in the event of a fault they ensure operation of the overcurrent device. Research at the Georgia Institute of Technology found that in the event of a ground fault, even when there is a grounding wire inside metallic tubing or conduit, by far the most fault current flows in and on the raceway, so long as the connectors and couplings are secure. Other than during fault-clearing, raceways have never been intended to serve as current-carrying conductors.

Only in submarine cabling is the surrounding medium intended to serve as a return conductor. A common practice adopted to some extent in this book is to use "conductor" variously to mean either the wire with its insulation, or just the wire inside the insulation. You can tell by context. "Conductor" sometimes is also used to mean "current-carrying conductor" (see NECA Article 100) as opposed to "grounding conductor." Here again you need to pay attention to context. Sometimes, of course, "conductor" may be used in the most general sense of "anything that, intentionally or not, conducts electricity," as in the sentence, "Dry skin is a relatively poor conductor; this is why lucky electricians often survive shocks that could kill."

Conduit body See NEC Article 100.

Condulet Conduit body. There are various designations, which apprentices used to be expected to absorb through their skin, as it were.

☐ An "E" condulet has one conduit entry;

☐ "C" has two, one at each end;

☐ "T" has three, one at each end and one in back, opposite the access plate;

☐ and the rest have "L-something" designations and two conduit entries, one in-line, and one at 90°. Holding one of those like a pistol, by the 90° conduit entry, the second part of the designation indicates where the access plate is located:

☐ "LB" for back,(See Figure 11-5(b));

□ "LL" for left,

□ "LR" for right,

□ and "LLR" for both right and left.

Convenience outlet The term now is used to mean a receptacle; until the late 1920s, the term used to include Edison-base sockets as well.

■ **G-1** *Another, outdated, kind of conduit body you might encounter.*

Coordinated protection There are a number of overlapping meanings. The term can be used EITHER for any system arranged so that only one protective device fails, rather than a larger portion of the system crashing; OR for the multilevel approach explicitly specified for health-care systems, wherein there are at least two levels of protection, so that, for instance, a subpanel will become disconnected if one circuit and its protection fail, rather than the function of the entire system being jeopardized; OR for achieving a necessary interrupting rating by using a Listed combination of devices, possibly including fuses and circuit breakers.

The first two purposes can be achieved by the study of tripping curves supplied by manufacturers. The last only can be achieved by using specified devices that have been investigated in a series of combinations for the purpose of providing the interrupting rating required, and appropriately marked.

With old wiring, the issue of coordinated protection comes up in two ways. First, when circuit breakers are installed that are not listed on the panelboard directory, or explicitly Classified for the use through the aftermarket, it is not merely a violation of NEC Section 110-3 (b). There is also the chance that the main breaker that is supposed to provide coordinated protection will be unable to do so. This is the problem that electricians most commonly encounter. Second, even though it is a Code violation, installing a subpanel without coordinating tripping curves can result in the same problem—or worse. A service panel normally will have a main breaker rated at 60 amperes or more. Since a subpanel could be fed from, for instance, a 30- or 40-ampere two-pole breaker in the originating service panel, it is even more likely that the tripping curve of the circuit breaker protecting the feeder will overlap the curves of the subpanel's branch circuit breakers.

Copperclad Steel-core ground rod with thick copper coating metallurgically bonded to it.

Copperweld™ Copperclad (see entry).

Correlating Committee NFPA group riding herd on CMPs (see entry) to ensure that articles do not conflict and do conform to general guidelines.

Cowboy (slang) EITHER someone doing unlicensed (and often ignorant) work OR someone taking unnecessary chances on the job.

CRAMPS (acronym) Construction, Renovation, And Modernization ProjectS.

Crossed neutral(slang) EITHER an identified conductor connected to a hot leg OR "crossing the neutral," (see entry).

Crossing the neutral (slang) Taking or offering bribes.

Crowfoot EITHER a crowfoot receptacle or cord connector, meaning a device having three straight slots, two of them angled, like a miniature of a NEMA 50 design, but categorized as NEMA 10-20 and rated 250/125 volts, 20 amps OR a support for fixture hickeys, screwed to the box or other base via two to four of its "feet." In old wiring, a crowfoot support was screwed through the holes in the back of a box so as to bite into wood. Less commonly, it was bolted to the box. In very old installations, or those by the very ignorant or indifferent, it was used without a box. See Fixture block.

CPSC (acronym) Consumer Product Safety Commission. An active voice in home safety, including electrical safety.

CSA (acronym) The Canadian Standards Association. Canada's standard-writing, testing, and inspecting agency. CSA also tests to UL standards, and marks products passing those tests, "CSA-NRTL."

Current-carrying conductor See NEC Article 100.

Cut-in notice The AHJ's go-ahead to the serving utility to provide power.

Cutout An overcurrent and disconnecting device.

Cutout box A box with a hinged door and a latch.

Daisy-chain EITHER a wiring layout that involves running from one enclosure to the next and splicing conductors, as opposed to radial wiring, OR a method of wiring the terminals of multiple devices within one enclosure using a single unbroken wire. The latter is a potentially risky system, for several reasons:

☐ loops sometimes are badly made;

☐ excess copper often shows;

☐ insulation often is pinched;

☐ inadequate free conductor length is the rule.

OR, more generally, "daisy-chaining" may be used to describe looping from device to device using their terminal screws or backwiring entries or both instead of splicing devices; NEC Section 300-13(b) makes this illegal for the grounded, neutral conductor of a multiwire circuit. The previous meaning, using a single, unbroken wire, although chancy, is not a violation of that section.

Decora™ A rectangular flush device in Leviton's line—and the most common shape for a GFCI. Bryant has, and has had, similarly-shaped devices on the market. A Bryant engineer reported that Bryant originated the rectangular-face design in the 1960s, but it did not gain ready market acceptance.

Denzar (also Baby Denzar) A type of ceiling fixture with a reflector and a glass bowl.

Despard A catch-all consisting of a fixture yoke into which you can fit one to three interchangeable and replaceable devices—any combination of switches, receptacles, and indicator lights.

Device ring A fitting to adapt a junction box to enable it to serve as an outlet, and also, commonly, to bring it flush to the surface. Different rings were made for 11B, 1900, and 8B boxes (see entries), for single gang and for two gang in the first two categories, and for the square boxes at different depths from flush to 2" recess. See entry for "Mud ring."

Dielectric Nonconductor or nonconducting, for electricians' purposes.

Dike Containment to prevent transformer fluid from spreading freely in the event of a rupture.

Dikes (slang) Diagonal cutters.

Donut (slang) Doughnut (see entry).

Double-break switch A switch that breaks a circuit in two places.

Double breasted (slang) A contractor having both union and nonunion divisions under one ownership. For example, a contractor might have fairly steady residential construction work, paid not just below union "A" scale but below the lowest wages the IBEW would be willing to negotiate. The contractor's separate commercial division might steadily employ a core group, staunch union members, and call on the IBEW's hiring hall upon winning major contracts. That division will pay the "prevailing wage," which is considerably higher than what they offer the nonunion workers in their other division.

361

Double pole A switch to break (the double-pole, single-throw version) or transfer (the double-pole, double-throw version) both legs of a two-wire circuit, or two legs of a three-wire circuit. The term double pole, double throw normally is not applied to four-way switches (see entry). A double-pole, double-throw switch has six terminals, in addition to any terminal used for grounding. Both a double-pole, single-throw switch and a four-way switch have four terminals in addition to any terminal used for grounding. Fortunately, you can distinguish the latter type without relying on your continuity tester, which when applied to a defective switch may not help you determine what should be in place. Unlike a four-way switch, a double-pole, single-throw switch has a handle with marked "On" and "Off" positions.

Double-tapping Attaching two circuits to one overcurrent device or, more generally, to one terminal.

Sometimes use of this term implies doing so in violation of manufacturers' instructions. Lacking instructions to the contrary, you must assume that a single terminal is intended for a single conductor.

Doughnut (slang) A stamped bushing, a reducing washer. Pairs are used, concave sides facing each other, for reducing the opening size of a knockout. Doughnuts are NOT Listed for use as part of the bonding path between metal enclosures and, say, armored cable connectors or metal raceways.

Downstream Further from, or heading away from, the source of power, measuring along the conductors. Compare Upstream (see entry).

DPDT (acronym) Double Pole, Double Throw (see entry under Double Pole).

DPST (acronym) Double Pole, Single Throw (see entry under Double Pole).

Drain hole An opening left—or, as CMP members have confirmed is intended, drilled—in the bottom of an enclosure or even a fitting such as a conduit body, literally to let condensation or infiltrating water drain out. The only guidance offered by the 1996 NEC for sizing drain holes is in Section 600-9(d). Unfortunately, just about anywhere water can drain out, vermin can crawl in. Sometimes installers deliberately leave extra openings in the bottom of enclosures, for instance by using indoor, nonsealing, connectors, and justify the violation of NEC Section 110-3(b) or 110-12(a) by saying that their purpose is to arrange the equipment to drain. This practice does not appear to match the CMP's intent.

Dress To prepare, usually in the sense of making tidy. One common use of the term is making wires in an electrical panel or trough line up tidily parallel and ensuring that they sweep through smooth, concentric bends, fulfilling the "neat and workmanlike" requirement of NEC Section 110-12. A far more important type of dressing, in terms of safe function, has to do with prebending or folding conductors at terminals before pushing devices back into enclosures. See Training.

Drift A solid steel cylinder, usually with some taper, but with flat rather than pointed or rounded ends. A drift is used to transmit driving force from a hammer. A drift is a relatively uncommon electrician's tool, but a drift of $\frac{3}{4}$" diameter or thereabout can be mighty handy on occasion.

Drive-it gun An explosive anchoring device (see entry).

Drop cord Extension cord, normally with a cage-enclosed lamp holder. A metal cage should be grounded; a plastic cage should be capable of withstanding the heat of the lamp without softening or charring.

Dryvin™ An expansion ·anchor with integral nail. Some find these far less satisfactory than other types of masonry anchor.

Duck seal, Duct seal, Duxseal™ A nonhardening sealing compound used to prevent building infiltration by air, moisture, and vermin.

Edison base Medium base, as in incandescent lamps.

Edison cells Nickel-iron rechargeable cells putting out 1.2 volts or so—once used, in series, to power bell circuits.

EGC Equipment Grounding Conductor. When you run a new EGC to existing equipment, you have to comply with the current—and ever-changing—rules on grounding and bonding. When a separate grounding wire is used inside a conduit, if at all possible the raceway and the grounding conductor should be bonded together at maximum intervals of 100' or less. While normally one must evaluate the impedance of one's grounding systems, a 1999 change in Article 110 will confirm and clarify that where a Listed system such as armored cable is installed as intended, one can rely on the manufacturer's assurances and need not oneself calculate impedance. Flexible metal conduit, on the other hand, does not carry the same assurance, though UL has evaluated it in lengths up to 100' for, and only for, bonding the secondaries of neon-type transformers.

EIA (acronym) Electronics Industry Association. Responsible in part for telecommunications standards.

El or Ell Elbow. A "factory el" is purchased prebent. A condulet (see entry) may also be referred to as an el: this certainly is true of the larger version known as a service elbow or service el. The 1999 Code cycle may clarify the fact that when Code panels say that a raceway should be arranged to drain, they intend for the installer to drill a hole at the lowest point of the run, normally in an el.

Electragist (archaic) Electrician—a term used quite a ways into this century.

Electrician's hammer (jocular slang) Lineman's pliers, from the practice of using them as substitutes for a hammer, for instance to hit a screwdriver against a locknut.

Electrician's pliers See under pliers.

Elexit™ A lighting outlet with integral fixture support, installed in the 1920s.

EMI (acronym) ElectroMagnetic Interference. EMI can be responsible for significant standing leakage (see entry). Therefore, GFCI tripping in older buildings can be laid to the presence of modern electronic equipment lacking EMI filters, as well as to faults, misconnecting, and old, deteriorated wiring.

Enamel Refers to paint, normally, not to ceramics. Enameled conduit still is mentioned in the NEC. As of the late 1980s, it still was the most commonly installed indoor raceway in Scotland.

Enclosure Something that prevents accidental contact with wires or devices. Very old versions were not necessarily nonflammable.

EMT (acronym) Electrical Metallic Tubing, or thinwall conduit. Covered in NEC Article 348. EMT has used many kinds of connectors and couplings:

☐ The modern weatherproof threadless, gland-ring type;

☐ the modern setscrew type;

☐ threaded ones integral from the factory;

☐ and some that were installed by indenting with a special tool.

So yes, that odd EMT fitting you may run into was a legitimate installation. Incidentally, the term, "thinwall conduit," while commonly understood, is strictly slang. Code panels have emphasized that EMT is not, technically, conduit. One never would make that mistake with ENT, Electrical Nonmetallic Tubing, whose use is far more restricted, ENT being more delicate, usually not sunlight-resistant, and furthermore subject to shattering if struck when cold. ENT, or "Smurf™" tube, looks completely different anyway.

E.P. Equipotential Plane. Should you do any work around substations or farms, establishing equipotential planes can be very important, if costly. The presence of a functioning, grandfathered system does not change that necessity. At substations, an equipotential plane is a matter of life safety. On farms where livestock are kept (not including poultry), stray voltage can impair the health and productivity of the animals. The 1999 NEC should also stipulate that where livestock regularly cross the edges of an equipotential plane, as when they are brought in and out of a barn daily, a gradually ramped potential change is needed.

Erickson™ A three-part coupling for RMC or IMC allowing two lengths to be connected without rotating either one. An alternative system serving the same purpose relies on set screws and has flexible seals.

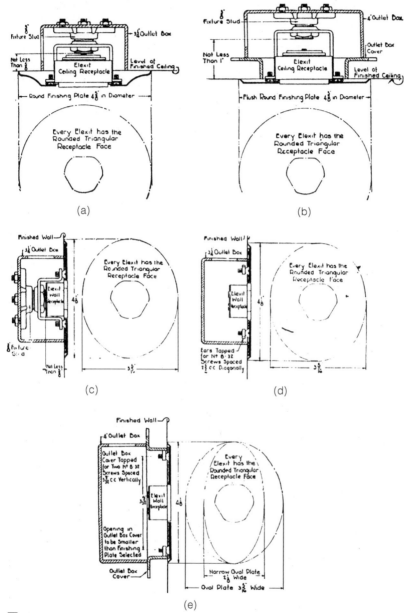

(a)

Fixture Stud — 3¼ Outlet Box

Not Less Than ⅜

Elexit Ceiling Receptacle

Level of Finished Ceiling

← Round Finishing Plate 4⅞ in Diameter →

Every Elexit has the Rounded Triangular Receptacle Face

(b)

Fixture Stud — 4 Outlet Box

Outlet Box Cover

Not Less Than 1"

Elexit Ceiling Receptacle

Level of Finished Ceiling

← Flush Round Finishing Plate 4⅞ in Diameter →

Every Elexit has the Rounded Triangular Receptacle Face

(c)

Finished Wall

3¼ Outlet Box

Elexit Wall Receptacle

4⅞

⅜ Fixture Stud

Not Less Than ⅜

3½

Every Elexit has the Rounded Triangular Receptacle Face

(d)

Finished Wall

3¼ Outlet Box

Elexit Wall Receptacle

4⅞

Ears Tapped for N° 8-32 Screws Spaced 3⅝ cc Diagonally

3⁵⁄₁₆

Every Elexit has the Rounded Triangular Receptacle Face

(e)

Finished Wall

4 Outlet Box

Outlet Box Cover Tapped for Two N° 8-32 Screws Spaced 3⅞ cc Vertically

Opening in Outlet Box Cover to be Smaller than Finishing Plate Selected

Outlet Box Cover

3¹⁄₃₂

4⅛

Every Elexit has the Rounded Triangular Receptacle Face

Narrow Oval Plate 2⅛ Wide

Oval Plate 3⅝ Wide

■ **G-2** *Elexit.*

365

Glossary

Exothermic welding Welding by a process that generates net heat. This is the only system of welding explicitly acknowledged in the NEC. It is one of the few systems that is recognized as acceptable for extending grounding electrode conductors.

Explosive anchoring device Using 22 caliber cartridges, such a device drives a short "nail," or an anchor from which a threaded stud protrudes. This system can create a very secure attachment to masonry, or a very unreliable one. There are several dangers associated with its use:

☐ It can damage hearing—90% of construction workers have significant hearing loss;

☐ The anchors can ricochet when the tool is not used with protective plates;

☐ and the anchors can pass through walls or floors to injure people on the other side.

FD A rectangular cast weatherproof box.

FR (acronym) Fire Resistant.

Finish rating How well a wall resists fire.

Fire resistance rating How well a wall keeps fire from affecting equipment on the other side. Fire resistance rating is the main characteristic you must attempt to restore after penetrating a fire-rated wall.

Fish tape A coiled line, most commonly of spring steel, or, more rarely, of nonmetallic and thus nonconductive material, for shoving through raceways and then pulling out with conductors attached. Fish tapes are of relatively little use for fishing in spaces such as walls.

Fixture block A block of hardwood to which a fixture was mounted via a crowfoot (see entry) or similar support. A fixture block was required in most cases where switches were mounted to wiring run in wooden raceways. It also was used with knob-and-tube wiring, in which case, at least later on, the conductors had to be protected by loom or porcelain tubes coming up flush to, or at least to within an inch of, the wall surface.

Fixture joint A mechanical transition fitting from a box or fixture stud (see entry) to a hanging fixture. A fixture joint often is designed to swivel; sometimes it incorporates a hickey (see entry).

Fixture stud A fixture support, originally used in open wiring, held in place by two bolts. The stud was prevented from rotating by those same two bolts. A no-bolt version patented in 1919 replaced the bolts with protrusions. In later work, a stud was inserted through the center back knockout of a box, and secured with a locknut.

Flex (slang) Flexible metal conduit, also known as greenfield; covered in NEC Article 350.

FLA (acronym) Full Load Amperes. Maximum normal current draw, other than during starting or stalling.

Flying splice (slang) An unenclosed splice.

Foot rule Wooden folding rule, commonly 6' to 8' when extended, sometimes manufactured with a hook on one end for the convenience of someone working alone who wants to take outside measurements, or a sliding extension for inside measurements, or both. "Six foot rule," "$4\frac{1}{2}$ foot rule" and so forth refer to various NEC spacing requirements.

Football clamp (slang) An acorn clamp (see entry).

Foundry switch A limit switch used in steel foundries. Foundry switches were manufactured with rubber boots to keep out the sand from foundry molds. These foundry switches found use in additional locations besides foundries.

Four-way switch A switch designed for multipoint switching of a single outlet or a single commonly switched set of outlets. See Chapter 4 for details. Compare Double pole (see entry).

FPE (acronym) Federal Pacific Electric, manufacturers.

French hook A fixture-hanging hook.

Friction tape Cloth tape imbued with pitch. In itself, friction tape is a good insulator, but its use is and, more to the point, was, to protect more delicate, better-insulating tape such as rubber tape.

FS A cast rectangular weatherproof box.

Gang In describing enclosures and box adapters, gang refers to the number of yokes that can be supported. In cover plates, correspondingly, gang refers to the number of device yokes or equivalent spaces in the box that can be covered.

Garden hose (slang) Liquid-tight flexible nonmetallic conduit, covered in NEC Article 351, Part B.

GEC (acronym) Grounding Electrode Conductor. The 1999 Code cycle clarified the fact that where bare conductors legally may be employed for grounding, there is no need to bring them into enclosures through clamps or connectors. An unused mounting hole or drain hole may serve. The GEC is the most common separately-run bare grounding conductor.

Gem box A flush-mount steel device box. The term now is used pretty much only for sectional switch boxes; originally, this was the name for the Chicago Fuse Manufacturing company's line of boxes.

GFI (acronym) Technically a misnomer, since the device is a Ground Fault *Circuit* Interrupter or GF*C*I—it interrupts the circuit, but the fault remains.

GFP (acronym) Ground Fault Protector, intended for protection of equipment rather than for life safety. Also GFPE, Ground Fault Protection for Equipment.

Gimme (slang) EITHER (Noun) an easily-solved challenge, known to some as a "no-brainer," OR (adjective) greedy, as in "a gimme (crooked) attitude towards performing inspections."; causing an unscrupulous third-party inspector working on commission to race through inspections, or any inspector to "cross the neutral" (see entry).

Goof plate (slang) A flush cover plate designed to hide patchy repairs or otherwise-unattractive surfaces. Goof plates usually are a bit longer and wider than standard cover plates; sometimes they are deeper as well, to give a flush finished appearance despite a protruding enclosure. Note that in a plaster or plasterboard wall, NEC Section 370-21 requires that damage around the enclosure be repaired to within $\frac{1}{8}$" of its edge; in such surfaces, the goof plate cannot legally be used to hide a big hole. It can, however, legally hide a very rough job of patching. Moreover, in a wood, cement, or other type of wall or ceiling, it can be used to hide a substantial hole without violating the NEC. It is possible that this disparity may be changed in a future edition of the NEC. Contrast a goof plate with a "Narrow" (see entry).

Gooseneck EITHER a flexible stem that is part of an old-fashioned desk lamp or wall-mounted fixture OR a similarly-shaped extension on the cover of a round hazardous location box cover. This latter type of gooseneck is used to convert the entire box into a sealing fitting, eliminating the need for sealing fittings at the conduits entering the box. One simply pours the sealing compound into the gooseneck. Its advantage over sealing raceways individually is that there is plenty of room in the box to separate the wires of, say, jacketed communication cable, so that sealing compound fills the space between them. This creates a better barrier.

Greenfield (slang) Flex (see entry).

Grounded conductor See NEC Articles 100 and 200. This is commonly known as the neutral (see entry), and as the identified conductor. It refers to the conductor which is grounded at the source of power. In three-wire, three-phase delta circuits this is not a neutral; in the very oldest wiring, it is not necessarily identified by being bare or natural gray. In modern switch legs, a white wire in cable is not a grounded conductor.

Grounding conductor See NEC Articles 100 and 250. A conductor that is not intended to carry any current except (temporarily) in the event of a fault. If a wire, it should be bare, green, or green with stripes.

Ground loop Current travelling, often dangerously, through the ground, as when an earth return is illegally employed. Because of the relatively high impedance of the earth, compared to grounding conductors, and especially compared to the grounded conductors of services, a ground loop may present a shock hazard for an indefinite period, rather than causing the overcurrent device to operate .

Gut (slang) A remodeling or renovation job in which walls and usually ceilings are removed, so that minimal fishing is required. This makes the work much more predictable.

Gutta Percha Malaysian tree sap used as underwater electrical insulation in early wiring.

Gypsum block A superb older system of very solid interior construction. Blocks of gypsum—the mineral that makes up plaster—were used to construct interior walls as one would use bricks or cinder blocks for exterior, and then plastered over. The resulting walls were wonderfully fireproof, sound deadening, and long lasting. Of course, if you need to run wiring through them, you may have to do quite a bit of channeling.

Hand drill EITHER a hand-held drill-motor OR a muscle-powered drill operated "eggbeater"-fashion via a crank rather than direct-driven like a brace (see entry).

Handi box, Handy box(slang) A utility box (see entry).

Hanger bolt A bolt threaded at both ends. One version combines a wood screw thread at one end with a machine screw at the other. Using this can make it easier to mount equipment on a wood joist or stud. Instead of holding the equipment up while tightening a screw or lag bolt, you first install the bolt. Then you stick the equipment onto the hanger bolt, over the protruding threads; as soon as a few threads of the nut catch on the bolt (with or without a washer), the equipment is held in place and you no longer need to support it by hand.

Harp A lampshade holder rising from the lamp base just before the lamp socket.

Heavy-up (slang) Service upgrade.

HEPA (acronym) High Efficiency Particulate Air (filter). HEPA filters used to be the most effective for negative pressure (pull air basically through what sits over your face) respirators. New "R100" particulate filters are quite an improvement over them. So are most "P95"s; even "N95"s.

Hickey, fixture support Different versions exist, screwing onto a pipe (or a crowfoot) and either being cast or pressed metal, either solid or (in more common usage) with openings, either with equal radii at both ends or reducing from, say, $\frac{1}{2}$" trade-size pipe to $\frac{3}{8}$" or smaller trade size.

Hickey, pipe A bender that requires more skill and permits more leeway than a "pipe bender," as its shoe is just long enough to permit the application of leverage against pipe (or tubing) without indenting or kinking it. The skill of the person doing the bending determines the smoothness and radius of the curve. Compare "Bender" (see entry).

Holophane™ A clear glass lamp shade with ridges/grooves for diffusion/scattering of the point light source.

Hot gutter (slang) The gutter or wireway between the point of service and the main disconnect.

Hot lazy (slang) Carter System (see entry).

HRC (acronym) High Rupturing Capacity, foreign (for example, New Zealand usage) for high Ampere Interrupting Capacity.

Hub EITHER, in more standard usage, a transition, normally threaded, between conduit and an enclosure, OR, sometimes, fixture joint (see entry).

ICBO (acronym) International Congress of Building Officials, publishers of some codes. See also BOCA, CABO, and SBCCI.

ICC International Code Council, an organization consisting of BOCA, ICBO, and SBCCI, but not CABO (see entries).

Ice pick Awl (see entry).

IDCI (acronym) Immersion Detection Circuit Interrupter.

Identified conductor Refers to a conductor with a white or "slate gray" (not just any gray) color. See NEC Article 200. The identified conductor normally is the neutral or grounded conductor (see entry). In old wiring, though, the presence of that "identification" is chancy.

IG (acronym) Isolated Ground.

IEC (acronym) EITHER Independent Electrical Contractors, Inc., the nonunion contractors' group most closely equivalent to NECA in its membership OR International Electrotechnical Commission, the organization that works for standardization of symbols, nomenclature, and to some degree safety standards.

IES (acronym) Illuminating Engineering Society.

Impedance Resistance, reactance, and capacitance.

Indicating Showing its state, normally visually. An example is a fuse designed so that it changes appearance when blown. See Figure G-1 for an example involving a switch.

Inlet A recessed male connector. When equipment is designed without an integral attachment cord, or provision for adding one, the cord set that you use to bring power to the equipment needs something to plug onto. In the event that a cord set, essentially an extension cord, used to bring in power is partly pulled out, those male prongs should not be exposed, lest they create a shock hazard. For that reason, the equipment's male connector is recessed as an inlet—the opposite of an outlet.

Intercommunicating In the case of alarms such as smoke detectors, intercommunicating means that when one is activated, all sound or flash the alarm.

Intermediate switch British for four-way switch.

Inverse-time An overcurrent device that opens proportionally faster, the higher the overload. Some circuit breakers rely on a thermal response for this characteristic, while others use a hydraulic feature, which makes their response less dependent on ambient temperature. The 1980 NEMA standard, AB-2, requires a circuit breaker operating in an ambient temperature of 25 degrees centigrade to open in no more than 900 seconds at 250% of rated current and 100 seconds at 350% of rated current. In other words, in a warm (77° fahrenheit) room, or at least a warm enclosure, a 20-ampere circuit breaker that takes 15 minutes (900 seconds, in more familiar terms) to trip under a 50 ampere load may or may not have lost its factory calibration, but it does meet minimum requirements! A maximum of a quarter-minute or a half-minute delay would seem more appropriate for such overloads, and indeed a reputable manufacturer may so specify. The *IEEE Soares Grounding Book* reproduces a table from Square D, Inc., giving such figures.

While an overcurrent device is waiting to operate, it is not limiting the energy passing through it. Therefore, inverse time devices are current-limiting in a sense only within their instantaneous ranges, where the current is high enough that they operate as fast as they are mechanically capable of opening. That instantaneous range may begin at five times or at twenty times rated current.

Considerations such as these make different Listed brands of breakers very much unequal. Unfortunately, we have no Consumers' Union to publish comparisions.

IPS Iron Pipe Size. There is more than one IPS specification for a particular trade size. Half-inch trade size PVC, for example, can be Schedule 40 IPS, Schedule 80 IPS, or the thinner Schedule A IPS. But the designations hearken back to the black iron that was being installed a century ago.

IR (acronym) InfraRed. IR detectors can in some cases be harder to fool than motion detectors that rely, actively or passively, on sound. For instance, they will not shut off the lights in a room merely because the occupant has not moved for a while.

IrDA (acronym) Infrared Data Association. Somewhat comparable to the EIA/TIA industry groups supporting voluntary standards.

Jackleg (slang) Unscrupulous—for electricians, specifically, the term is used to mean unlicensed or without a permit. Originally the word was "blackleg," and meant strike-breaking.

Japanned Varnished (or enamelled).

J-box Junction box.

Jelly Jar An outdoor light fixture, traditionally employing one 60-watt incandescent lamp. It consists of a rounded metal canopy with two cap nuts for 8-32 screws; a rounded metal cap section attached to the center of the canopy, through which two fixture wires enter; and a replaceable lampholder descending from the cap section via a short fixture nipple and cap. A circumferentially ridged, cylindrical, glass diffuser and lamp protector either threads into the cap section, or, more commonly, is held to it by three setscrews.

Jiffy box (slang) Utility box (see entry).

Jockey leg (slang) Traveller (see entry).

Jog To apply power for just a moment, to confirm direction of rotation or to make parts align.

Joint An old term for connections, mainly splices; used also for terminations.

JB (acronym) Junction box.

Kearneys Split bolt connectors, after the brand of that name.

Kellem™ A cord connector providing strain relief, originally a private brand (now owned by Hubbell). A "romex™ connector" is not designed or listed for use on flexible cords. More generally and colloquially, "kellem" is used to refer to a wire cage gripper, whether used for strain relief on a cord, or as part of wire- and cable-pulling equipment, or elsewhere.

Kett™ A brand of tools that included a drill motor manufactured with a remote switch, whose body mounts on 3/4" trade-size pipe. Together with the remote switch, this characteristic enables it to

---

BRYANT

ECONOMY CABLE GRIPS
Quick Reference Selection Guide
for Grip Attachments

STRAIN RELIEF GRIPS

WIDE RANGE BUS DROP
Used indoors for cable support where flexible cable connects electrical equipment to bus duct. Support or restrain air hose and water hose. See pg. 11.

WIDE RANGE STRAIN RELIEF
Indoor use only for wiring of electrical enclosures, machine tools, portable power tools, bus drop cable systems. See pg. 8.

DELUXE CORD
Indoor or outdoor use where subject to moisture, splash, or washdown. Examples are crane hoist and pendant drop stations, hand tools, pumps, and processing equipment. See pg. 9.

LIQUID-TIGHT, FLEXIBLE METAL CONDUIT
Wiring of machine tools, electrical enclosures, motors, and systems subjected to vibration, flexure, motion, or strain. Also available in straight, 90°, or 45° configurations. See pg. 9.

SUPPORT GRIPS

SINGLE U EYE
For single hook attachment of permanent indoor/outdoor cable. Available on heavy-duty, standard duty, and service drop grips. See pgs. 12, 13, 14, 16, 18, 20, & 21.

DOUBLE U EYE
For double hook attachment of permanent indoor/outdoor cable. Available on heavy-duty and standard duty grips. See pgs. 12, 13, 14, 16, & 18.

LOOPED BALE
For wrap-around attachment to an existing fastener in permanent indoor/outdoor cable. Available on standard duty, heavy-duty and light-duty service drop grips. See pgs. 15, 17, 19, 20, & 21.

SINGLE OFFSET EYE
For offset hook attachment of permanent indoor/outdoor cable. Available on heavy-duty, standard duty and light-duty support grips. See pgs. 15, 17, & 19.

PULLING GRIPS

HEAVY-DUTY REVOLVING EYE
For underground wiring and overhead heavy-duty pulling of service lines and new construction cable. See pg. 25.

HEAVY-DUTY FLEXIBLE EYE
For overhead transmission and distribution line stringing. See pg. 26.

LIGHT-DUTY FLEXIBLE EYE
For light industrial pulling of electrical cable and for underground and industrial plant wiring and re-wiring. See pg. 28.

SLACK GRIPS
For removing underground cable and pulling slack in existing cable and new installations and when end of cable is not available. See pgs. 29 & 30.

OTHER SPECIALTY GRIPS

SPLICING GRIPS
Used as temporary splice for rope, cable, and wire rope or as reinforcement to protect cables and hoses. Available with flexible tube for pulling grips. See pg. 32.

CONDUIT SUPPORT GRIPS
Ideal for supporting electrical cable inside rigid conduit via a supporting ring. See pgs. 22, 23, & 24.

■ **G-3** *Kellem-type grips. Courtesy Jim Noonan and Bryant.*

be operated by somebody standing on the floor, rather than on a ladder or stilts, to drill joists.

Key EITHER, more or less as in common usage, the metal operator of a locking mechanism (as in the removable operator of a key switch (see entry) OR the integral switch operating a light fixture (as opposed to, for instance, a keyless porcelain) OR that protrusion of plaster or, more commonly, sand coat (see entry) through and behind lath which holds it in place, "locking" it to the lath (see entry).

Key switch A switch that required inserting a "key" (actually, a fairly simple metal shape) to flip it on or off.

Keyhole saw A hand saw with a narrow triangular blade. In recent years, this has largely been supplanted by the reciprocating saw, and to a lesser extent the jigsaw.

Keyless An unswitched or remotely switched fixture, as opposed to "keyed," having a switch as an integral part of the fixture.

Kick A short offset (see entry), normally one raceway diameter or less.

Kindorff™ A brand of mounting hardware consisting of squared U-shaped rigid steel channels with openings for bolts and threaded rod. The lips at the U-opening support special clips that provide additional means of equipment mounting.

Kleins™ Lineman's pliers (see entry). After Klein tools, in business since the mid-1800s.

Klieg An arc lamp (named for the Kliegl Company of Long Island, N.Y.).

Kludge (slang) Kluge (see entry).

Kluge(slang) A lash-up or make-do solution. Also kludge.

Knife switch An open switch, normally live-front, with no quick-make or quick-break characteristics. (See Figure 5-7.)

Knob A support for "knob-and-tube" wiring, consisting of a nonporous porcelain knob, glazed or unglazed, with a hole through the central axis. A 10d or larger nail, or equivalent wood screw, went through this hole to secure it to the structure, and conductors were pinched between its two parts or tie-wired to it. Besides keeping them in place, knobs held conductors away from the building surface.

Knockout EITHER an embossing that can be removed to create an opening OR the opening created thus or by drilling with a hole saw or or using a hole punch, or the equivalent threaded opening in a cast box. There are various types of knockout:

(a)

(b)

■ *Knob (a). Knob-and-tube wiring emerging from loom (b).*

☐ circular openings in standard trade sizes for entry of conductors into enclosures or wiring compartments—these are the most common ones;

☐ rectangular openings in panel covers for circuit breakers;

☐ rectangular openings in the backs of fixtures for access to the enclosures over which they may be mounted, and of surface boxes to convert them into extension boxes. (See Figure 7-4(a));

☐ variously-shaped openings embossed in boxes containing integral cable clamps, for cable entry;

☐ in some old boxes, semicircular openings going back from the front edge to enable them to be mounted over existing gas pipe;

☐ odd-shaped openings in the backs of boxes intended to be suspended between joists, for connectors used with box-mounting straps.

Knockout seal or plug The closure device required when a knockout has been removed and no cable and pipe installed, or when cable or pipe was installed and later removed. Most commonly, these are press-fit, but they are also available with screw anchors that go on the enclosure interiors. A threaded version is used for cast, weatherproof boxes.

K.O. (acronym) KnockOut (see entry).

K.O. seal Knockout seal or plug (see entry).

Kwikchange™ Decora-type device made by Leviton.

Labeled See NEC Article 100. Equipment can be Listed without there being a label on the individual item as installed, so while an item's Labeling demonstrates the fact that it is Listed, albeit not necessarily used within the scope of its listing, the lack of a Label does not prove that it is not Listed. By the time equipment qualifies as old wiring, labels have dried up and fallen of.

Lamp EITHER properly used as a noun, a light bulb plus its base or terminals, OR, as a verb, to install one, OR, more loosely, as a noun, a light fixture. Used with an adjective, as in "desk lamp," the latter meaning is clear and proper. This technicality is not something to fuss about, so long as all parties understand each other.

Lamp socket A brass-shell lampholder. Be alert for deteriorated cardboard separator/insulators in old ones.

Lath Either wood strips (wood lath) or expanded metal mesh (metal lath) nailed across studs and joists to hold plaster in constructing walls and ceilings. Lath has nothing to do with lathes, outside ancient history. Towards the end of the period when plaster walls were common, plasterboard underlayment was introduced in place of lath. (See Figure 8-2.)

Lay In stranded wire, the angle that the strands make with the central axis.

Lazy neutral (slang) Carter System (see entry).

LB A type of conduit body. See entry under Condulet.

Light bulb Technically, this refers to the glass itself rather than the entire lamp including the filament(s) and base. This technicality is not something to fuss about, as long as all parties understand each other.

Lineman's (also linemen's) pliers Pliers designed for both grabbing and cutting. Sometimes they are made with a crimping die behind the pivot. Other times, the area behind the pivot is intended for use in grabbing fish tape, so that the tape can be pulled using one hand over the plier handles and the other over the jaws.

Listed See NEC Article 100.

LL A type of conduit body. See entry under Condulet.

Loom A semi-rigid cylindrical cloth covering for conductors, used to bring open wiring or knob-and-tube wiring into outlet boxes or otherwise protect it from damage. At one point, it was manufactured by the Wiremold company! Ettco was another manufacturer, calling it "nonmetallic conduit." See illustration G-4(b).

Loop, looping EITHER daisy-chain (see entry) OR simply the curve into which the end of a conductor must be shaped so as to fit snugly around the shaft and under the head of a terminal screw OR ground loop (see entry).

Los Angeles Switching (slang) Carter System (see entry).

Luminaire Light fixture.

Lump hammer A short, one-hand, sledge hammer.

Made Factory- (or, occasionally, shop-) prepared. Made 90s are common, made 45s less so except in RNMC (see entry).

Make Establish contact. Compare Break (see entry).

Make up Complete an installation; tighten; secure; prepare ahead of time.

Mare's Tail An attachment used for pulling the middle of a cable. See entry for Bird Cage.

Medallion A usually ornate plaster (or imitation plaster) ornament surrounding a ceiling outlet. A medallion can be quite useful in concealing, say, a modern, deep box that has replaced an inadequately sized one, when the fixture or canopy is too small to do so. See entry for Rose.

Megger™ (brand, and also generic slang) (verb, noun) (Employ a) megohmmeter (see entry).

Megohmmeter A battery-or crank-operated device for impressing 1,000 volts or so on wiring, once you have determined with a normal ohmmeter that its insulation appears intact. A megohmmeter sometimes but only sometimes will detect incipient insulation failure. Using a megohmmeter can be hell on any electronic equipment, or even on capacitors, that inadvertently were left in the circuit you are checking.

Mercury switch A very long-lasting switch whose contacts tended not to corrode because they made and broke in a small pool of mercury. A mercury switch will not work if installed upside-down.

Michigan 3-way (slang) Carter system (see entry).

Minibreaker A circuit breaker in the shape of an Edison-base plug fuse. It is indicating and resettable, but its operation is magnetic only. Tripping is automatic only, but it can be unscrewed. It will not fit in a no-tamper fuseholder.

MB (acronym) Main Breaker.

MLO (acronym) Main Lug Only.

Modular In telephone wiring, a modern jack that permits quick-release connection and disconnection, without use of a screwdriver or punch-down tool.

Molding (slang) EITHER moisture-proofed wooden raceway OR metal or plastic surface raceway OR a plastic wire cover that is not Listed as a raceway.

Monument EITHER a pedestal box (see entry) OR a larger, ground-mounted splice box such as is used by telephone or cable companies.

■ **G-5** *Wooden molding installation.*

Motor generator A rotary converter that can be used to change voltage and phase, to electrically isolate, and sometimes to provide short-term uninterruptible power by virtue of its rotational inertia.

Mouse hole A hole drilled for a temporary feeder, often through a fire-rated wall in a location such as a theater. Older theaters, and buildings used as theaters, may have insufficient power to serve broadcast equipment or shows such as rock concerts. Firestopped or not, conduit-sleeved or not, mouse holes drilled to accommodate temporary feeders can violate the local building code. For this reason, a proposal for the 1999 NEC to authorize mouse holes on a blanket basis was rejected in committee.

Mud box (slang) A concrete box (see entry).

Mud ring (slang) Used loosely, a device ring. Used more precisely, only the type of device ring with a round opening for support of a fixture as opposed to a rectangular opening for support of a device or devices.

Multihat A combination inspector (see entry).

Multimeter A volt-ohm-ammeter, normally.

Multipoint switch, switching Three-way or four-way (see entries).

Multiwire A circuit with two or more "hot" conductors, each of which powers loads connected between it and the one neutral grounded conductor shared by all the "hot" conductors. In residential multiwire circuits, the disconnects must be operated simultaneously, as by multipole circuit breakers, circuit breakers with Listed handle ties, or pull-out fuse blocks. In commercial and industrial locations, this requirement is absent.

Regardless of legalities, the practice is one that will enhance safety. Polyphase circuits (see entry) do not have that exemption, nor do single-phase circuits connected between two hot legs. In those latter two cases, circuit breakers must be of the common-trip design.

Mungo (slang) EITHER stolen material OR, more commonly, scrap copper.

Myers Hub™ A self-bonding and self-sealing two-piece hub for connecting RMC or IMC to an enclosure, for instance in service work. It is a screw-down rather than a bolted hub.

NAHB (acronym) National Association of Home Builders. The NAHB is primarily of service to small-scale builders, remodelers, handypersons, and developers. They do have programs of interest to electricians, including Smart House™ and photovoltaics demonstration projects.

379

Nalox™ The trade name of one brand of wire-pulling compound.

Narrow (synecdoche) A cover plate for enclosures within partitions 2" wide or narrower (Think of a wired office partition.)

NBFU (acronym) National Board of Fire Underwriters, early publishers of the NEC.

NEC (acronym) National Electrical Code, NFPA 70.

NESC (acronym) National Electrical Safety Code, utilities' equivalent to the NEC. The two documents have been separate since 1915.

Needlenose See under Pliers.

Neutral The term would be identical with "identified conductor," (see entry) except that in many older wiring systems, the identification is nonexistent, or at best rather muddy. Instead of finding "all cats are gray at night" you will often find "all insulation is gray or black in old boxes." The precise term commonly intended by "neutral" is "grounded conductor." Strictly speaking, the term "neutral" is not very meaningful with respect to a two-wire circuit. Therefore, this book uses "grounded conductor" for more accuracy, unless referring to a wire that is definitely a neutral. This is not a point to belabor in normal conversation with a customer.

New work Electrical installation undertaken as a building is being constructed.

NFPA (acronym) National Fire Protection Association. The NFPA is the author of the NEC and of many, many other standards covering topics such as lightning protection, sprinklers, hospitals, and fire alarms; see Code.

NIOSH (acronym) National Institute for Occupational Safety and Health. The Federal agency that does research to set standards for manufacturers of safety gear, and also to form bases for OSHA (see entry) to set their standards.

Nipple, close nipple, chase nipple, offset nipple Threaded pipe of considerably less than full (10') length.

A close nipple is a very short length with running thread. The 1999 NEC should make it clear that running threads are forbidden on IMC and RMC *at couplings*. This does not forbid your using a single close nipple, with four locknuts and a couple of bushings, to connect adjacent *enclosures*.

A chase nipple also is all thread, up to its integral bushing.

An offset nipple has a kick in it of about one diameter, maybe $\frac{7}{8}$" for a $\frac{1}{2}$" trade size nipple.

NIST (acronym) National Institute for Standards and Technology; formerly the National Bureau of Standards or NBS. NIST is responsible for calibrations, and for some international coordination of standards.

NM (acronym) EITHER NonMetallic, when used as an adjective, OR NonMetallic sheathed cable, covered in NEC Article 336, and informally known as Romex™.

NMRC (acronym) Nonmetallic Rigid Conduit, also known as PVC conduit, although fiberglass versions, for instance, also fit this classification.

Nonmetallic conduit a term formerly used for loom (see entry). Now refers to either NMRC or nonmetallic flex/"sealtite."

Nose The junction of the conductors or cable and the fish line or tape or rope. A nose normally is made up of many wraps of electrical tape covering the folded-over junction, and slathered with pulling compound to facilitate a smooth pull.

NRTL (acronym) Nationally Recognized Testing Laboratory. An organization such as Underwriters Laboratories, Inc., Factory Mutual, Inc., and about a dozen others, recognized by the Department of Labor as qualified to List devices and equipment as complying with ANSI standards. The acronym often is pronounced to rhyme with "Myrtle" or "turtle."

OC, OCP (acronym) OverCurrent, OverCurrent Protection (or Protective).

Octopus (slang) A receptacle adapter permitting the attachment of additional cord connectors beyond the number the receptacle is designed to accommodate.

Offset Two identical bends, back to back, usually near the end of a raceway, so that on either side of them the lengths of raceway are parallel but shifted.

Offset nipple See Nipple.

Okie rope (slang) Nonmetallic-sheathed cable, at least in parts of California.

OL (acronym) OverLoad.

Open An open connection; an unintended interruption in a circuit.

OSHA (acronym) Occupation Health and Safety Administration. This U.S. agency, part of the Department of Labor, sets safety standards for workplaces and employees, except for government employees. Even government agencies, while exempt from OSHA record-keeping requirements, inspections, and fines, are bound by the fact that OSHA rules represent generally accepted standards of safe practice.

Outlet Loosely, a receptacle outlet. More precisely, any place where electricity leaves the fixed wiring, including lighting outlets and, arguably, hard-wired appliances, but not, for instance, switches or other control devices. This book uses the more precise meaning.

Outlet plate A type of equipment that used to be permitted as a base for device mounting before the NEC required the installation of devices in boxes with sufficient room to enclose conductors.

Ovalduct™ An oval raceway once produced (and which you may still encounter in the field) that was intended for use in under-plaster extensions. It used correspondingly shaped fittings, which fit in correspondingly elongated knockouts in boxes.

Ovalflex™ Armored cable corresponding to Ovalduct™ (see entry).

Pan (slang) EITHER Pancake box (see entry), OR panelboard OR Ceiling panel (see entry).

Pancake (slang, but common usage) A shallow outlet box, 3" to 4" round, $\frac{1}{2}$" or, nowadays, $\frac{5}{8}$" deep. At one point, they were manufactured without fixture mounting ears. (See Illustration 5-3).

Panel blank A knockout seal for an opening in a panel cover. A panel blank for brand A may not be suitable for use with brand B. See Knockout Seal.

PC (acronym) Pull Chain. There is little consistency, jurisdiction to jurisdiction, as to whether a pull-chain fixture simply can be replaced where the current Code requires a light fixture controlled by a wall switch or equivalent means.

PE (acronym) Professional Engineer.

Peanut butter jar (slang) Jelly jar (see entry).

Pedestal box An enclosure that is mounted above the floor, or above a counter, on an integral equivalent of a conduit stub (see entry).

Penetrox™ Trade name of an antioxidant compound—noncombustible.

Persuader (humorous slang) A large, high-leverage, or heavy tool used to accomplish a job with less sweat—for instance, a sledge hammer used where a lump hammer has had difficulty, or a 24" pipe wrench to turn a pipe which your pocket channels have not managed to budge.

Piggyback (slang) A tandem circuit breaker (see entry).

Pigtail EITHER (noun) a short length of conductor attached to enable termination OR (verb) to attach a pigtail.

Plaster ring (slang) Device ring (see entry).

Pliers

> **Diagonal cutters** Pliers that can be used to grip, but really only have cutting edges.
>
> **Electrician's pliers (slang)** The term can mean lineman's pliers (see entry), or can refer to other pliers considered particularly suitable for electricians, such as needle-nose pliers with a crimp die
>
> **Needle nose** For the purposes of this book, long-nose pliers. Some restrict their use of the term to long-nose pliers that come to quite narrow, sharp points. Needle noses are mainly useful for grabbing anything in an enclosed space too small to admit lineman's pliers. An example is turning—though not completely tightening—a locknut in the back of a deep, narrow box. Some use a needle nose for bending wire into hooks to go under terminal screws. Lineman's pliers work quite as well for that purpose.
>
> **Polynose** Channels (see entry).
>
> **Pump** Water pump pliers (see entry).
>
> **Side cutters (slang)** EITHER lineman's pliers OR, less properly, diagonal cutters.
>
> **Snipe nose** A type of long-nose pliers.
>
> **Stork beak** A type of long-nose pliers.
>
> **Telephone pliers** Long-nose pliers without a cutting jaw.

Polynose See under pliers.

Polyphase A circuit with at least one load connected between the hot conductors fed by separate phases of a system of two, three, or more phases. Used more loosely (and, technically, incorrectly), to mean a load connected between two or more hot conductors, even of a single phase system.

Porcelain (synecdoche) A single lampholder designed to cover an 8B box or an equivalent opening.

Post-and-wire Knob, as in knob-and-tube (see entry).

Pots (slang) EITHER transformers OR potentiometers OR dashpots.

Presenting problem What the customer complains about or asks for, or the symptoms you initially observe, as opposed to the root of the difficulties.

Pressure connector A wire connector that does not rely on soldering, brazing, or welding. The term includes but is not limited to devices such as Wirenuts™.

Preprint The former term for ROP (see entry).

Pull el, pull elbow An extra-short conduit body (see entry under Condulet) creating a 90° turn. Basically the LB configuration, but with equal "legs" and the opening and cover on the outside of the angle between them

Pump pliers Channels (see entry).

Punch-down tool A tool for pushing a signal wire down between insulation-piercing teeth, and also to trim the end of the wire.

Pushmatic™ An obsolete brand of circuit breaker. Its handle was operated by a push to turn it on if it was off, or to turn it off if it was on, rather than by toggling.

PV (acronym) PhotoVoltaic.

PVC (acronym) EITHER PolyVinyl Chloride OR, used more loosely, nonmetallic rigid conduit of PVC.

Quad (slang) Four. So a "quad" receptacle has four outlets.

Qualified Competent to deal with a particular installation. This competence extends well beyond professional training and skill, to include familiarity with the particular circumstances—layout and design, for starters. A plant maintenance worker may be qualified, in this sense, while a master electrician, with a half-dozen licenses, is not.

Rabbitt (slang) Mungo (see entry). (Rabbitt is also used as a verb.) Not to be confused with rabbett.

Raceway A system for enclosing and protecting wiring, such as conduit. The term formerly was used to refer specifically to surface raceways such as Wiremold™ and wooden molding. Some still use it exclusively for surface raceway, NEC Article 352.

Radial wiring Wiring with a separate cable to each outlet, as opposed to looping. This design most commonly is employed for dedicated circuits. It rarely is used with multioutlet circuits because of the added expense. When used with multioutlet circuits, it does have the advantage that any of the outlets can easily be converted to a dedicated outlet with its own circuit—or subgroups of them can be separated out.

Rag wiring (slang) Conductors whose insulation includes vegetable fiber. Originally, this term was strictly used to refer to conductors insulated with varnished cambric, type VC. These literally were wrapped in cloth, which was impregnated with insulating varnish, much like the varnish used on magnet wire. For greater protection, some of them were coated with lead, making them type VCL. In recent decades, the term has been used more loosely. The most common version of so-called rag wiring now encountered is copper insulated with an inner layer of rubber, and that covered with braided cotton.

Range An electric range consists of a cooktop plus, normally, an oven. In existing installations it can be wired with a 240/120-volt, three-wire circuit with the frame bonded to the grounded conductor. The same is true of a wall-mounted electric oven, but technically not of a freestanding one, which somehow was not covered under the same exception and therefore must have separate grounded and grounding conductors if it lacks a cooktop.

Ramset™ One brand of explosive anchoring device (see entry).

Rational Easy to follow—taking few words to describe accurately, as when writing out a circuit directory.

Rawlplug™ A traditional brand of fiber-and-metal anchor for securing screws going into masonry or solid plaster. The term sometimes is used generically for soft anchors.

RCD (acronym) Residual Current Device—the European equivalent of a GFCI or GFP (see entry).

385

Reciprocating saw An electric saw with straight, narrow blades of various lengths, materials, and tooth designs, which has eliminated the need for many other saws in rough work.

Red hat (slang) An antishort bushing required with armored cable. Its use is optional in installations of MC cable and flex. Now plastic, it used to be fiber—sort of like cardboard. Before these red bushings were introduced, brass ones sometimes were used.

Red head EITHER (slang) a red hat OR, trademarked, a masonry anchor designed to drill its own hole, using a suitable hammer-drill.

Red lead Lead-based primer formerly used for corrosion protection of metals. The name was retained after the formulation was changed, removing the lead.

Red leg (slang) The phase or terminal, in 3-phase, 4-wire delta systems, that is furthest (has the greatest voltage difference) from the grounded midpoint. There are entries for this under many other names.

REN (acronym) Ringer Equivalence Number (see entry).

Renewable Rebuildable, as in replaceable contacts. Renewability used to be common in fuses, too, and once upon a time in receptacles. In some foreign usage, "renew" simply means to replace

Return The grounded conductor (see entry), in grounded systems.

Rheostat A variable resistor; the term sometimes is still used, imprecisely, to refer to dimmers, although they now work on a different principle. Rheostat-type dimmers were tougher, but inefficient.

Rigid (synecdoche) Rigid metal conduit, covered by NEC Article 346. The term sometimes is loosely used for Intermediate Metal Conduit as well.

Ring As a verb, EITHER to score conductors, especially stranded conductors, so that subsequent bending easily causes a fatigue break, OR to run your knife in a circle through insulation. This is far more likely to nick, or, worse, "ring" the conductors than the "pencil-sharpening" method of cutting back insulation. As a noun, in old-style four-wire telephone station wiring, the ring conductor was the red wire. It corresponded to the circular, concentric conductor in old switchboard phone plugs, roughly equivalent to the shell in a plug fuse, fuseholder, lamp, or lampholder. The green wire was called the tip and corresponded to the central conductor in the plug. The yellow wire was the ground, and the black wire powered the light. Modern replacements use red and green as one pair, with black and yellow serving for a second telephone line. See Chapter 9, "Relatively Rare Situations."

Ring out To test for continuity and identify circuiting. This job used to be performed with a battery-operated bell or buzzer, hence the name.

Ringer equivalence number The classification of a telephone device such as a handset in terms of how heavy a load it constitutes.

Ripper A tool used to remove cable sheath, or at least prepare it for removal by slitting it at a controlled depth, avoiding harm to the insulation of the enclosed conductors.

Rise In choosing device rings or adapters to comply, for example, with the requirement for flush mounting in NEC Section 370-20, rise refers to the distance between the base of the ring or adapter and the surface against which equipment will be mounted. The thickness of that base may or may not be figured in.

ROC (acronym) Report On Comments. It is a listing of critiques offered, by both the public and NFPA panels, regarding proposed amendments to the NEC that have been published in the ROP (see entry). It includes panel responses to those critiques. None are final, though, until after the vote at the annual meeting of the NFPA.

Rocker A switch whose operation is rather like a toggle switch (see entry), except that you push in the top or bottom to operate it rather than flipping a handle up or down.

Romex™ Nonmetallic sheathed cable. The name is used generically, after Rome Cable Company's product.

ROP (acronym) Report On Proposals. A listing of proposed amendments to the NEC. The ROP includes panel actions, to be confirmed or overturned by vote of the full membership or the correlating committee (see entry).

Rose EITHER a ceiling ornament, like a smaller version of a Medallion (see entry), OR a Rosette (see entry).

Rosette An old type of ceiling canopy used for pendants or for pull chain switches. Some rosettes were fused, using solder wire. Some could be twisted directly into wood molding. Rosettes also were used on walls, with turn switches (see entry) installed in them.

Rotary converter A three-phase motor whose idling creates or smoothes out three-phase power that has been approximated through use of capacitors. Converters are used where the serving utility provides only single-phase but loads require three-phase.

Rotary switch See turn switch.

RTV (acronym) Room-Temperature Vulcanizing (compound). Silicone bathtub caulk-and-seal is an example. Other versions, such as butyl rubber compound, are available. An indoor connector plus silicone seal is not a legal substitute for a Listed outdoor waterproof connector. That said, should you find indoor connectors, of long standing, in place outdoors, it would be going overboard to take apart, say, the service entrance. Adding RTV is a reasonable compromise. It is also excellent sealant for when you mount something with a rear opening, such as an old-style FS box, against a damp or wet or uneven surface, even if the screw heads more or less close the openings. Most importantly, some manufacturers' instructions explicitly require you to apply sealant between their equipment and the mounting surface, at least at the edges. This will do the job. Some solderless wire connectors are Listed for wet locations or direct burial, thanks in part to being filled with RTV. Two cautions: first, this does not mean you can take a regular splicing device and legally convert it for burial by adding caulk; second, a Listed wet location connector may not be reusable, because the RTV has been disturbed.

Rubber tape An insulating tape of stretchy, self-annealing, natural or artificial rubber, with relatively poor resistance to mechanical damage.

387

Run EITHER (verb) to install wiring OR (noun, slang) circuit, more or less. As a noun, a "run" refers to a sequence of outlets or enclosures fed from one to another to another. Thus, an "end-of-run" outlet has one cable entering its enclosure, but a "mid-run" outlet has at least two—one coming from upstream and one or more proceeding downstream. One circuit may feed runs going in two directions.

Runner (slang) Traveller (see entry).

Running thread Pipe that has been threaded continuously from one end to the other. This weakens it considerably, as well as—if field-threaded—destroying the surface corrosion protection. See also Close Nipple, under Nipple.

RX (acronym, slang) Romex™, meaning nonmetallic-sheathed cable.

Saddle, saddle bend Back-to-back offsets deep enough and spaced far enough apart to create a sort of U or inverted U shape to accommodate an obstruction the raceway needs to get past. Some consider a saddle bend to "use up" the four 90° bends allowed in a raceway between enclosures or conduit bodies or pull els.

Salting (slang) Union members, often salaried union organizers, hiring on to a nonunion crew for the specific purpose of bringing the shop into the union.

Sand coat In plaster walls, most of the thickness of the wall finish actually is made of a mix with a high proportion of sand, that being covered with a relatively thin coat of plaster of paris. In repair, it usually is simplest to cut a piece of drywall to fill most of the space, and plaster over that.

Sawzall™ Milwaukee brand reciprocating saw (see entry). Also used generically.

SBCCI (acronym) Southern Building Code Congress, International. Also see BOCA, CABO, and ICBO.

Scratch awl See Awl.

Screwit™ A Wirenut™-style solderless connector used in the United Kingdom.

Sealtite™ Liquid-tight flexible metal conduit, covered in NEC Article 351A. The brand name is used generically.

Selfeed An auger bit.

Service el or ell Large conduit body (see entry). Normally LB style.

Shunt conductor EITHER a conductor used to divert dangerous voltage OR a traveller (see entry) OR a conductor illegally installed behind a meter to bypass it and steal power.

Sidewinder (slang) A circular saw with a nonworm drive. The term also is used for a circular saw with its blade extended past the blade guard. Potentially dangerous due to bypassing the guard, the latter is used for making straight flush cuts. An electrician is more likely to rely on a reciprocating saw with a flexible blade, and merely approximate "flush" and "straight."

Shell The threaded cylindrical portion of a lamp base and lamp holder.

Sherardized Galvanized.

Ship auger, Ship's auger A self-pulling wood drill bit, generally quite long.

Side cutters See under pliers.

Side job (slang) Work contracted independently, normally without a permit, by an electrician or apprentice lacking an electrical contractor's license—sometimes in competition with his employer, and sometimes using the employer's tools or even materials.

Skin (slang) To remove insulation from conductors.

Skin effect The increase in impedance with frequency resulting from the fact that, especially at high frequencies, most of the current flows toward the outside of a conductor due to electrons' mutual repulsion. This is far more pronounced with large conductors than with smaller conductors run in parallel.

Slip conduit (slang) The British equivalent of EMT, formally known as Class B conduit.

Slip-in connectors Connectors that secure conduit or cable to an enclosure without a locknut, and in some cases without a setscrew. Thomas and Betts made slip-in armored cable connectors as far back as 1926.

Slugin™ A stud-type masonry anchor that is installed using its own driving tool.

Smurf™ **tube, tubing (slang)** Nonmetallic (commonly blue) tubing, covered in NEC Article 331.

Snap switch A spring-loaded toggle switch, which is consequently fast-break and fast-make, as opposed to a knife switch or mercury switch. Used loosely to mean any SPST (see entry).

Sneaky Pete (slang) A dummy pole-top transformer, hiding a supplemental electric meter emplaced to detect power theft.

SNI (acronym) Standard Network Interface (see entry).

Snipe nose See under pliers.

Soap (slang) To apply pulling lubricant. Loosely, the lubricant itself.

Socket EITHER a receptacle outlet, OR a brass-shell lamp holder—more commonly referred to as a lamp socket (see entry), OR, more rarely, a porcelain (or plastic) lampholder that mounts directly to an octagon box or a mud ring.

Socket outlet A wall receptacle, in European parlance.

Spade A flat drill bit; it does not pull itself as does an auger, but is cheaper, is easier to sharpen, and does not get hung up as easily.

Spark, Sparkie, Sparky, Sparks (slang) British, mainly, for electrician.

SPDT (acronym) Single Pole Double Throw EITHER a three-way switch (which see) OR, if it has a center-off position, a switch for use when one wishes to supply power either to one of two loads, or to neither load, and never to both loads at once.

Speaking tube Before the development of intercoms, thin brass tubing was run through walls to carry the voice. It is probably little more effective than a string with cans tied at each end, but it represented the state of the art for many years. Later, those same tubes served as convenient wiring chases. Some used them as raceways, before rules for the latter were formalized. At one point in the early 1920s, in some areas, electricians were not permitted to embed armored cable in plaster, even though its armor was galvanized. When a wall's construction was plaster on brick, the electricians ran one or two cables inside a speaking tube, which was then plastered over.

Specular Mirrorlike. The adjective is used to refer to luminaire reflectors that are designed to throw more light out and waste less light and hence electricity by conversion, inside the fixture, to heat than do standard fixture reflectors. A similar or greater improvement can be achieved by cleaning fixture lenses and diffusers, and even lamps and reflectors, in establishments which do not do so as part of their regular maintenance program. Adding reflectors plus cleaning yields even better results.

Split bus panel A loadcenter with separate busbars serving the main section and the branch circuit section. Incoming power is connected to lugs on the main busbars, rather than to a main circuit breaker or main fuse holder. Therefore, so long as power enters on the service conductors or feeders, everything attached to the MAIN section is live. One circuit breaker or set of fuse holders in the MAIN section is permanently connected to cables that feed the BRANCH section's busbars.

Split main A loadcenter with more than one main overcurrent device or set of overcurrent devices. Split bus panels (see entry) are of this type.

Split-wire To feed a receptacle so that the (normally two) outlets are independent or semi-independent. Sometimes this means that they are on separate circuits, with or without a shared grounded conductor. Other times, they just are separately controlled, one switched and the other always live, or both switched separately.

SPST (acronym) Single Pole Single Throw, the most common simple "snap switch."

Square D Switch (slang) At one point this was used to refer generically to an enclosed safety switch.

Stacking die A compact, handheld apparatus which held $\frac{1}{2}$" and $\frac{3}{4}$" trade size dies for threading the very end of pipe stubs emerging from the ceiling. In old wiring, when an existing gas fixture was to be replaced with an electric light, a stacking die was used as follows. Gas service to the location was shut off, the fixture was removed, and the gas pipe cut back far enough to leave just sufficient length to enter an outlet box that would be flush with the ceiling. Then the stacking die would be used to thread the cut-back gas pipe just enough to later get a fixture support hickey, and perhaps a locknut, onto it. Then wiring would be brought to the location, and the box and then the fixture hickey mounted. Last, the box would be wired, the ceiling patched, and the fixture hung. If gas service needed to be restored in that pipe, the fixture hickey would be of the type that also served as a pipe cap.

Stakon™/stakon tool A crimp terminal/ a crimper. The crimper also has insulation-stripping dies and cutters and thread-clearers for a few small bolt sizes such as 6-32 and 10-24. The term is used generically.

Standard network interface A telephone box used as a demarcation point where premises wiring is connected to the serving telephone company's wiring. There is a test jack, and the two systems can be separated in order to establish whether problems originate in the premises wiring or in the incoming service.

Standing leakage Ground leakage current normally produced at certain equipment, despite the absence of any fault.

Star drill A steel bit for hand-drilling through masonry. It is hit with a lump hammer and turned, hit and turned, repeatedly. It serves as the manual equivalent of a hammer drill or electric hammer.

Station wire The telephone cable leading to a jack.

Stinger (slang) The high leg of three-phase, center-ground delta systems—for example, in 240-volt delta, the leg at 208 volts to ground. Also red leg, bastard leg, etc.

Stork beak See under pliers.

Structure Anything stationary that is human-built, so far as the NEC is concerned. This includes skyscrapers, sheds, fences, even poles.

Stub (verb, noun) (To bring up) a short length of conduit, either as a nipple or as a bend at the end of a fuller length.

Stubby (slang) An extra-short screwdriver.

Stud EITHER, more commonly, a minor vertical structural member OR a fixture stud (see entry).

Stud box, or shallow stud box An armored cable outlet plate also known simply as an outlet plate (see entry).

Sweat EITHER to solder or more rarely to braze, as in "to sweat a seam" (early electricians used to fabricate enclosures by sweating together pieces of sheet lead); OR to have beads of moisture appear. For example, an enclosure in a moist environment may "sweat" because it is slightly cooler than its surroundings, kept cooler by virtue of the fact that the conduits attached to it serve as heat sinks. Corrosion or, in the extreme case, shorting can result.

Sweat equity Permitting a customer to put in labor to reduce the cost of a job. Even when this takes place as part of a "Christmas in April" type charity effort, the ramifications in terms of liability can be complex. See Chapter 7, "Setting Limits and Avoiding Snares."

Switch leg When hot and grounded conductors both come directly to a lighting outlet, and two wires go from the light to a switch which controls that outlet, the wiring between light and switch consists of a hot conductor descending and a switched hot conductor returning. Either of these two wires or the combination may be referred to as the switch leg.

T&B (acronym) Thomas and Betts, manufacturers.

T&M (acronym) Tests and Measurement.

Tampin™ An anchor equivalent to the AJ (see entry).

Tandem breaker Two independently operating single pole circuit breakers—as opposed to a two-pole breaker—incorporated in one body, hence attached to the same busbar and system pole or leg. Also "piggyback."

Tap EITHER a device for creating or restoring threads in holes OR a bug (see entry) OR a tap conductor (NEC Section 240-21, etc.)

TCD (acronym) NEC Technical Committee Documentation, former name for ROC (see entry).

TCR (acronym) NEC Technical Committee Report, former name for ROP (see entry).

TEFC (acronym) Totally Enclosed, Fan-Cooled. A motor design that, like oil-cooled as compared to air-cooled vehicle engines, allows tighter control of operating parameters. This potentially permits a greater range of operating conditions and greater efficiency.

Telco (slang) Telephone company.

Telephone pliers See under pliers.

Thermoplastic A material that softens when heated. One result of that characteristic is that when used above its temperature range, thermoplastic insulation that is under pressure can deform and even flow enough for the wires that it was protecting to short out. Some badly-constructed portable hand lamps have shocked users when the thermoplastic around the lamp shell softened and exposed metal. Another result of that characteristic is that modern plastic insulations tend to get hard and brittle when cold. Other shortcomings are discussed in the Fine Print Note to NEC Section 310-13.

Thermoset A material that hardens, at least during its manufacture, when energy is added. Generally, this means that it hardens when heated, but some thermoset materials are hardened through radiation or other means of adding energy. The fact that a material hardens when heated does not usually mean it softens when cooled. Some thermoset materials actually soften when heated—but do not become fully plastic (malleable). Natural and artificial rubbers fall into this category. The shortcoming of at least natural rubber and comparable materials is that they do dry out, embrittle, and eventually crack over time and also with exposure to heat.

Thinwall (slang) EMT (see entry).

Three-way A switch required for multipoint switching. See Chapter 4.

Throwaway boxes (slang) Junction boxes tossed into the space above the finished ceiling, used for the transition from fixture whips to branch circuit wiring. (The NEC requires boxes to be secured, making this usage illegal.)

TIA (acronym) EITHER the Telecommunications Industry Association, which is partly responsible for telecommunications standards OR Tentative Interim Amendment, an NEC change deemed so urgent that it could not wait for the next regular Code change cycle.

Tie wire Wire used for mechanical support, usually made of mild, as opposed to hard-drawn, steel.

Tie wrap (slang) Ty-Rap™; (see entry) used generically.

Tiger Saw™ The Porter-Cable company's reciprocating saw (see entry).

Tip The green wire in old, four-wire telephone systems. See entry for "ring" for further discussion.

Toggle EITHER a bolt anchor that turns or spreads after inserting into a hollow wall OR a switch that is operated not by pressing a button or contact, nor by pulling a lever, but by flipping a handle (see entry for rocker).

Tombstone (slang) Pedestal box (see entry).

Torchiere An open-top floor lamp, shining against the ceiling and thus primarily illuminating the room indirectly. Usually, a torchiere is supplied with a semitransparent shade below the lampholder. According to a CPSC safety alert, 500-watt halogen lamps in torchieres should be replaced with 300-watt lamps, and covers added.

Torpedo level A pocket-sized or at least tool pouch-sized level 6" long.

Training Preshaping conductors before pushing a device into an enclosure. This important practice ensures that they will NOT fold up in response to resistance in such a way as to push or pull on terminal screws and loosen the terminations. In addition, conductors can be shaped so that they end up where there is minimal chance of their being pinched or crushed by a device or nicked by a screw.

Traveller Either pole on three-way switches other than the common (see entry). The term is also used to refer to the conductors attached to these poles. See Chapter 4, "An Esoteric Switching Layout."

Triple tap A screwdriver handle with a blade containing three steps of tapping threads: either 6-32, 8-32, and 10-32; or 6-32, 8-32, and 10-24. Some triple-tap screwdrivers have removable, reversible blades with different threads on the two ends, giving six options.

Triplen Third-order harmonics. At a basic frequency of 60 Hz, this means the superimposed frequencies of 180-Hz, 540-Hz and so forth. Many of the problems associated with using solid-state equipment on multiwire circuits are due to triplen harmonics. They add in the neutral, rather than cancelling.

Trunking British for wireway.

Tube EITHER (used in the term "knob-and-tube wiring"—see entry) a porcelain cylinder inserted in a bored hole to protect an open conductor passing through OR a Speaking tube (see entry).

Tubing Circular raceway with thinner walls than conduit, thus offering conductors less protection.

Tungar A type of vacuum tube used as a rectifier.

Turn switch A switch that is operated by twisting its handle. There are both on-off multiposition versions. The latter is used to operate a three-way lamp, or, as part of a three-wire socket, for the operation of two separate lamps.

Twin (slang) Tandem Breaker (see entry).

Two phase EITHER an antique polyphase system consisting of two hot wires, with current 90° out of phase, and either separate or combined return conductors, plus possibly a separate neutral conductor, making for 3-, 4-, or 5-wire two-phase systems OR, to some electrical engineers and lay persons, any system with two hot conductors, including one with current 180 degrees out of phase, which electricians call "single phase." See Chapter 9, "Relatively Rare Situations."

Two-way switch British for three-way switch (see entry).

Ty-rap™ A self-locking cable tie, usually plastic but sometimes stainless steel, named after the well-known brand but used generically.

Uncle (slang) Mungo (see entry).

Undertakers' wire (slang) Underwriters' wire.

Underwriters' wire Early insulated conductors, with insulation consisting of multiple wraps of cotton, each wrap individually lead-sheathed. It served well in dry areas, but...it quickly got renamed, "undertakers' wire." Still, it was better than the more highly-flammable paraffin-dipped cotton insulation it replaced.

Union A coupling with special characteristics, such as threading onto the ends of two conduits to connect them without the need to turn the conduit, or else such as flexibility, allowing for thermal expansion.

Upstream Closer to or heading towards the source of power as measured along the conductors. Compare Downstream (see entry).

Utility box A surface metal device or junction box, normally 2" wide by 4" tall, in varying depths. Utility boxes are mostly from $1\frac{1}{2}$"

to 2⅛" deep; 3" deep versions have been made, and extra shallow ones, too.

Often, they have rounded corners, although some are tack-welded of flat components. Utility boxes normally have ½" or perhaps ¾" trade size knockouts, and less commonly other sizes. When surface-mounted, they require covers listed for use with utility boxes, rather than the wider ones designed for flush installations. Utility boxes also can be embedded in dry masonry walls, in which case they do take flush covers. Utility boxes have nothing particularly to do with utility companies.

VOM (acronym) Volt-OhmMeter. The term also often is used to mean volt-ohm-ammeter.

VTVM (acronym) Vacuum Tube VoltMeter.

Washer, reducing Doughnut (see entry).

Water pump pliers Channels (see entry). The grooves in water pump pliers often are slighter than those in channels, making them functionally "equivalent" but actually a poor cousin.

Weatherproof wire A fabric-covered, but not properly insulated, wire. It was used for overhead lines and also in some early cloth-covered, nongrounded nonmetallic-sheathed cable.

Western Union splice An in-line soldered splice. The conductors coming from opposite directions are wrapped around each other, with the ends in a tighter lay.

Wiggy™ Wigginton™ tester. A solenoidal voltmeter. The term is used generically.

Wild leg (slang) Red leg (see entry). (Also stinger, bastard leg, etc.).

Withstand rating The former term for short-term current rating.

Wire Skinny round flexible metal, nothing more specific. It could be for tying, it could be for support, it could be for grounding, or it could be a circuit conductor, solid or stranded, or part of a flexible cord. The term also is used loosely to mean cord or cable. Compare building wire (see entry).

Wireman Electrician.

Wiremold™ Surface metal raceway, perhaps the best-known brand; hence the term is used generically. It used to also refer to wood electrical raceways.

Wirenut™ A (brand name of) solderless twist-on connector, often used generically.

WN (acronym) WireNut™.

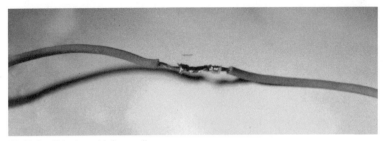

■ **G-6** *Western Union splice.*

Wrench tight Mechanically secured to the maximum extent possible, so as to minimize impedance in a path, as well as ensuring mechanical support and minimizing the chance of foreign matter—even gases—entering, or heat or sparks leaving. With threaded connections, this means literally tightening with a wrench or groove-joint pliers. Added security can be achieved by using a conductive thread sealant, which also serves as a lubricant so that the connections can be threaded a bit tighter using the same applied force. Some types of hardware relying on setscrew connections are Listed for uses that are required to be "wrench-tight." A wrench is NOT applied to a setscrew connection. While few inspectors will demand that connections in a mechanical system, as opposed to, say, connections at conductor terminals, be tightened with a torque wrench, installers still need to use reason. All couplings and connections, especially those using setscrews or compression rings, can be stripped by applying too much force.

XS A round or octagonal box $3\frac{1}{2}'$ diameter by $\frac{3}{4}''$ deep.

Yoke The part of a device used to support its body at each end from the enclosure.

Zinsco A brand of quite noninterchangeable circuit breaker.

Zip cord Two-wire, hence nongrounding, flexible cord (as opposed to building wire) designed so the insulated conductors can be pulled—"zipped"—apart, retaining their insulation. Should you zip apart zip cord and find that one of the conductors leaves its insulation, that zip cord is bad. Zip cord's days are numbered, because cords with an overall covering tend to be slightly safer due to the added layer of protection.

Fluorescents

Many old fluorescent fixtures remain in use, and they can present special problems that you should be prepared to handle if you plan to work on old wiring systems. For instance, it is worth your while to stock fluorescent starters. This does not mean you need to maintain preheat-type fixtures as such. Such a fixture can be reconfigured easily into functioning as the rapid start type, eliminating the need for a starter. The modification is not so very great a departure from the design with which the fixture earned its Listing. Of course, if your customer is perfectly happy with the fixture as it was, and replacing the starter will restore it to health, you have no business spending the time and money to rework it.

Any electrician who has worked on fluorescents will be familiar with certain problems. Besides the usual malfunctions stemming from lamps (including incorrect lamp insertion), ballasts, and starters, you occasionally will find fluorescents acting up due to two sometimes-nonobvious problems. One is that, in the fullness of time, the end terminals make poor contact with the lamp pins. This frequently is evident upon examination, especially when the source of the problem is cracked porcelain. Incidentally, some electricians swear by coating the pins on a fluorescent tube with antioxidant before inserting them. The other reason is bad contact, or no contact, with an internal lead. The leads from one terminal to another commonly are secured by spring pressure, the same way many switches and receptacles can be backwired. Sticking a wire into a spring clip is far from the most secure way to make an electrical connection, especially over the long term.

If none of these turns out to be the problem, you are back to looking at the circuit wiring. That is beyond the scope of this appendix, which is devoted to one special type of fixture. Consider these additional concerns specific to working on fluorescents.

☐ While the lamps run very much cooler than incandescents, even thermally protected ballasts generate a fair amount of heat. Wires in the vicinity of a dying ballast that was not type P, thermally protected, can get quite thoroughly cooked. This can, on rare occasion, damage splices to the point that they fail, affecting downstream outlets. Type TW wire is forbidden within 3 inches of a ballast; older insulation types are even more vulnerable.

☐ Older ballasts that rupture or smoke sometimes put out toxic, or carcinogenic, substances. Your fingers are not going to fall off because you handle a filthy, blackened ballast—but do at least give them a good washing before you eat. And if a ballast is smoking, avoid breathing the filthy stuff.

☐ Even if the fixture is not terribly old, beware of relamping with energy-saving lamps unless you have installed an energy-saving ballast. A 40-watt ballast can be damaged by the use of 34-watt lamps. One still can buy 40-watt lamps in premium versions, triphosphor or deluxe types, and these are what you must use if you need 40-watt output. New 25-watt 48" lamps are designed to be compatible with older 40-watt ballasts.

☐ Energy-saving electronic ballasts have some disadvantages compared to magnetic ballasts. According to a utility study, high-frequency ballasts cause more emissions (line disturbances) and also are more vulnerable to them. This problem manifests both in terms of operating characteristics, specifically hum and variations in light output resulting from line disturbances, and of premature failure due to that susceptibility.

☐ Excessive hum sometimes warns of a dying ballast. More often, though, when there are no other symptoms, hum simply means a mounting screw or nut needs tightening. When tightening has failed to do the trick, wedging in a metal shim sometimes has worked.

☐ Although you are very unlikely to encounter the substance in fluorescent fixtures, asbestos still is a health hazard potentially associated with them. At one point, well before the advent of thermally protected ballasts, fluorescent ballasts were notorious for damaging fixtures and ceilings. The solution was to install asbestos washers, about $\frac{3}{16}$" thick, under the ballasts, as standoffs, around the mounting screws. The only way you would encounter these—which by now could well be crumbly and thus hazardous, at least in quantity—would be if the electricians previously replacing ballasts left the washers in place.

☐ Some straight-line fluorescents will have been installed as though they were Circlines-covering enclosures. When you come across old standard, noncircular fluorescents installed over outlet boxes, they should serve as warnings that you are dealing with creative wiring: nonstandard, probably illegal, and, most to the point, potentially hazardous. Incidentally, while you may find those providing access to the outlet box through a ½-inch knockout quite annoying, such fixtures do not threaten to slice your fingers as badly as do fixtures with freely-chopped, jagged, hand access holes.

☐ Grounding is another problematical issue with older fluorescent fixtures. As mentioned at the end of Chapter 2, "Troubleshooting," when you work on an outlet you should do your best to leave it properly bonded and grounded. While the NEC may not demand grounding of light fixtures that are more than 7 feet up and not otherwise accessible, the instructions that come with fluorescent ballasts usually explicitly demand that they be installed in the vicinity of grounded metal.

What makes this a problem is that older fluorescent fixtures often lacked a ground screw or grounding lead. When they are fed by a grounded metal cable or raceway, entering a single knockout (as opposed to one removed from a set of concentric or eccentric knockouts) grounding is accomplished automatically. Where knock-outs are not single, or, especially, where the wiring method is non-metallic, what can you do? Some inspectors will accept the use of a ground clip, although that is really intended to grab onto the rather thicker wall of a metal box. Some will accept the use of a sheet metal screw, although that is definitely not up to Code.

401

Here are two answers. The metal is too thin to drill and tap, but in-stalling a nut and bolt to (scraped) bare metal, especially installing them with a lock washer or a toothed washer, is a fine solution. A less obvious solution, in the case of nonmetallic-sheathed cable entering a single knockout, would be securing the cable's ground wire to a bonding bushing installed on the cable connector, after the locknut. The path would be ballast to housing to locknut to bushing to ground wire. So long as all those elements are made up tight, that path will have reasonably low impedance.

With that, the digression into the special problems associated with nonincandescent lights is complete. This is not the place to talk about high-intensity discharge fixtures.

Old Splices—Different Than Modern Ones, Not Necessarily Worse

Almost all branch circuit splices nowadays are made with twist-on solderless connectors such as Wirenuts™. Some of them, which as a professional you should not employ, are simply plastic shells. Any decent model contains an internal spring. (This is not to deny that there are functional installations relying on insulating wire caps that lack internal springs.)

Old splices (those installed before, say, World War II) were done up differently—some better, a few worse. They were almost always bulkier, and they took longer to create. Splicing, or "making up joints," was done by first twisting the wires together (not necessarily clockwise), then soldering them, then taping them with self-annealing rubber tape, and over that applying friction tape. Talk about solid!

These splices are easy to recognize, and at worst inconvenient to work with. For those unfamiliar with it, friction tape is pitch-impregnated cloth, which may either unwind or tear when you attack it. Old installations using rubber tape probably will not look like they are covered with tape. The rubber will be a solid black mass under the friction tape, and it will adhere to the wires quite well. It hardens enough over the years that it often will break up when squeezed with pliers or groove-joint pliers such as Channelocks™. You can pick it off with your fingernails, after you absolutely confirm that the spliced conductors are dead and will stay dead.

The main exceptions to the rule that old, old splices were good and bulky black masses are those few splices that used porcelain caps. These look like modern solderless connectors, in white, but lack a modern connector's internal spring. Electricians began to use them to cover their splices in the 1940s. They were trusted because porcelain

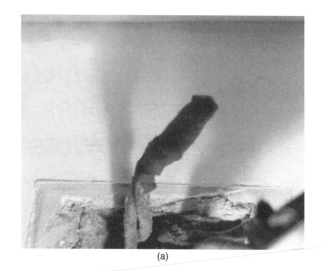

(a)

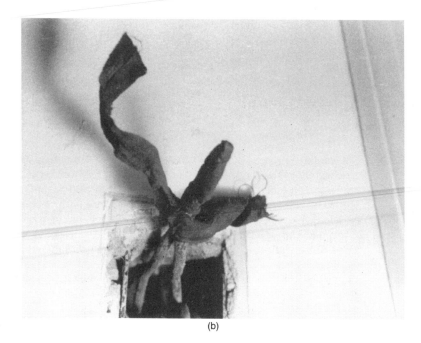

(b)

(c)

■ **A2-1a,b,c** *Old-timey splices. Unwrap the friction tape and you find a solid mass of rubber. Remove that and you find conductors that were twisted and soldered, and maybe folded over for compactness.*

was known to work perfectly fine as a protector in knob-and-tube installations. Given that they serve basically as insulators, using them did not make for very good connections—unless the wires inside first were soldered, a practice that would have been quite unusual. Their use was one reason old-timers used to put a wrap of tape around the skirt of a solderless twist-on connector. It was not that they necessarily thought the wires had been stripped back too far; it was not necessarily that they were afraid something metallic would poke under the connector's skirt and short against the wires; it was simply that they were afraid the connector would fall off.

Having said that, take warning: DO NOT assume that a screw-on connector was taped due to habit. Instead, assume that when you remove the tape, the connector could fall off and the wires could spring apart. Assume that as you unwind the tape, you may come across bare copper—or at least nicked insulation.

The first plastic splicing caps had a big problem: if an installer kept twisting them onto the wires, the wires eventually broke through, out the back. This is why 3M's Scotchlok™ incorporated an internal metal sleeve or cap enclosing the spring: it restrained the wires from breaking out of the back of the shell. It did not actually keep

Old Splices—Different Than Modern Ones, Not Necessarily Worse

the wires rigidly constrained, as it was bigger than even the expanded spring, which was "live" in that it expanded and contracted to some degree with the microscopic thermal changes in
the contained wires.

There were a few other exceptions to the practice of soldering and
two-layer taping as a means of splicing.

☐ Some soldered-and-taped connections did not incorporate
rubber tape, but trusted friction tape to both insulate and
protect.

☐ The first solderless connectors incorporating springs were
introduced in the early 1950s by 3M. The "L Scotchlok™" was
a spring such as you now find inside a modern solderless
twist-on connector. The spring came with an integral handle
to use in winding it onto the wires. Once it was on, you teased
the handle by bending it back and forth until metal fatigue
broke it off. Then you taped your splice.

☐ Setscrew and crimp-on connectors also were used since before
the 1920s. Some lovely setscrew connectors were designed to be
used in combination with a screw-on insulating cap, a combination that looks not unlike a modern solderless twist-on
connector.

■ **A2-2** *An early setscrew splicing device. Taping still was required.*

Electrical Myths

The following document was prepared some years ago to help bridge gaps between electricians and laypersons. The list of misunderstandings comes from research that went into the test, Safety Awareness For Electricity-Residential (SAFER). SAFER was tested (validated) against the responses of laypersons, electricians, and inspectors. Validation meant confirming that there are marked differences between the responses of people in the first category and of the professionals in the latter two, whose responses can be trusted to be more knowledgeable.

Your preconceptions can make you oblivious to other people's assumptions. Electricians' difficulties in communication with laypersons can stem from three elements. One is lack of good faith, which is reflected in a cluster of behaviors that are hard—many would say futile—to fight: willful misunderstanding, intentional obscuring, and indifference. A second is ignorance of terminology (jargon) or different understandings of the same language. A third is different preconceptions—what you might think of as laypersons' myths about electricity. Education can help there.

The following statements are genuine quotes. Each myth is followed by the reality, a reality that often is not obvious to customers, to electricians' and inspectors' surprise. The reality is described in a way that might help you to explain electrical concepts and jargon to customers. None of the suggestions are guaranteed to work with all customers, any more than the boilerplate specifications that a lazy engineer patches in from his or her last job are guaranteed to make sense in terms of today's blueprint. But they offer something you can start with, carefully observing whether they help you and the customer make sense to each other.

Electrical panel

"A circuit breaker will trip and prevent an electric fire, unless the breaker is bad."

☐ A standard circuit breaker cannot tell whether the 5 amperes it is sensing are being drawn by a correctly operating appliance or burning across the electric arc at a bad connection.

"Circuit breakers mean modern wiring."

☐ Circuit breakers have been around for a long time (and could have been put in with rag wiring, old rubber and cotton insulated conductors, whose insulation probably is deteriorated).

☐ A service upgrade that converts a house from fuses to circuit breakers need not have touched anything other than the panel. To know the condition of wiring, you must look at the wires.

"Needing more circuits means that you require a service upgrade."

☐ A subpanel is a viable alternative, unless you're adding significant loads.

"Having empty spaces in a panel means that circuits can be added."

☐ The busbars are what count: coordinated protection is important, so panels with room in their Main sections may not accommodate more branch circuits. Also, one cover can fit several panels, so blank spaces in a panel cover do not always correspond to available spaces on the busbars inside the panel. Read the schematic.

"Having the right size fuse or circuit breaker means you're protected from fire and shock."

☐ GFCIs and grounding are also needed. The right brand, as well as the right size, is needed. Surge protection and arcing fault protection are useful—not to mention smoke detectors, carbon monoxide detectors, and sprinklers.

"To determine service size, add up circuit breakers' or fuses' ratings."

☐ The incoming cable, and then the busbars, give service capacity.

"Circuit breakers are much better than fuses."

☐ They can fail "on." They ARE more convenient, of course.

"If a circuit breaker trips, you should be able to flip it back on, unless there's still a problem."

☐ With some brands, you need to turn the breaker to a full off position first, before it can be reset.

"If power is off and I test all my circuit breakers, and none are out, I need an electrician."

☐ A blackout could be the problem—check with neighbors.

☐ You could have a subpanel elsewhere feeding everything you find off.

☐ Alternately, perhaps the panel in which you looked is a sub-panel; the main might be in an electric room or outside.

☐ If power is off to only part of the house, you might have tripped a GFCI.

"A "heavy-up" means changing the panel to circuit breakers and providing more room."

☐ In an upgrade, service cable is changed first of all. If someone upgraded your electric panel without involving the utility, the service almost certainly was not changed, and the work may well have been done on the sly.

"A "heavy-up" means you can handle bigger loads on old circuits."

☐ In itself, a heavy-up rarely helps with problems such as flickering or dimming lights; a circuit's capacity is limited by its wire size.

General

"If it works it's probably okay. Corollary: anybody can wire by going "monkey see, monkey do" or by following installation instructions."

☐ Cars work fine without brakes—until you have to slow down.

"As long as you don't cut a wire and blow the fuse, doing carpentry, painting, siding, plumbing and so forth doesn't affect wiring."

☐ Damage can occur unnoticed.

☐ Recessing outlets is a violation and creates some degree of hazard.

☐ Losing grounding continuity is common in cases where people are careless of wiring.

☐ Shorts can degrade electrical equipment or otherwise create hazards, even when they are small enough that they fail to blow fuses. Paint, dust, moisture, and other foreign material can interfere with electrical equipment.

Electrical Myths

Wires and cables

"That grey stuff is aluminum wiring, which could cause a fire."

☐ Scrape it before making assumptions. If it is old, rubber-insulated wire, the silvery stuff is probably tinned (solder-covered) copper. Yes, in that case it may cause a fire, but if it does so that will be because the insulation has deteriorated from age, not because an aluminum connection has developed high resistance.

"Any aluminum wiring is dangerous."

☐ Aluminum branch circuit wiring, which goes to a switch, a light, or a receptacle, is relatively dangerous compared to copper, especially when not installed in accordance with the best practices. (There are expensive intermediate measures available, other than replacement, to reduce its hazard.) Heavy cables of aluminum alloy, the type that bring power in to the house, are safe when correctly installed with antioxidant in connectors designed for them. Aluminum cable feeding heavy appliances such as stoves and central air conditioners are an in-between case. Some jurisdictions forbid them entirely; where they are allowed, they often are installed incorrectly.

"To splice wires, twist or fold them together, then wrap with tape to insulate."

☐ That approach is not satisfactory even with low-voltage wires.

"To splice wires, twist, solder, and tape."

☐ Never solder the ground wire.

"To splice wires, use a wirenut."

☐ That is okay ONLY with the right size connectors, appropriate wire preparation, proper twisting, correct pretwisting if pretwisting is used, and, if you are judicious, only if the connector has an internal spring.

"To get more power, change the fuse or circuit breaker."

☐ If the problem is not bad connections or too much equipment on one circuit, change wire size first; only then may you change the protective device where legal for the loads you are feeding.

"To get more power, pull new wires into old cable (or new cable through the holes in studs, inside the walls, in place of the old)."

☐ That option is only viable with pipe, rarely with flexible conduit. Cable that is in place normally does not work as a drag line, being stapled or otherwise secured.

"Tape is tape."

☐ Would you use cellophane tape to hold a bandage on?

"There's no ground in this system; it's just two wires."

☐ If it consists of two wires in armor or pipe, the steel is probably a ground.

"There's no neutral wire; I can just use the ground wire for the return."

☐ You chance electrocution if you do so.

"That skinny aluminum wire in the armored cable is the ground; I can splice the Romex™ ground wire to it."

☐ If you do not understand the difference between grounding and bonding, Romex™ and BX, you probably are using the wrong clamps or connectors as well.

Legality

411

"If an electrician did it, it's legal."

☐ Would that it were so; not all are scrupulous about permits and inspection.

"An electrician is an electrician."

☐ If so, you would not see evidence of illegalities such as when a "licensed journeyman or journeywoman," lacking master and contractor licenses, advertises his or her contracting services in the newspaper serving an area where such licenses are required.

"If it's in your own home, you don't need a permit."

☐ This may be true if you live in the backwoods; generally, where you have adjoining neighbors, fire departments, and other amenities, you have to play by the rules.

"You need an electrician to make it legal."

☐ Homeowners' permits usually are an option—provided that the homeowner actually does the work.

"You can make it legal by pulling a permit for the installer."

☐ That constitutes fraud.

☐ Doing so means that you lose the protection of a legitimate contractor's insurance, bonding, and licensure. Of course, a contractor who suggests you pull the permit may well not have the license, insurance, and bonding.

Receptacles

"Three prong means grounded."

☐ Well, it should. It does when the receptacle is installed correctly.

"Heavier, 20 ampere, is better than 15 ampere."

☐ Most appliances draw 1 to 5 amperes. Significant excess capacity is not doing any good (or harm). Still, the closer the fuse or breaker size is matched to the actual need, the better the protection. A heavier-duty switch or receptacle, though, where applied legally, conceivably might last longer without deterioration, and thus might enhance safety.

"If an appliance cord connector sparks when withdrawn, there's a problem."

☐ It probably means nothing more than that the appliance was not turned off.

"If there's insufficient power for a particular load at a receptacle, power to it can be beefed up."

☐ Usually not.

"Well, then a new line can be run to the receptacle."

☐ Theoretically true, but, practically speaking, this ignores the usually easier choice of running a new line to a new receptacle for the appliance.

"If there's a working receptacle, you have a source of power from which to run additional outlets."

☐ Not necessarily; there are restrictions. Some circuits must remain limited to small appliance or laundry or bathroom use.

☐ There also are limitations based on circuit capacity.

Switches

"Plastic cover plates can be changed to metal—or other plastic, or wood."

☐ Yes, but they need to be Listed. Also, metal plates need to be grounded, normally by a metal screw or screws connecting them to devices bonded to ground.

"Up is on, unless the switch was installed the other way around, in which case down is on."

☐ Not with 3-way or 4-way switches.

"Dimmers can handle anything—low voltage lights, fluorescents, receptacles, motors."

☐ Dimmers and timers and sensors come with various restrictions.

"Where there's a switch, we have power available to power additional lights or appliances."

☐ Not if it is fed by a "switch leg," without a grounded conductor.

Lights

"You can simply replace a surface incandescent fixture with a fluorescent."

☐ Not with any fluorescent. The fluorescent probably requires an opening in back, and probably cannot be dimmed.

"Except for different-diameter bases, a light bulb is a light bulb."

☐ Fixtures, especially recessed ones, specify light bulb types, including wattages, shapes, diameters, and sometimes reflectors.

"A recessed light is a recessed light."

☐ Some must be kept away from thermal insulation; only some can legally be set in concrete.

"A plumber is trained to properly wire a disposal—or heater."

☐ Some are; some are not.

"An electrician is trained to plumb a disposal—or an ice maker."

☐ Most are not.

"An engineer is at least as good as an electrician."

☐ Not at wiring buildings—the training, even of an electrical engineer or an electronics technician, has very, very little overlap with that of an electrician.

"If a newspaper accepts advertising from an electrician, he or she must be a licensed contractor."

☐ Newspapers can be pretty careless about the advertising they run. After all, they are not endorsing—or patronizing—their advertisers.

Index

1920s-type installations, 267–272
1930s-type installations, 272–274
1940s-type installations, 267, 274
1950s-type installations, 265–266

25/50 percent rule, 156, 209

3M Corp., 191

A

AC cable (*See* armored cable)
access to electrical outlets, 168
accessibility for handicapped, 195
adapters, 304
  Edison-base, 303
  Fustat adapters, 50
adapting old systems, 4
adding new circuits, 118, 122
AFC/Nortek, 311
age vs. condition, 261–262
air conditioners, 292
Al/Cu devices, 190
Allied Tubing, 278
Aluminum Assoc., 192
aluminum wiring, 110, 185–192
  branch circuits, 186
  copper-clad aluminum, 186–189
  large cables, 185–186
  repairing, 189
  solderless connectors, 190–191
  swapping in, 189–190
  systemic repair, 190
  terminal strips, 191–192
  variations on solid aluminum wiring,
    187–189
ammeters, 301–302
Amp Corporation, 190
Anderson, J., 313

Andre, J., 147
ANSI standards, 251
antique fixtures
  historic buildings and safety, 211–213
  reinstallation after upgrade, 251
apartment complexes (*See* commercial
    buildings)
Aptiva computers, 196
arcing fault circuit interrupters (AFCIs), 119,
    121–122
arcing or sparking, 12, 18–19, 121–122
  ground fault circuit interrupters' (GFCIs')
    response to, 19
ARCO, 253
armored cable (*See* cable, armored)
Aronstein, J., 189, 191
Arrow Hart, 287
Arrow-Hart circuit breakers, 202
artistic or antique fixtures (*See also* antique
    fixtures), 211–213
asbestos, 203–204, 276, 344
attic fans, recalled, 337
autotransformers, **2**, 219–221, 220
Ayrton, H., 291

B

backfeeds, 22, 181
backwiring, 16, 46–47
Bakelite, 276, 283
ball chain, **148**
Barrett, D., 176
Barry, Dave, 119
Becquerel, A-E., 263
Bell bits, 298
Bezane, N., 312
Black, D., 293
blanking off outlets, 47–48
blown fuses, 12
  overfusing, 50–51, 109
BOCA code, 208–210

Illustrations are indicated in **boldface**.

417

418

421

R

raceway systems, 146, 242, 277–279
  conduit, 278
  fill, 278–279
  plastic (Wiremold), 332
  supports for, 161
  surface wiring, 158–161
  wooden, 277–278
radio controllers, 195
Radio Shack®, 194
ranges, grounding, 177
rare situations, 5, 165–205
  access to electrical outlets, 168
  aluminum wiring, 185–192
  backfeeds, 22, **23**, 181
  bonding cable, 200
  breaker boxes, 200–202
  Class 2 wired relays, 193
  crowfoot receptacles, 178
  fuses and fuseboxes, 202–204
  ground fault circuit interrupters (GFCI),
    204–205
  grounds, 176–178
  knob-and-tube wiring, 179–180
  loadcenters in odd locations, 165–169
  meters inside house, 184–185
  multiwire circuits, 169–175, **171, 172, 173**
  new fixtures at hazard from old, 204–205
  photovoltaic systems, 175, 181
  relay switching, 192–193
  Smart House™ systems, 195–196
  sound activated devices, 193
  splices, buried, 179–180
  taps and jumpers, 181
  telephones, 199
  thermostats, 196–199, **198, 199**
  unexpected power, 180–183
  unlisted (non-UL) equipment, 183–184,
    211–213
rebuildable devices, 92
recall of products, 7, 329–338
  attic ventilators, 337
  ceiling fans, 335–336
  chandeliers, 332
  circuit breakers, 333
  consumer goods, 329–331
  electric heaters, 333–334
  fluorescent lights, 330, 331
  furnace igniters, 337
  ground fault circuit interrupters (GFCIs), 330
  halogen torchiere lamps, 330

recall of products (*Cont.*)
  holiday light strings, 330
  lighting, 331–332
  loadcenters, 333
  pool and spa motors, 336
  portable lamps, 330
  raceways, plastic (Wiremold), 332
  reporting recalls to customers, 338
  service panels, 333
  smoke detectors, 334–335
  sprinkler controllers, 337
  stereos, permanently installed, 338
  sump pumps, 337
  surge protectors, 332–333
  switches, 332
  well pumps, 336–337
  wire and cable, 330–331
receptacles, 27–28, 35–36
  blanking off, 47–48, **47**
  bonding (*See also* grounds), 36–37
  crowfoot receptacles, 89, 178
  fatigue, wear and tear, 15–17
  grounds, 35–40, 94–95
  multiwire circuits, 169–175, **171, 172, 173**
  myths and misconceptions, 412
  oddly-shaped, 89, **90**
  ranges and dryers, 177
  replacing, 35–36
  spacing regulations, 266
  two-prong replaced with three-prong, 113
recessed boxes, 108
"red leg," 220, 227
referrals, 318
relay-operated switching, 192–193
remote-control devices, 69, 192–195
  low-voltage systems, 194–195
  Smart House™ systems, 193, 195–196
  X-10™ systems, 194–195
renewable fuses, 117
researchers, expert help in dating, 307
resources guide, 6, 297–318
return conductors, 11–12
rewiring, 5, 119–120, 143–161
  25/50 percent rule, 156, 209
  aluminum wiring, 185–192
  blanking off old outlets, 151–152
  cabling layout, 152–154
  choosing to rewire, 143–144
  cooperating with other trades, 145–146
  copper-clad aluminum, 186–189, **188**
  crawlspaces, 154–155
  critical situations, 144

426

About the Author

David E. Shapiro is a master electrician and consultant, certified as an inspector and plan reviewer who practices in the Washington, D.C. metropolitan area. He is a former contributing editor to *Electrical Contractor* magazine and author of many articles on electrical work which have appeared in *The Washington Post*, *Practical Homeowner*, *New England Builder*, *EC & M*, *CEE News*, and *IEC Quarterly*.

429